Technology and the Contested Meanings of Sustainability

Technology
AND THE CONTESTED MEANINGS OF
Sustainability

AIDAN DAVISON

STATE UNIVERSITY OF NEW YORK PRESS

Excerpts from "Eve to Her Daughters," "Habitat," and "Geology Lecture" from *A Human Pattern: Selected Poems* by Judith Writh. Copyright © 1966, 1973 Judith Wright. Reprinted by permission of ETT Imprint.

Excerpt from "Mysticeti Testament" from *The Trumpeter* 6, no. 3 (1989): 119–120 by Freya Mathews. Copyright © 1989 Freya Mathews. Reprinted by permission of the author.

Excerpt from "Healing" from *What are People For?* by Wendell Berry. Copyright © 1990 by Wendell Berry. Reprinted by permission of North Point Press, a division of Farrar, Staus and Giroux, LLC.

Excerpt from "The Way of Pain" from *A Part* by Wendell Berry. Copyright © 1980 by Wendell Berry. Reprinted by permission of North Point Press, a division of Farrar, Straus and Giroux, LLC.

The epigraphs to the three parts are excerpted from "Song of the Open Road" in *Leaves of Grass: The 1892 Edition* by Walt Whitman, published by Bantum Books, a division of Random House Inc.

Published by
STATE UNIVERSITY OF NEW YORK PRESS
ALBANY

Printed in the United States of America

For information, address
State University of New York Press,
90 State Street, Suite 700, Albany, NY 12207

Production, Laurie Searl
Marketing, Anne M. Valentine

Library of Congress Cataloging-in-Publication Data

Davison, Aidan, 1966—
Technology and the contested meanings of sustainability / Aidan Davison.
p. cm.
ISBN 0-7914-4979-3 (hc. : alk. paper) — ISBN 0-7914-4980-7)pbk. : alk. paper)
1. Technology—Philosophy. 2. Technology—Moral and ethical aspects. 3. Sustainable development. I. Title.

T14.D29 2001
303.48'3—dc21

00-045622

10 9 8 7 6 5 4 3 2 1

for Mara, Angus and Dale

anchors to a world worth caring for

CONTENTS

Preface ix

Introduction 1

PART I

SUSTAINABLE DEVELOPMENT, AN ECOMODERNIST PROJECT

1. Agenda: Toward Ecoefficiency 11
2. Politics: Confusion, Cooptation, and Dissipation 37
3. Metaphysics: Making Nature Secure 63

PART II

THE TECHNOLOGICAL WORLD

4. Building a Deformed World 93
5. Revealing an Inhospitable Reality 115
6. Disorienting Moral Life 141
7. Recovering Practical Possibilities 159

PART III

THE SEARCH FOR SUSTENANCE

8. A World Worth Caring For 181
9. Sustaining Technology 199

Notes 215

Index 275

PREFACE

I ACCEPT AS SELF-EVIDENT the claim that current crises in ecological systems and in development strategies demand fundamental technological change. Yet the inquiry I offer in this book opens up a path of thinking about sustainability and technology that takes us well beyond the now deeply familiar preoccupation with efficiency, a preoccupation being imposed on environmental thought by dominant sustainable development discourses. I consider questions about sustainability to be essentially normative and any answers they prompt essentially contested. The engineer's search for the design specifications of Sustainable Technology—a technology promised to produce efficiency and equity in equal measure—is therefore profoundly misleading. So too is the claim that we have no alternative but to persist with the forms of technological skill that presently define our practices.

The nub of my argument is that the ideal of sustainability—the focus of so much thinking about our ecological predicament—beckons us toward the now unfamiliar yet still morally resonant question of what sustains us. In so doing, this ideal offers to move our understanding fluidly back and forth between moral and technical questions and between our moral experience and our technological practices. The contested moral meanings of sustainability call upon us to hold product and producer together in our thinking, opening up a space within which our understanding of technology can move into the aspirations that animate our moral lives, aspirations about what we most want to sustain in our experience. For to suggest that there is a need for technologies that are sustaining is to beg the question, Sustaining of what? And to suggest that there is a need to sustain our world through human skill is necessarily to ask, How are we to develop in the craft of giving and receiving sustenance?

In these pages I thus happily confound those contemporary habits of thought that associate the idea of technology with an array of neutral objects that are theoretically interesting only to the extent that they embody scientific knowledge. Similarly, I confound claims that technology is an autonomous, transhuman force in social affairs. In my view, to canonize or to demonize technology is to misunderstand it. Technology, like any other intrinsic, constitutive facet of our humanity, bears witness to the saint and the devil within us all.

I challenge the assumption that technology stands external to the human condition. I necessarily challenge the corollary of this assumption, that our ideas stand external to our experience. Whether walking alone along river cliffs at dawnsong or seated with others within the brick walls of a university seminar room, deliberation is embodied activity. To deliberate is to engage with our reality. I have no interest therefore in conforming to the pretence that this book stands outside of my own limitations, or that it treats ideas dispassionately. My favorite Australian poet, Judith Wright, wrote *Eve to Her Daughters* the year I was born, 1966. This poem speaks directly to my experience, disclosing the world into which I was entering, the world of my father:

> So he set to work.
> The earth must be made a new Eden
> with central heating, domesticated animals,
> mechanical harvesters, combustion engines,
> escalators, refrigerators
> and modern means of communication
> and multiplied opportunities for safe investment
> and higher education for Abel and Cain
> and the rest of the family.
> You can see how his pride had been hurt.
>
> In the process he had to unravel everything,
> because he believed that mechanism
> was the whole secret—he was always mechanical-minded.
> He got to the very inside of the whole machine
> exclaiming as he went, So this is how it works!
> And now that I know how it works, why, I must have invented it.
> As for God and the Other, they cannot be demonstrated,
> and what cannot be demonstrated,
> doesn't exist.
> You see, he had always been jealous.
>
> Yes, he got to the centre
> where nothing at all can be demonstrated.
> And clearly he doesn't exist; but he refuses
> to accept the conclusion.
> You see, he was always an egotist.[1]

The story told in this poem has matured and gained strength in my experience over the last three decades. Over this time, the anxiety of which it tells has spread from the margins toward the mainstream of my affluent society. Many now see that the New Eden of "central heating, domesticated animals, mechanical harvesters, combustion engines" often tears down more than it builds up. The warmth of our homes is recklessly drawn from the carbonized essence of ancient life-forms. Our animal factories are barbaric. Industrial agriculture mines the ancient wealth of our soils. Urban life is dominated by the pollution, torn bodies, and social dislocation imposed by our car-dependency. These and many other practices are unfit to be sustained by any world that I would wish my children to become parents in.

Of course, my deliberations about unsustainability also incorporate within them the vast affluence, abundance, and security of my technological world. I recoil from the New Eden built by my fathers yet take as my due its comfort. Even my most mundane activities are embedded within practices that exacerbate the life-long suffering, poverty and oppression of nontechnological societies, degrading further their culture and ecology.

Wright's poem spans and holds within it my deliberations about what sort of life to live. As a child in the late 1970s, I was surrounded with images of the horrific famines of Ethiopia. A Tasmanian, I sought to make sense of the irrevocably complete slaughter of my island's original peoples and tigers. Living alongside Sydney's famous beaches, I was unwilling to swim in their sewer-stained tides. Entering the laboratory intent upon my daily guillotining of rats, my body ached in rebellion. Depleted by the Sisyphean labours of consumption, I have sought resourcefulness at home. Travelling up-river inside the air-conditioned cabin of a motor boat in Borneo, for want of any other mode of transport, I have watched a video of *World Championship Wrestling* through the ritually distended ear lobe of a Penan tribesman, and I have recoiled from the New Eden.

If the present work has a precise beginning, it is not to be found on a page but in those strange months more than a decade ago, when I was a diabetes researcher by day—ministering with delicate touch to a fickle High Performance Liquid Chromatograph—and a masseur by night—using my fingers to practice a very different kind of healing skill. In my movement from biochemistry to cultural studies, I have drawn strength from Arne Naess's insight that a "philosopher, in contrast to a professor of philosophy, is one whose philosophy is expressed in his or her life."[2] It is in this spirit that I offer this book as a contribution to the emerging philosophy of sustainability and in the knowledge that there is no substitute for the lived expression of sustainability's contested meanings. Through it I seek, in an unavoidably small way, to nurture the many wonderful practical experiments that are underway in the building of more sustaining worlds.

In completing this study, the guidance and friendship of Ian Barns and Patsy Hallen has been an inexhaustible source of sustenance. Their unflagging commitment to practical theory and their care for the living earth holds this book together as strongly as does the binder's glue. I must especially thank Patsy for the gift of her extraordinary eye, reflected in the photograph that graces the cover with a view of her *Jarrah* hut, nestled at the edge of Jarrah *(Eucalyptus marginata)* forest in the southwest corner of Western Australia. This image resonates with contested meanings that invite us to probe matters as diverse as the interplay of image and reality, the paradoxical power of the pastoral ideal in technological culture, and the ability of nature and technology (forest and building) to sustain each other. The irrepressible corridor life of Murdoch University's Institute for Sustainability and Technology Policy, in which we teach, has also kept me in touch with other inspiring people who have taken on the good work of building more sustaining worlds. I warmly thank Jim Cheney, Deborah Kennedy, Brad Pettitt, Val Plumwood, Steven Rodgers, Wendy Sarkissian, Laura Stocker, Tabatha Wallington, Karen Warren, and Julie Webb for comments on parts of this manuscript. I wish I were further able to acknowledge countless conversations with staff and students of the Institute, the gems of which are now inextricably embedded in my work. And I am fortunate to have received encouragement and suggestions from Albert Borgmann, Freya Mathews, and Landgon Winner, as well as from the reviewers enlisted by SUNY Press. If only I were as good a listener as the counsel of all these kind folk was wise. In addition, I have leaned heavily on the sturdy editorial support provided by Kate Barnett, Tabatha Wallington, Cathy Hunter, and the staff at SUNY Press.

This book is more than anything else a dialogue between my everyday world and my theoretical confusions. I thank those from whom I have learned so much as we joined together to heal some of the wounds in our shared world. I am indebted to Roma and Jack Brown for unstinting help at home, as I am to my mother, Mary Anne Davison, for regular visits and shared memories of Tasmania's wildness. And to Dale Brown, my debt is beyond expression and beyond redress. Unimpressed by clever theories, Dale just knows, and knows much that I have long striven to learn. Through her, I have the joy of acknowledging the wisdom that our children, Angus and Mara, have been able to provoke out of my (often reluctant) soul.

INTRODUCTION

THE TRIUMPHANT HISTORY of industrialization is shadowed by a history of social oppression and ecological degradation. The vast, unprecedented affluence that has concentrated in highly technological societies is shadowed by poverty and pollution, the extent of which is also vast and unprecedented. This shadow is cast over every aspect of our glamorous present. This is so because much of our technology persistently lacks the ability to sustain ecological flourishing and social well-being.

Contemporary technological systems, or latemodern technosystems, as I shall call them, are lacking in sustain-ability.[1] While they are unsurpassed in their productivity, they are, characteristically, not nourishing of their world. They are manifestly unsustainable. It is not difficult for many of us, especially those of us who have grown up with the environmental movement, to imagine the ways in which industrial factories may be unsustainable. The toxic black clouds billowing from their smokestacks, the sterile sea around their outfalls, their voracious appetite for nonrenewable resources, the demeaning forms of work they create, are often evidence enough. Wendell Berry reminds us, however, that the lavish latemodern home is no less a factory of waste and destruction:

> With its array of gadgets and machines, all powered by energies that are destructive of land or air or water, and connected to work, market, school, recreation, etc., by gasoline engines, the modern home is a veritable factory of waste and destruction. It is the mainstay of the economy of money. But within the economies of energy and nature, it is a catastrophe. It takes in the world's goods and converts them into garbage, sewage, and noxious fumes—for none of which we have found a use.

What Berry would have us recognise is that our affluent latemodern lives, the life-styles of the "industrial conquistador," are systematically, constitutionally unsustainable.[2]

Acknowledgment of systemic ecological degradation and systematic economic injustice is now widespread. Since 1980, this acknowledgment has increasingly gained expression through the language of sustainable development.[3] Guided by this language, environmental issues have entered the mainstream of political debate. Sustainable development is now an immensely powerful institutional, theoretical, and grassroots framework for shaping practical solutions to the problems of industrial and postindustrial unsustainability. It was "the developmental paradigm of the 1990s," the principal achievement of which was, so we are told, to unite economic and environmental objectives through fostering awareness of how humanity is united in sharing its common home, the fragile earth.[4] As we shall see in chapter 1, it is thought by its most optimistic advocates to be lighting the path toward a Golden Age of enduring prosperity.

A great diversity of views are accommodated within the language of sustainable development. There is much disagreement about what the root causes of unsustainability are and thus about what policies for sustainability should look like. There is nonetheless broad agreement that the symptoms of unsustainability include many if not all of the following:[5]

- persistent destitution of more than half of the human population
- sustained population growth within impoverished communities
- disempowerment of women and indigenous peoples by development strategies
- neocolonial, patriarchal, and militaristic bias within the global economy
- widening economic disparity within and between nations
- overconsumption of resources by wealthy nations
- social, physiological, and psychological pathologies of overconsumption
- crippling economic burdens arising from side-effects of past development
- depletion and contamination of nonrenewable natural resources
- political and military tensions arising from resource scarcity
- escalation of technologically mediated risks to ecological and human health
- rapid and irrecoverable loss, both local and global, of biological and cultural diversity
- alteration and degradation of global ecological life-support processes

Inextricably intertwined, these symptoms constitute evidence of a truly global crisis in social and ecological stability.[6]

Evidence of global instability is now recorded in countless publications. Everything from rates of species extinction to rates of infant mortality are proclaimed almost daily in classrooms and television documentaries. Most of us in affluent latemodern societies—what I shall refer to, loosely in part I and then with tighter focus in part II, as technological societies—are well aware of the alarming evidence of unsustainability. Many of us know, for instance, that the populations of two thirds of all recorded bird species are in decline; that the average global temperature has been steadily increasing since 1950; that half of the world's languages have been lost during the modern era, with half of those remaining predicted to disappear early this century; and that the number of people living in poverty has increased in the postwar period, with the gap in income between humanity's richest twenty percent and poorest twenty percent more than doubling to a ratio of over sixty, despite the fact that the Gross World Product has increased six-fold during this time.[7] The litany of empirical representations of unsustainability goes on and on. But such statistics rarely portray the systemic, cultural character of our global instability.

Perhaps more revealing is the juxtaposition of Mexico City, a contemporary monument to unsustaining development, on the site that once sustained the Aztec city of Tenochtitlán.[8] Built on and around Lake Texcoco, largely out of stone, Tenochtitlán was founded upon a sophisticated system of urban agriculture and aquaculture called Chinampa. It housed a population of 200,000, five times that of Renaissance London or Seville, within a complex system of confederated clan politics. The lofty river basin now dominated by Mexico City then supported a population of between one and two million people.[9]

The diary of the conquistador Bernal Diaz records his amazement on entering this valley in 1519: "With such wonderful sights to gaze on we did not know what to say, or if this was real that we saw before our eyes." Diaz, a footsoldier in Hernán Cortés' army, goes on to describe the

> gardens with their many varieties of flowers and sweet-scented trees planted in order, and their ponds and tanks of fresh water into which a stream flowed at one end and out of which it flowed at the other, . . . and the variety of small birds nested in the branches and the medicinal and useful herbs that grew there . . . I may add that on all the roads they have shelters made of reeds or straw or grass so that they can . . . purge their bowels unseen by passers by, and also in order that their excrement shall not be lost.[10]

Tragically, within only fifty years of the arrival of the Spanish, the Aztec cities of the Valley of Mexico were "pale shadows of their former selves."[11]

Today, the mountain-locked Valley of Mexico is dominated by Mexico City, a deeply unstable urban structure in which 70 percent of infants have dangerously high levels of lead in their blood.[12] This is a megalopolis of almost twenty million people—grown from two million in 1940—in which two and a half billion (US) dollars has recently been spent in an effort to control the city's extreme air pollution, regarded as the worst in the world, while slot machines have been installed on city streets to dispense oxygen.[13] Depletion of groundwater has caused large parts of this city to sink, in some areas by up to nine metres—this, the same place where once Tenochtitlán perched atop a maze of canals.[14] Most of the Mexico City's freshwater is now pumped, using vast amounts of energy, from dams more than 1,000 metres below.[15] Equally vast are the amounts of money being sought to combat the city's debilitating water scarcity by buying and piping water from the United States of America (U.S.).

This information provides only the merest indication of the unsustainability of Mexico City. It provides enough, however, for us to glimpse what lies behind the Mexican activist Gustavo Esteva's claim that "development as a social experiment has failed miserably."[16] Yet the juxtaposition of Tenochtitlán and Mexico City is not the juxtaposition of an idyllic, sustainable past and a dystopian, unsustainable present. Matters are much more complicated and more ambiguous than this. Human societies have always had to struggle to sustain themselves. At the time the Spanish arrived, Tenochtitlán was not much more than one hundred and fifty years old. Further, this city was itself at the center of a military empire set upon, in fact, economically dependent upon, bloody conquest. It existed adjacent to the ruins of a former civilization. The city of Teotihuacán had dominated the valley for over seven centuries before the civilization that created it collapsed around 900 C.E. and was supplanted by the Toltec empire that itself collapsed only three centuries later.[17]

The juxtaposition of Tenochtitlán and Mexico City is more than anything else a simple yet powerful reminder of the diverse forms that human development can and has taken. It is a prompt for the simplest of questions, such as, What can *development* mean? What is truly sustainable in human affairs? What is it that nourishes the ability of a society to sustain itself? The juxtaposition of Tenochtitlán and Mexico City prises open the rigid rhetoric of progress, laying it bare before the ambiguity that inheres in all social achievements. It juxtaposes the experience of an Aztec citizen walking around Tenochtitlán's once awesome town centre, replete with temples, with the experience of Mexico City's equally awesome latemodern vastness and technological bustle. It holds before us the dissonance that lives on in the contemporary experience of Mexico's indigenous peoples.[18] What is on display in the history of the Valley of Mexico is certainly great flux, but not linear, unequivocal movement toward or away from human betterment. What is displayed in this long-enchanted place is the inescapable struggle of human communities to sustain their world. What is displayed is the communal and individual search for sustenance that begins anew with every birth.

In considering the idea of sustainable development, we need to hold the essential ambiguity of notions like 'development' and 'sustainability' before us. We need to uphold this ambiguity publicly, or so it seems to me, because only in this way are we likely to be stimulated to think deeply and openly about the nature of our latemodern crisis. The slavish pursuit of absolute certainty has long fed the roots of modernity's predicament. It is extremely concerning, therefore, that the discourses of sustainable development are becoming increasingly dogmatic, technocratic, and hegemonic. The ideal of sustainability is being steadily redefined by the modernist orthodoxy of technological progress. Latemodern evidence of unsustainability is being taken as proof of the fact that technological progress has not yet gone far enough and that ecological efficiency is first and foremost economic efficiency. As a result, a great diversity of cultural concerns about sustainability are being usurped by the global agenda of sustainable development. The appearance of either consensus or intellectual clarity in embracing rhetoric about "our common future" is superficial and deceptive. Those who see the ideal of sustainability as a prompt for challenging political and moral questions, those who see it as latent with contested moral meanings, are finding themselves either enclosed within the stifling language of efficiency or pushed toward the margins of global environmental politics.

Through the present inquiry I seek to display one of the paths along which the ideal of sustainability can be recovered from the presently dominant rhetoric of sustained progress and can thus be placed squarely in the range of basic normative questions about how we are to live. I do not attempt—indeed, believe it impossible—to provide an analytical description, a catalogue, of the meanings of sustainability, for they are essentially contestable. Indeed, they extend beyond the reach of abstract, theoretical articulation and into the heart of our embodied experience of a sustaining world. My approach, then, is to reconfigure the meanings of sustainability as prompts for open-ended questions that focus our attention on the nature of moral experience in our technological world. The pivot on which I shift the ideal of sustainability from technocratic to cultural discourses about development is the philosophical acknowledgment that technology is constitutive of, and not external to, our humanity. My discussion is in three parts, each of which is quite different in content and disciplinary focus. Further, while each chapter is structured so that it can be, at least partially, understood in isolation, the nature of my inquiry unfolds only through the transdisciplinary breadth of scope and movement of the whole.

Part I, Sustainable Development, an Ecomodernist Project, provides a critique of the agenda of sustainable development and the modernist project that informs it. Chapter 1 offers a descriptive review of sustainable development discourses. This review illustrates how "the environment" has been transformed into a mainstream political issue, and how the resultant ecomodernist agenda is founded on the pursuit of ecoefficiency. This pursuit is encouraged by a technological optimism that extrapolates existing trajectories in resource management, corporate liberalization and planetary science toward the ethical goal of an ecologically stable, equitable regime of global governance.

The impression that the language of sustainable development has fostered broad agreement about how the challenge of global ecological problems may best be met is refuted in chapter 2. The essentially contested language of sustainability has been fundamentally coopted and confused by ecomodernism. I show that concerns about cultural, in contrast to technocratic, sustainability have been rendered either inarticulate or unheard within the discourses of global environmental politics, reinforcing the neocolonial oppression of humanity's poor majority.

In chapter 3, moving from politics to metaphysics, I look at the cultural underpinnings of the language of sustainable development that support its claim to provide a universally applicable, culturally neutral account of both development and sustainability. We see that ecomodernist interpretations of ecological notions such as 'limits', 'scarcity', and 'holism' are actually reinvigorating the cultural project of mastering nature. Drawing from the literature of radical ecology, I explore ecomodernism's dualistic metaphysics, encountering in the process the fact that many ecological philosophers become bogged down in abstraction through neglecting the interplay between the world-views that define our thinking and the technological world that defines our experience.

I discuss this coevolution of ideas and experience under the heading of The Technological World in part II. I develop ontological, moral, and epistemological paths for understanding technology in chapters 5, 6, and 7, respectively, on the basis of the description of the technological world I set out in chapter 4. In this description I counter instrumentalist and determinist explanations of technology by drawing attention to technology not as object or knowledge but as practice. Technologies are constitutive of, not external to, our humanity, and they express, shape and perpetuate our philosophical commitments. Through them we build worlds of practice. And through modern technologies, I shall argue, we have built an unsustaining world that is fundamentally deformed because it dislocates means and ends in our experience.

In chapter 5, by way of a critical engagement with Martin Heidegger's postwar writings about technology, I argue that the material deformation of our world is simultaneously a deformation of our understanding of reality. The ecomodernist project of mastery is informed by a *technological ontology* that portrays the moral purpose of human life as sustained productivity in an inhospitable reality. In the technological world, sustainability is inevitably construed as the managerial demand to produce a less perilous future. Yet, unlike Heidegger, I do not think that our practices are implacably enframed by the ontological imperatives of production. Exploring my position through a rejection of his claim that technology has obliterated the conditions of genuine moral *praxis*, I affirm the rich and irreducible ambiguity that inheres in our technological experience.

I explore how we might go about recovering the ambiguity of our moral experience of technology from the confines of modern moral theory in chapter 6. Rejecting the attempt to found moral discourse on scientific reason, an endeavor that seeks to strip ontological narratives about what is good in a world from rational prescriptions about what is right to do, I argue that technology has become destructive of its own ennobling impulses. The reduction of practical reason to procedures of technical control in latemodern forms of life has thoroughly disoriented moral life.

In chapter 7, I offer a substantive, in contrast to a procedural, explanation of the nature of practical reasoning. Through an account of a world-disclosive 'rationality of relationality', I seek to recover sight of those moral horizons within which our visions of 'the good' might begin to make practical sense of the technological possibilities placed before us.

Part III is entitled The Search for Sustenance, and it offers a tentative, yet hopefully emblematic, example of practical normative discourse capable of holding together the problems of technology with the ideal of sustainability. I enter into a form of discourse in chapter 8 that holds together my experience of technology with my concerns about the systemically unsustaining practices of our shared technological world. Reflecting on the place in which I live and on the people, beings, stories, and things with which I share this place, I present a highly specific but, I hope, culturally resonant expression of the yearning to develop in the craft of giving and receiving sustenance. I draw on these personal reflections in chapter 9 to compose a modestly but irrepressibly optimistic cultural vision of the latemodern possibilities for the practice of sustainability. I give voice to my conviction that we can be informed and sustained by the particularity of our worldly relationships in the political task of joining with others to care skillfully for our shared world.

I

Sustainable Development, an Ecomodernist Project

From this hour I ordain myself loos'd of limits and imaginary lines,

Going where I list, my own master total and absolute,

Listening to others, considering well what they say,

Pausing, searching, receiving, contemplating,

Gently, but with undeniable will, divesting myself of the holds that would hold me.

—Walt Whitman

CHAPTER ONE

AGENDA

Toward Ecoefficiency

INTRODUCTION

> Humanity's relationship with the biosphere will continue to deteriorate until a new international economic order is achieved, a new environmental ethic adopted, human populations stabilized, and sustainable modes of development become the rule rather than the exception.
>
> —International Union for the Conservation of Nature, United Nations Environment Programme, World Wide Fund for Nature, 1980

> Sustainable development is development that meets the needs of the present without compromising the ability of future generations to meet their own needs.
>
> —World Commission on Environment and Development, 1987

> [The] integration of environment and development concerns and greater attention to them will lead to the fulfilment of basic needs, improved living standards for all, better protected and managed ecosystems, and a safer, more prosperous future. No nation can achieve this on its own; but together we can—in a global partnership for sustainable development.
>
> —*Agenda 21*, UN Conference on Environment and Development 1992

> Population, consumption, technology, development, and the environment are linked in complex relationships that bear closely on human welfare in the global neighborhood. Their effective and equitable management calls for a systemic, long-term, global approach guided by the principle of sustainable development.
>
> —Commission on Global Governance, 1995

AS THE FOREGOING quotations illustrate,[1] the language of sustainable development has evolved into a powerful and international policy framework. The term first appeared in the UN *Stockholm Declaration* (1972) and the consequent *Cocoyoc Declaration on Environment and Development* (1974).[2] Yet it was not until the *World Conservation Strategy* (1980), prepared jointly by the International Union for the Conservation of Nature (IUCN), United Nations Environment Programme (UNEP) and the World Wide Fund for Nature (WWF), that the ideal of sustainable development became central to development discourses.[3] By the time of the IUCN's 1991 report, *Caring for the Earth*, this ideal had entered the mainstream of political debate in most Northern, highly industrialized nations and in many Southern, industrializing nations.[4]

The influence of the language of sustainable development was ensured through its adoption in the mid-1980s by the UN-sponsored World Commission on Environment and Development (WCED), commonly known as the Brundtland Commission.[5] As I detail in section II, sustainable development is now the dominant policy framework for guiding technological responses to global and regional and increasingly to local ecological problems. This framework differs from prior postwar development strategies in its claim that the aims of economic growth and ecological well-being are complementary. According to the Brundtland Commission, the notion of sustainable development challenges policy makers to revive while changing the quality of economic growth.[6] By way of its success in reframing anxiety into a peculiar form of techno-ethical optimism, and in appealing to powerful political and economic interests, the Commission's influential 1987 report, *Our Common Future*, arguably remains the manifesto for sustainable development policy.[7]

The language of sustainable development has guided environmental issues from the margins toward the core of political debate, although this movement is far from complete.[8] It is "a banner under which transformed environmentalism has marched into the public consciousness."[9] Under this banner, environmental issues have gained much scientific and economic credibility. Environmental concerns are no longer solely the province of radical activists. Politicians of all stripes, business leaders, advertising executives, and bureaucrats share this language with grassroots environmentalists. More importantly, perhaps, this language has also helped to professionalize environmental concern. Sustainable development discourses are now everyday fare for architects, demographers, educationalists, engineers, economists, futurists, lawyers, managers, philosophers, political theorists, scientists, and sociologists.

In its adoption by the core institutions of the North—the institutions of the technological society—the language of sustainable development would appear to represent a remarkable triumph for the environmental movement. But it has been a problematic victory at best. This language has not simply introduced environmental concerns into mainstream policy; it has enabled a fundamental redefinition of these concerns. By the late 1980s, this redefinition had shaped a distinct second wave of postwar environmental concern.

The first wave of postwar concern was directed against the current of the technological society, arguing that modern forms of technology and economic organization are inherently anti-ecological and socially iniquitous. The second wave, in contrast, moves with the current of the technological society, thereby rapidly gaining political momentum. It has not only discredited earlier calls for substantial political reform to technology and economics, it has reinvented the environmental crisis as an opportunity for an even more urgent, unflinching pursuit of economic growth, global managerialism, corporate liberalization, and technological globalization. Although both waves share a concern for ecological sustainability, this concern is, I will argue in chapter 2, essentially contestable. But so that we can contest the ideal of sustainability without becoming lost in abstraction, that is, without being trapped in elegant theoretical constructions of either utopian or dystopian form, there is a need to understand the practical ways in which this ideal is being actualized as a techno-economic agenda that itself lacks genuine openness and cultural self-awareness.

This chapter establishes the subject around which the critical and constructive dimensions of this book are subsequently focused. It locates sustainable development in the history of postwar environmental concern (section I) and in contemporary international policy (section II), before going on to explicate the technological (section III) and ethical (section IV) dimensions of the optimism that infuses the emerging ecomodernist agenda.

I A BRIEF HISTORY

The first wave of environmental concern, as I apply this metaphor, describes the ecological awareness arising with Northern countercultural discontent in the 1950s to the late-1970s.[10] This concern was bound up with a broad platform of political reform encompassing peace, civil rights, feminist, New Left, and neo-Marxist movements. Beginning with three books published in 1948–1949—Fairfield Osborn's *Our Plundered Planet*, William Vogt's *Road to Survival*, and Aldo Leopold's still resonant *A Sand County Almanac*—and in more mature form with the appearance of Rachel Carson's *Silent Spring* (1962), Paul Ehrlich's *The Population Bomb* (1968), Theodore Roszak's *The Making of a Counter Culture* (1969), Barry Commoner's *The Closing Circle* (1971), Donella Meadows's et al. *Limits to Growth* (1972), and Herman Daly's *The Steady State Economy* (1974), ecological arguments became a familiar feature of popular debate.[11] By the early-1970s, scholars such as Murray Bookchin, Arne Naess, Rosemary Radford Ruether, and Richard Sylvan were establishing the foundations of a new academic subdiscipline, an "environmental philosophy."[12]

The First Wave of Environmental Concern

The first wave of concern about environmental issues was deeply skeptical of the modernist model of progress and called for far-reaching spiritual, moral, and economic change in technological societies. It adopted a characteristically antigrowth position with respect to the orthodoxy of unlimited economic growth and technological globalization. The 1972 *Blueprint for Survival*, from the editors of *The Ecologist*, was typical in its claim that the "principal defect of the industrial way of life with its ethos of expansion is that it is not sustainable."[13] *Limits to Growth*, a study commissioned by the Club of Rome, was of particular importance in conferring empirical credibility on the argument that the biophysical and political limits to human population and human avarice had been breached. Talk of an impending cultural paradigm shift was common. Leading theorists of the first wave, such as Ivan Illich, Herbert Marcuse, and Fritz Schumacher, provided trenchant and influential critiques of neoclassical assumptions about the complementarity of technological proliferation and human or ecological well-being.[14]

First wave discontent led to the founding of an identifiable environmental movement comprised of a great many activist groups, mostly in North America, such as Greenpeace, the Environmental Defense Fund, and Friends of the Earth, and a Green political movement, mostly in Western Europe. In its extreme forms, this discontent manifested in a number of doomsday accounts of impending global "ecotastrophe."[15]

The nascent environmental movement was largely restricted to the North and was "met with suspicion in developing countries."[16] In particular, Garrett Hardin rightly attracted strong criticism for his neo-Malthusian descriptions of the "tragedy of the commons" and "lifeboat ethics."[17] Hardin's arguments placed the heaviest burden of responsibility for ecological problems on the burgeoning populations of the South. In so doing, he ignored the historical role of the North in the creation and maintenance of the conditions in which poverty and population growth become uncontrollable. Hardin, like many others, also chose to ignore the fact that the bulk of the world's resource consumption and the bulk of waste and pollution generation takes place in the North.

The political naivete and downright elitism of much Northern environmentalism, combined with a growing awareness of the severe environmental problems confronting poor communities, contributed to the rise of Southern grassroots environmentalism in the 1970s. Conservation for Southern groups like Kenya's Green Belt, India's Chipko, and, more recently, Mexico's *Zapatista* movements, to take three of the best-known examples, rests not on the preservation of scenic wilderness, but on the struggle of local people, and especially women, to reclaim political control of their environments, resources, and livelihoods.[18] It is in the struggles of the rural South during this period that we can see the first real practical attempts to found an explicit political agenda on the interdependence of social justice and ecological well-being.

Not surprisingly, however, claims that immutable ecological and social limits were being breached by technological development, and with devastating consequences, brought a strong reaction from those within mainstream institutions. Books by John Maddox (1972), Wilfred Beckerman (1974), Perry Pascarella

(1977), and Julian Simon (1981) hotly denied that there was any crisis, ecological or otherwise, and argued that the opportunities for human prosperity had never been greater.[19] The 1970s also saw the beginnings of "corporate activism." Business leaders, especially those in the U.S., began "getting together to support a conservative antiregulatory agenda and financing a vast public relations effort."[20] Efforts by theorists and practitioners to debunk the cultural critiques of the first wave as negative, regressive, and impractical were so successful that limits to growth arguments "were no longer found in mainstream discourse by the end of the 1970s."[21] The 1979 OPEC oil crisis was also vital in re-establishing economic growth as the preeminent concern of global politics.[22]

The Second Wave of Environmental Concern

Despite the weakening credibility of limits to growth critiques, environmental problems drew increasing media coverage and political acknowledgment in the early 1980s. A range of different terms competed for the role of articulating the environmental challenge.[23] Universities established environmental science departments, environmental legislation was enacted, the environment became a standard component of curricula, "green" products landed on supermarket shelves, and recycling entered the lexicon of local government. As broad awareness of ecological unsustainability grew, attention turned from cultural critiques to technical descriptions of problems and solutions. The environmentalist movement continued to grow but began to polarize into reform and radical, "light green" and "dark green," realist and fundamentalist factions.[24]

In the midst of all this flux the language of sustainable development bound together a distinct second wave of environmental concern, transforming it into policies for "ecological modernization," or what I shall call the cultural project of ecomodernism.[25] "Ecological modernization," according to Andrew Gouldson and Joseph Murphy, "proposes that policies for economic development and environmental protection can be combined to synergistic effect."[26]

John Dryzek's characterization of this second wave as Promethean is apt. His representation of the first wave as "survivalist" is problematic, however.[27] It obscures the way in which the Promethean, ecomodernist agenda is just as preoccupied with the end of human survival as are countercultural agendas. The reports of the Independent Commission on International Development Issues (1980), chaired by Willy Brandt, the Independent Commission on Disarmament and Security Issues (1982), chaired by Olav Palme, and the Brundtland Commission are indicative of a redefinition of, rather than an eclipse of, concern about survival.[28] As the Brundtland Commission put it, humanity must now be "ever mindful of the risk of endangering the survival of life on Earth."[29] However, unlike the first wave, the second wave of concern about planetary mortality affirmed not the existence of ecological and social limits, but the "need" to wrest control of our future from arbitrary ecological forces, placing our destiny squarely in the hands of progressive, efficient global managers. *Our Common Future* undermined limits to growth arguments by placing at the centre of the language of sustainable development the following question: How is economic growth and technological expansion to be sustained?[30] This shift in focus away from limits and toward growth continued throughout the 1980s with the result that the "environmental concerns of the 1990s have proved to be about the implications of profligacy, of plenty, rather than the limits to growth."[31]

Consistent with this shift, sustainable development has invigorated corporate activism. By the late 1970s the first wave of corporate activism had effectively nullified the wide appeal of radical environmentalist critiques. Nevertheless, corporations struggled throughout the 1980s to come to terms with the emergence of the environment as a mainstream electoral concern. By 1990, however, it was widely recognized among business leaders that, through the vehicle of sustainable development, there is "no reason why we can't make the environmental issue *our* issue."[32] In particular, the reactionary, anti-environmental "Wise-Use" movement that has been gaining ground in the USA since the late 1980s has begun "flocking to the banner of sustainable development" raising the possibility that it may be used to mask "a new and expanded environmental backlash."[33] This resurgence of capitalist fundamentalism has, of course, been given great momentum by the collapse of state socialism in Europe. The anxieties of the Cold War have been largely superseded by a triumphal optimism—even on the part of those communist regimes that remain, such as China—in the blessings of global trade liberalization.

The shift from the dire warnings of the 1972 report of the Club of Rome toward the optimism of the second wave—what Gregg Easterbrook heralds as the "coming age of environmental optimism"[34]—is clearly displayed in the Club of Rome's 1995 report: "Sustainable development is a process which necessarily involves all sectors of society, including those not yet born. The early phase—when key actors in this process felt they faced 'no win' . . . or 'zero sum' . . . outcomes—has been replaced by a phase in which key actors . . . have been looking for 'win-win' outcomes."[35] The 1997 report to the Club of Rome, *Factor Four*, provides much of the technical detail behind this optimism, claiming that "resource productivity can—and should—grow fourfold. . . . Thus we can live twice as well—yet use half as much." This is, the authors contend, a "novel and simple" message that "heralds nothing less than a new direction for technological 'progress'."[36]

Interpreted in cultural rather than in narrow engineering terms, this message is not novel; nor does it establish a new direction for technological progress, or so I seek to convince the reader in coming chapters. Promethean environmentalism has gathered the support of mainstream social institutions because it expounds a technocratic understanding of social development. In chapters 2 and 3, I offer political and metaphysical critiques, respectively, of the new ecomodernist optimism. But first, in the remainder of this chapter, I make clear what it is in institutional, practical, and ethical terms that I seek to contest.

II A GLOBAL LANGUAGE

The Brundtland Commission broadcast the appealing message that the imperatives of economic growth and environmental conservation converge. We heard that sustainable economic growth is the solution to endemic poverty in the South and more efficient resource management the key to the long-term well-being of the planet. The resultant strategy of sustainable development was presented as having six strategic aims:

1. overcoming poverty through sustained growth in the global economy
2. conservation and enhancement of the global resource base
3. ensuring a sustainable level of population
4. implementation of a precautionary principle in risk assessment
5. integration of environment and economics in policy making
6. strengthening the institutional basis of global governance[37]

The conceptual concern that weaves these aims together is the laudable one of 'equity'. Most of the principles ascribed to sustainable development arise out of an interest in intergenerational or intragenerational equity. So, for instance, the adoption of a precautionary approach to development, which prescribes that technological developments must proceed only where future, long-term impacts are sufficiently known so as to be judged to fall within acceptable limits, becomes unavoidable once a prior commitment to the equal rights of future generations is established.

On the basis of its strategic aims, *Our Common Future* effectively identified four sites in development strategies in need of urgent reform, two practical and two conceptual. On the practical side, there is a need for (1) international political, economic, and legal institutions capable of implementing strategies for realizing intergenerational and intragenerational equity and (2) industry to be founded on more environmentally efficient technologies. On the conceptual side, there is the need for (3) an extrapolation of neoclassical economic theory to incorporate the environment and (4) a vast increase in basic scientific knowledge about the functioning of global environment. I shall address the first of these areas of reform below, moving on to consider the mutually supporting relationships between the remaining three in the next section.

The Role of the United Nations

In line with the perceived need for transnational institutional reform, the UN General Assembly (UNGA) responded to the tabling of *Our Common Future* by initiating preparations for a UN Conference on Environment and Development (UNCED). Dubbed the Earth Summit, this meeting was held in Rio de Janeiro in June 1992, marking the twentieth anniversary of the Stockholm conference and a period during which environmental concerns moved from the margins to the centre of international politics.[38] The ten-day Earth Summit attracted the largest gathering of heads-of-state up to that point.[39] It was preceded over a two-year period by four, month-long preparatory conferences at which most of the detailed negotiations took place.[40] These negotiations finally yielded two global conventions (on Biodiversity and Climate Change), a statement of Forest Principles, a statement of the principles underlying sustainable development, and an 800-page action plan, *Agenda 21*, for the implementation of sustainable development into the twenty-first century.

The institutional momentum generated by the Earth Summit was reflected in the creation of the UN Commission on Sustainable Development (CSD) in 1993.[41] The CSD took the view that Agenda 21 was authorized to be "the principal framework for coordination of relevant activities within the United Nations system."[42] The CSD faces an awesome task in achieving this coordination. As would be expected, UNEP is closely involved in sustainable development initiatives. Yet its long-time rival, the UN Development Program (UNDP), has lost little of its extensive influence within the UN system and also oversees many sustainable development programs.[43] The greatest problem faced by the CSD, however, is not so much that of refereeing internecine disputes but that of securing the financial commitment of member states necessary to implement *Agenda 21* programs. For the principal role of the CSD is to oversee the financing of *Agenda 21* through the Global Environment Facility (GEF).[44]

In addition to the work of the CSD, a number of recent UN conferences have picked up on sustainable development themes, while the report of the Commission on Global Governance, *Our Global Neighbourhood* (1995), offers an upbeat sequel to the vision of global cooperation articulated by the Brundtland Commission.[45] This report makes the bold claim that the world community is "uniting around the idea that it should assume greater collective responsibility in a wide range of areas, including security, sustainable development, the promotion of democracy, equity and human rights, and humanitarian action."[46]

Progress in the implementation of *Agenda 21*, and thus the performance of the CSD, was reviewed by a special session of the UNGA in June 1997. This meeting, prosaically termed Earth Summit+5, was confronted with clear evidence that not only was *Agenda 21* not being implemented, but that ecological and social "maldevelopment" had worsened considerably since 1992.[47] The president of the UNGA, Razali Ismail, declared the postsummit period a tragedy: "Five years on from Rio, we face a major recession; not economic, but a recession of spirit, a recession of the very political will that is essential for catalyzing real change. The visionary ambition of Agenda 21 is tempered by some damning statistics which show that we are heading further away from, and not towards, sustainable development."[48]

As evidence of the lack of political will invested in sustainable development policies, several participants noted that financial development assistance, rather than increasing toward the target of 0.7 percent of Gross Domestic Product, had declined from 0.35 to 0.27 percent in the period 1992–1997, to reach its lowest level since 1970.[49] The sad result of this is that the GEF "is pitifully small, at $US2 billion, in proportion to the $US125 billion [annually] that is needed."[50] According to radical Southern critics such as Martin Khor, this miserable state of affairs reflects the way that "the liberalization free-market paradigm . . . has undermined the sustainable development partnership paradigm."[51]

It is certainly true that the central institutions of global economic liberalization have readily adopted the language of sustainable development. Indeed, the Bretton Woods institutions—the General Agreement on Tariffs and Trade (GATT), International Monetary Fund (IMF) and World Bank—are crucial to the further evolution of this language.[52] It is noteworthy, therefore, that the GATT's 1995 Marrakesh Agreement, establishing the World Trade Organization (WTO), adopts the idea of sustainable development. This document asserts that although international relations "should be conducted with a view to raising the standards of living, ensuring full employment and a large and steadily growing volume of real income and effective demand, and expanding the production of trade in goods and services," it should do so "while allowing for the optimal use of the world's resources in accordance with the objective of sustainable development."[53] What this means, the 1996 joint meeting of the CSD and the WTO made plain, is that these organizations see no inherent policy contradiction between sustainable development and an open, expanding, and deregulated trading system.[54]

Equally, the World Bank has apparently been undergoing a slow and painful process of restructuring since the publication of the Brundtland Report with the aim of becoming more "environmentally aware."[55] In its 1992 *World Development Report*, the bank rhetorically posed the question, "What is sustainable?", responding, with disarmingly circular logic, that "[s]ustainable development is development that lasts."[56] Sustainable development, strategic environmental assessment and environmental economics are now everyday terms within the World Bank, although many insist there have been, to date, only minor changes in the bank's lending policies.[57] Disturbingly, the Bank's 1999/2000 *World Development Report* avoids mention of sustainable development in any titles or subtitles, using it only peripherally in the text. The entry into the twenty-first century may well prove, with hindsight, to have been a convenient point at which to switch rhetorical strategies, the ascendant rhetoric in the World Bank seeming to be the abstruse one of development for "globalization and localization."[58]

NATIONS UNITED?

While nation states have largely failed to commit financially to the sustainable development programs of the UN, they have been eager to parade their domestic policy commitment to sustainability. A majority of nation states now have national commissions for implementing *Agenda 21*, although it is unclear whether this reflects genuinely new initiatives or merely a repackaging of existing policies in the glossy rhetoric of sustainability.[59]

We can take Australia's domestic approach, the National Strategy for Ecologically Sustainable Development (NS-ESD), as a case in point.[60] Initiated in 1990 by a federal government discussion paper, the NS-ESD was ratified by federal and state governments in 1992 and was based on the extensive deliberations of nine working groups drawn from federal and state bureaucracy, unions, academia, conservation groups, and industry.[61] This consultative process was an apparent acknowledgment of the UN's claim that one of the "fundamental prerequisites for the achievement of sustainable development is broad public participation in decision-making."[62] One observer of the working group process enthused that "the degree of consensus [achieved] was remarkable."[63] Nonetheless, the working groups and the subsequent strategy were strongly criticized by many as being unnecessarily bureaucratic and tokenistic.[64] Environmentalists were particularly critical. They argued that the resultant strategy offers, amongst other things, vacillating and contradictory policy recommendations, abrogates Australia's global responsibilities, and lacks a genuinely participatory decision-making process.[65] Conversely, business groups and unions cautioned, successfully as it turned out, against legally binding interpretations of ecologically sustainable development on the basis that these would undermine Australia's competitiveness in the global marketplace.[66]

In the end result, the NS-ESD is regarded by many on both sides of this debate as failing to produce any concrete results beyond an increased sophistication in political rhetoric on the environment. As with *Agenda 21*, momentum generated by the preparation of the NS-ESD has flagged since its release, while decision-making responsibility for sustainable development has been devolved informally throughout federal and state bureaucracies, making its implementation practically impossible to assess. Significantly, the Australian Government's position at the Kyoto International Summit on Climate Change in December 1997 avoided any reference to the NS-ESD and emphasised, above all, the need for a sustainable domestic coal industry.[67] The extraordinary permission granted to Australia to continue to increase its "greenhouse" emissions was unashamedly presented by the Government as a coup. It is difficult, perhaps impossible, to reconcile policy principles such as intragenerational equity and international responsibility within documents such as the NS-ESD.

Undeniably more successful than national approaches to sustainable development have been initiatives that take up *Agenda 21*'s hope that the "participation and cooperation of local authorities will be a determining factor" in fulfilling the objectives of sustainable development.[68] "It can certainly be argued," observes Michael Redclift, "that Agenda 21 has proved most useful as the basis for wide-ranging enquiry, on the part of local authorities themselves, into the best methods of increasing local-level sustainability."[69] The most notable conclusion being drawn by these authorities, to date, seems to be that sustainability demands much greater community participation in the decisions that affect them. Just as with international and national sustainable development strategies, however, we must note that while changes to the political rhetoric of local governments are certain, strategic change in policy is not.[70]

The Global Marketplace

The amorphous rhetoric of UN diplomats stands in stark contrast to the enthusiasm with which business has embraced sustainable development. In the view of the director of the UN Division for Sustainable Development, Joke Waller-Hunter, a "strong partnership between business, government, and international organizations is the foundation on which sustainable development will need to be built. . . . The challenge is for business to take the lead in this fundamental transformation of our economic life."[71] Business is having apparently little difficulty addressing this challenge. But where and how, exactly, is it that business wishes to lead us? Echoing the Bretton Woods institutions, business leaders tell us that "the cornerstone of sustainable development is a system of open, competitive markets in which prices are made to reflect the costs of environmental as well as other resources."[72] The key to such free markets is, apparently, that they allow the "best aspects of the human propensity to buy, sell, and produce" to "be an engine of change."[73]

If they did not overtly seek the media spotlight at the Earth Summit, business interests, in the form of the (World) Business Council for Sustainable Development, certainly shaped its outcomes. The Council's figurehead, Swiss billionaire Stephen Schmidheiny, was appointed as personal adviser to the Earth Summit's Secretary-General Maurice Strong, himself a wealthy businessman. It came as no surprise when Strong obligingly described this group as "a cadre of the world's leading practitioners of sustainable development."[74] Backing up this claim, the International Chamber of Commerce launched its Rotterdam Charter for Sustainable Development in 1991, providing sixteen principles for "policies to reconcile free market economics with sustainable development."[75] And what is the nature of this reconciliation? As I display in section III, the answer to this question is being worked out in the ecomodernist disciplines of environmental economics, environmental science, and environmental technology.

III A TECHNO-ECONOMIC OPTIMISM

The reconciliation of free-market economics and sustainable development, for which the UN hopes, depends on the capacity of economic policy to direct wealth generation into more environmentally efficient technology. Thus, just as sustainable development is promoted as being contingent on sustained economic growth, it is also promoted as requiring sustained technological proliferation. In chapter 3, I explore the Enlightenment traditions that so tightly join ideas of technical efficiency to ideas of human flourishing, thereby shaping broad notions of social progress into narrow policies for technological progress. To ensure that this exploration can work back from the reality of contemporary techno-economic structures, I want to display here the unfettered technological optimism that is at the core of the economics of ecomodernization.

THE ECONOMIC LEXICON OF SUSTAINED GROWTH

The primary aim of environmental economics is to internalize the environment within neoclassical models of economic growth. British economists David Pearce, Anil Markandya, Edward Barbier, and colleagues have now published five influential "blueprints" for the "green" neoclassical economy.[76] These authors claim that the key principles guiding the economics of sustainable development are futurity, equity and environment.[77] Futurity recognizes the importance of long-term concern; equity refers to the need for both intergenerational and intragenerational equity; environment calls for "increased emphasis on the value of natural, built, and cultural environments."[78]

Environmental economists reject the first wave of environmentalism. In Pearce's view, "[a]dvocates of lifestyle change involving wholesale rejection of economic growth have been losing ground for some considerable time. Fashionable in the early 1970s, it is not heard so much in the 1990s, a part reflection of its noncredibility."[79] This rejection of economic growth lacks credibility, apparently because sustainability can be achieved simply by maintaining (even increasing) the sum total of natural capital (that is, resources and life-support processes) and human capital (that is, technology, skills, and information).

The growth of financial capital combined with a nondeclining capital stock produces sustainable development, if this is broadly defined as "continuously rising, or at least nondeclining, consumption per capita, or GNP, or whatever the agreed indicator of development is."[80] Defined in this way as nondeclining welfare, sustainable development is not new to neoclassical theory.[81] The first UN Development Decade was, after all, launched in 1961 with the claim that development was synonymous with accelerating "progress toward *self-sustaining growth* of the economy," a claim developed in detail by the influential American economist Walt Rostow in the 1950s.[82]

Pearce and his colleagues contend that weak forms of sustainability are produced when no distinction is made between "natural capital" and the rest of the capital stock. In weak approaches, so long as capital stocks are constant, it is unimportant whether natural capital such as old-growth native forests are substituted with a corresponding value of human-generated capital such as genetically enhanced plantation trees. In contrast, a strong account of sustainability is generated if this distinction is made, thus requiring the maintenance of a constant "stock" of natural capital such as old-growth native forest.

Environmental economists profess to having an open mind about the relative merits of these two conflicting approaches.[83] Pearce admits that, in theory, the principle of futurity combined with the irreversibility and unpredictability of much anthropogenic ecological change should push the balance in favor of strong sustainability.[84] In practice, however, environmental economists routinely adopt the weak approach of substituting natural capital with human capital. Indeed, to a great many conventional economists, even weak approaches requiring that natural capital be given a monetary value at all are regressive and unnecessary.[85] With the peculiar otherworldliness of orthodox economic theory, these economists hold onto the illusion that natural resources are so abundant that losses in natural capital are statistically negligible.

In many other respects, however, environmental economics converges with neoclassical orthodoxy in its use of instruments such as Cost-Benefit Analysis and market-based incentives such as the Polluter Pays and User Pays principles.[86] Like conventional economists, Pearce and his colleagues support market mechanisms over government intervention and regulation. They champion global free trade. Pearce claims that "sustainable development requires the phasing out of selective tariff barriers and/or their replacement with low but relatively uniform ones."[87] "[T]he policy implication of a negative association between freer trade and environmental degradation is not," says Pearce, "that freer trade should be halted."[88] Rather, the implication is that domestic policies must fully internalise the environment within the costs of their goods and services. In this view, the trade produced by sustainable development takes place on an international playing field levelled by stripping all domestic markets of price manipulation, including that manipulation produced by the failure to reflect the true and full environmental costs of products.

The claim that internalizing the environment will ensure that market decisions will direct technological change toward more ecologically efficient paths is central to environmental economics. To paraphrase the corporate leader Schmidheiny, on the path to "ecoefficiency," profitability and environmental cleanliness will go hand-in-hand.[89] The Director of UNEP has correspondingly claimed that there is "a consensus on the need for an increase of a factor of 5 to 10 in eco-efficiency, if the goal of sustainable development is to be achieved."[90] This increase in ecoefficiency is of course attractive to the UN for more than reducing the pressure on scarce resources. The increased profitability it promises, and the microeconomic focus it offers, appears to be a much more realistic strategy for funding *Agenda 21* than does the prospect that national governments will match impressive rhetoric about global responsibility with genuine reform of the macroeconomic structures of the global economy.

Technological Optimism

The Brundtland Commission cemented into the foundations of sustainable development policy the conviction that technology is the neutral instrument of social institutions. *Our Common Future* recognized that the major obstacles to the goal of sustainable development are imposed by the present state of "technology and social organization."[91] We read, however, that with "careful management new and emerging technologies offer enormous opportunities for raising productivity and living standards, for improving health, and for conserving the natural resource base."[92] The ultimate cause of unsustainability according to this Commission is a passive, if devastating, carelessness. Indeed, as far back as 1968, the emerging discourse on sustainability has produced the observation that "human societies must begin learning how to halt the historically careless application of technology."[93]

The Brundtland Commission's call for careful management of technology is consonant with the view put forward by neoclassical economists, both conventional and environmental, that the function of policy is to deploy economic instruments so as to guide and focus technological innovation. The description of "wayward technology" offered by the technology policy theorist Ernest Braun is typical of this approach:

> Technology is tossed and buffeted by many forces—internal and social—and as a result is a wayward creature of human ingenuity. . . . Wayward creature that technology may be, it is yet amenable to conscious social policy measures as these constitute one of its many influences. The aim of policy is not perhaps to tame the beast, but to curb its excesses and guide its power to where society most needs it.[94]

The implicit message here is that if we are careless in handling a beast as powerful as technology, the consequences are likely to be severe. Conversely, it is suggested that with careful management this wayward "pet" can be tamed and reared to maturity by its "masters."[95]

Through its capacity to direct technology, under the whip of careful policy, the technological society is viewed by its mainstream institutions as being capable of self-correcting the aberration of unsustainability. This is a view on full display in the World Resources Institute's 1991 vision for transforming technology:

> Technological change has contributed most to the expansion of wealth and productivity. Properly channelled, it could hold the key to environmental sus-

> tainability as well. . . . Today the climate for innovation seems uniquely rich. . . . If environmental goals are integrated into these innovations, the transition to a sustainable future will happen faster, cost less, and have longer-lasting results.[96]

In similar although less breathless fashion, the Brundtland Report advocated a "reorientation of technology" built upon a "deeper" and "better" scientific understanding of natural systems.[97] And what, according to this report, are these better technologies likely to be?

> Information technology based chiefly on advances in micro-electronics and computer science is of particular importance. Coupled with rapidly advancing means of communication, it can help improve the productivity, energy and resource efficiency, and organizational structure of industry. . . . The products of genetic engineering could dramatically improve human and animal health. . . . Advances in space technology . . . also hold promise for the Third World.[98]

Exactly what makes these technologies new or reoriented in relation to the history of unsustainable technological development is not clear. The Brundtland Commission would seem simply to be describing the existing trajectories in latemodern society that are likely to develop with or without an overt ecological rationale. To confuse matters further, the Commission itself notes that these "new" technologies "are not intrinsically benign, nor will they have only positive impacts on the environment."[99] Yet, other than offering the orthodox proviso that new technology be a "mainspring of economic growth," we are handed no criteria by which to understand or assess the intrinsic character of technologies.[100]

The Brundtland Commission has prepared the way for win-win ecomodernists like Easterbrook who dream of a complete technological restructuring of the biosphere. Easterbrook laments nature's "structural flaws and physical limitations" and inverts the conventional environmentalist maxim that human activity poses a threat to the planet, arguing that "nature needs us—perhaps, needs us desperately."[101] Extrapolating from current trajectories in high technology, mainly in gene technology, he asks us to imagine with him a New Nature in which technological ingenuity has ended no less than predation, war, extinctions, disease, aging, collisions between earth and celestial bodies, radiant "waste" of the sun's output, and even human mortality itself.

The technological optimism running through *Our Common Future* gained further momentum through *Agenda 21.* Pratap Chatterjee and Matthias Finger propose that it is "probably not exaggerating to say that technology is the biggest hope that emerges from UNCED in general and Agenda 21 in particular."[102] And hopefulness rather than clarity indeed characterizes the understanding of technology that emerged in Rio. The word *technology* appears only once in its forty chapter titles, while the text makes only a few fleeting references to the concept of 'technology' itself. Environmentally sound technologies were defined in "motherhood" style as those that "protect the environment, are less polluting, use all resources in a more sustainable manner, recycle more of their wastes and products and handle residual wastes in a more acceptable manner than the technology for which they are substitutes."[103] But, after all, who is going to advocate more polluting and less acceptable technologies?

During the cut and thrust of preparatory negotiations, matters were blurred further. One bureaucrat resorted to the observation that *environmentally sound technology* is a term that "depends on the technical and economic conditions of a specific country," with "'soundness' defined on a case-by-case basis," whilst another document suggested that "there is no universally agreed upon concept of environmentally sound technology. The environmental soundness of technology is a relative rather than an absolute term."[104]

With the proviso that this term expresses no tangible commitments, *Agenda 21* did however acknowledge that the notion of environmentally sound technology cannot be uncoupled from the need to alleviate poverty and human suffering.[105] Nonetheless, in the end, this action plan ceded responsibility for making connections between technology and equity to the motive of ecoefficiency within the global marketplace.

Ecoefficiency

So what exactly is ecoefficiency? According to the World Business Council on Sustainable Development, it is "the delivery of competitively priced goods and services that satisfy human needs and bring quality of life while progressively reducing ecological impacts and resource intensity, through the life-cycle, to a level at least in line with the Earth's estimated carrying capacity."[106] This tortuous definition at least serves to remind us of the ways in which modernism fuses ideas of economic progress with those of technological efficiency. Examples of this fusion abound in the sustainable development literature. There is, for instance, the World Resources Institute's claim that sustainable development "will be possible only through a transformation of technology . . . to new technologies that dramatically reduce environmental impact per unit of prosperity."[107] Similarly, the Council of Academies of Engineering and Technological Sciences enthuses that while there are "many obstacles to the transition to sustainable development, technology provides a means to overcome them."[108]

This urgent technological optimism undergirds the Promethean hope that "we can live twice as well yet use half as much." The reluctance of Northern leaders to make binding commitments to reduce resource consumption at the Earth Summit is being replaced by growing enthusiasm as these leaders become convinced that advances in ecoefficiency make the maintenance of economic growth and reduced consumption compatible.[109] The ecoefficiency or "cradle-to-grave" product management strategies currently being put in place include:

- reducing the amount of material and energy inputs for production
- replacing hazardous or polluting materials and fuels with less damaging alternatives
- utilising waste from one stage of manufacturing as raw materials for another
- producing products that require less energy during use, recycling, and disposal
- designing products that can be reused or upgraded and easily recycled or biodegraded
- implementing product stewardship for the whole product life cycle
- shifting from providing products to supplying services[110]

Technologies and technosystems (that is, technological systems) that make possible gains in ecoefficiency are referred to generically as "environmental technology."[111] Just as there is an important distinction to be drawn between environmental (reformist) and ecological (radical) economic theory, a matter I take up in chapter 3, I reserve the term *ecological technology* and related terms, such as *ecotechnology*, *ecotechnics*, *ecological engineering* and *ecological design*, to refer to practical alternatives to sustainable development that directly relate reform of technology to political and moral reform of the technological society.[112] The cultural critiques that inform these alternatives are explored in political terms (especially in relation to the South) in chapter 2, and in more ethical terms in chapter 6.

Environmental Technology: Three Ecomodernist Trajectories

Although international bureaucrats are deliberately obtuse in their descriptions of what constitutes an environmentally sound technology, the ecomodernist disciplines of environmental science and environmental engineering have been much more forthcoming in their attempts to design and produce more ecoefficient technologies. According to the environmental engineer Jesse Ausubel, there are three major technological trajectories within environmental technology: industrial ecology, decarbonization, and dematerialization.[113] Of these, industrial ecology appears to be the primary vector of change, defining the other two as it is concerned with the overall design of modern technosystems. As Robert Socolow explains it, industrial ecology "explores reconfigurations of industrial activity in response to knowledge of environmental consequences."[114] It represents, according to its practitioners, a "New Paradigm" of technology in which "industrial processes . . . interact with each other and live off each other, economically and also in the sense of direct use of each other's material and energy wastes."[115]

Industrial ecology and the related notion of "industrial metabolism" came to prominence only during the 1990s.[116] They describe, apparently, the "way

21st century engineering will be done—the only way."[117] The attempt to draw analogies between ecological and industrial systems reflects the hope that industrial systems can mimic the interconnectedness characteristic of ecological systems and establishes the claim that industrial systems, like ecological systems, are evolving.[118] Past industrial practices are thus represented as the vulnerable and unavoidable early stages in the evolution of diverse and sustainable "climax" industrial ecologies. "A crucial goal of evolutionary (as opposed to revolutionary) industrial ecology," says one industrial ecologist, "is to develop a sustained dynamic of technological improvement."[119] And the nature of this dynamic? It arises, so it seems, out of the search for "superior efficiency and productivity in serving human needs with reduced environmental impact, reduced consumption of raw materials, and inclusion of wastes and products at the end of their lives in the industrial food web as both material and energy."[120] This dynamic thus gives rise to trajectories of decarbonization and dematerialization.[121]

Decarbonization refers to decreases in technological dependence on the nonrenewable, finite, and diminishing stock of fossil fuels. As an indicator of this trajectory, Gross National Product in Northern countries increased by fifty percent from the mid-1970s through the 1980s, while total energy consumption increased by only fourteen percent.[122] This relative reduction (though still actual increase) has been achieved through the overlapping strategies of increasing efficiency of use, that is, through conserving more power and utilizing power more effectively for each unit of carbon emitted, and through reducing overall carbon production, that is, through phasing out of the most inefficient fossils fuels through to the use of alternative energy sources, as well through reductions in energy demand. Better forms of electricity grid insulation and design, use of 'waste heat' from energy systems, more energy-efficient appliances and thermal design in buildings are examples of the former. Switching from coal and oil to "cleaner coal" and natural gas, as well as nuclear, hydroelectric, photovoltaic, geothermal, wind, wave, biomass, and other renewable energy systems are examples of the latter. Nuclear power, for instance, is now being promoted as an environmentally sound response to climate change, with the Club of Rome reported to have "evolved from nuclear critic to nuclear promoter."[123]

While, strictly speaking, decarbonization could refer to a radical agenda of dramatic reductions in energy demand, as well as a move to decentralised and locally-managed renewable generation systems, ecomodernism advocates decarbonization based on very technically complex systems of "large spatial scale, centralization, and management by elites."[124] So ecomodernism selects coal cleaning technology over education programs encouraging less reliance on frivolous domestic appliances. It selects electric cars over bicycle-centerd reform of cities. It selects nuclear power over backyard solar ovens. The end point of this trajectory is a "hydrogen economy" in which, Ausubel tells us, the "transition to natural gas and hydrogen can be quite painless. . . . Oil companies can benefit as much from natural gas as oil, and in fact several of the 'majors' have already redefined their business toward . . . natural gas and hydrogen."[125]

The trajectory of dematerialization refers to a decoupling of materials and affluence: that is, the decline in size, weight, and overall resource demands of industrial products. As shaped by ecomodernism, in contrast to possible broader interpretations, this trajectory refers to attempts to reduce the use of resources while maintaining or increasing productivity and efficiency. Examples of this trajectory include the trend toward miniaturization of computers, fiber-optic cables that are one-fortieth of the weight of copper cables, and a human heart micropump implant. The present transition from an Industrial Age to an Information Age, whereby intensive resource use is giving way to an intensification in the production and movement of information, is also accelerating this trajectory. And lest information be seen as too intangible a product, some in the World Bank contend that information technology "can be deployed even in regions that lack adequate water, food, and power. . . . In fact, this technology is often indispensable for meeting basic human needs."[126]

Ecomodernist trajectories do not thus require an overall reduction in resource use. Indeed, this is unlikely if the market base of consumers and per capita consumer demand continue to increase, as it seems they must if neoclassical economic growth is to be sustained. An evolutionary theory of industry is consistent with an agenda for incremental rather than fundamental change, and industrial ecology proposes design changes to existing industrial processes, rather than calling for alternative processes *per se*. For example, cradle-to-grave product management is careful not to provide a critique of the social character of products, or of their consumption. We need not be surprised then when we find that, rather than responding to the ecological and social injustices of car dependency, ecomodernists celebrate the life cycle approach of BMW, whose ZI sports car has an all-plastic skin of "recyclable thermoplastic" that can be completely disassembled in twenty minutes, allowing it to be "shed" at regular intervals as accidents, maintenance, or aesthetics dictate.[127]

Discussion of the trajectories of environmental technology appears skewed toward the concerns of industrial production and away from concerns about natural production, especially the production of food and timber.[128] This can be explained, however, once we recognize that at the heart of the ecomodernist enterprise is an attempt to subsume forms of natural production within ecoindustrial systems. "Agriculture has become virtually an 'industry' in developed countries," asserted the Brundtland Commission, an observation even more telling now, fifteen years on, with the mushrooming of biotechnology industries.[129] That is, the trajectories of ecoindustry will affect agriculture, fisheries, and forestry in the production of everything from fertilizer, herbicides and pesticides; agricultural and aquacultural pharmaceuticals; heavy machinery;

irrigation, refrigeration, filtration, sterilization, and navigation systems; food processing processes; to transport and marketing systems. Those aspects of natural production not subsumed within the hardware of ecoindustrial production are relegated to the software of resource management and are often swamped by the latest attention-grabbing, skin-shedding toy. Yet even this software depends upon the mapping, monitoring, and measuring of the "Earth system" performed by the latest astounding hardware of planetary science, such as satellite, remote-sensing, remote-robotic, and simulation devices.

Business Aspirations in the Age of Ecoefficiency

Sustainable development is spreading the message that environmental technology is good for business. Green management is the "next competitive weapon."[130] Consumers are making their green preferences felt. Companies seem to be willingly taking up the burden of "manufacturing a sustainable future," as a Unilever executive recently put it.[131] As soon as the clumsiness of today's technologies "penetrates the consciousness of a significant number of captains of industry and politicians," declares Ernst von Weizsäcker, "the race will be on to overcome the dinosaur technologies and to gain the lead in creating the efficiency revolution in the use of resources."[132]

The vast majority of environmental technology development is taking place in the North. Inevitably, the objectives of Northern governments and corporations involve the South becoming significant consumers of these technologies. In this vein, Marcello Colitti, Chairman of the Chemical Company of the European ENI Group, suggests that sustainable development can help overcome "economic stagnation" in the North, "mainly due to the market saturation resulting from the decline of consumer demand."[133] That is to say, it is difficult create new markets within static populations composed of the "three-car, two-house, five-television-set and countless-gadget family," no matter how capital rich they may be.[134] The only viable strategy for generating continued economic growth that Colitti can see is that of "enlarging the market and creating sustainable development in non-industrialized countries, where demand is far from saturation."[135]

Of course, many advocates of Northern capitalist interests make this claim without bothering to resort to the label of sustainable development. Take *Business Week* magazine's gleeful celebration, in the wake of Eastern European socialism's collapse, that "capitalism is flourishing . . . affluence is lifting millions out of poverty, giving many the chance to purchase their first Fiats and Toyotas as well as their first Apple computers and Panasonic VCRs."[136] We should be delighted, apparently, that the newly affluent will push annual car production from forty-four to sixty-one million units, while annual international air travel will treble between 1995 and 2010. Fortunately, most Northern governments are more circumspect in their celebration of consumerism, conceding that such trends meet neither the challenge of intragenerational or intergenerational equity. Nonetheless, their desire to take full advantage of the expanding markets of the South has not weakened in the rhetoric of sustainable development. Consider the U.S. Government's 1994 report *Technologies for a Sustainable Future*: "Today, environmental technology offers a win-win opportunity for our nation and the world as a whole: economic growth through the development and diffusion of environmental technology will result in more jobs, and a clean environment will mean a higher standard of living for ourselves and the generations that follow." The nature of the win-win opportunity for the rest of the world becomes clearer as this report continues: "Being the world leader in clean technologies and processes will make our companies more competitive, increase our exports, and help us lead the world toward sustainable development."[137] To some, including physics Nobel laureate Abdus Salam, it is just this kind of Promethean leadership that the "poor and socially backward" South needs, "because the economies of its countries are not knowledge-based and cannot be in step with the change that comes as a corollary to rapid development in science and technology."[138]

Claims about the cultural need of the South to move "forward" dovetail neatly with the claims of U.S. bureaucrats about the need for the U.S. to beat other technologically advanced countries in the rush to make a profit out of environmental problems in poor countries. Nonetheless, in the face of this all-too-obvious motive, says this vice president of the World Bank, the Northern ecoefficiency agenda is essential to the future economic sustainability of the South: "Together, the technology revolution and the economic revolution are producing a completely new world economy that is high-speed, knowledge-driven, global, and disciplinarian. For 4 billion of the world's people it could be the birth of a Golden Age of sorts; for the first time they will have a serious opportunity to catch up with the rest of the world."[139] And so we are taken full circle, back to the plaudit with which Harry Truman launched the era of development in 1949: "We must embark on a bold new program for making the benefits of our scientific advances and industrial progress available for the improvement and growth of underdeveloped areas. . . . The old imperialism—exploitation for foreign profit—has no place in our plans. . . . Greater production is the key to prosperity and peace."[140]

IV A GLOBAL ETHIC?

The ecomodernist pursuit of sustained growth in the global marketplace gives the language of sustainable development a hard, sharp edge. This edge is wrapped, however, in the softer, alluring vision that sustainable development is gathering humanity together as one cooperative, caring community. Despite the weight of their technical arguments, ecomodernists thus resort to normative claims to push home their case for an environment-led revitalization of economic activity.

Moral burdens were placed on the language of sustainable development as early as the UN's 1972 Stockholm Declaration.[141] By the time of the 1980 IUCN meeting, this vision was firmly linked with "a new ethic embracing plants and animals as well as people."[142] Only two years later, the UN's *World Charter for Nature* stated that "every form of life is unique, warranting respect regardless of its worth to man."[143] At the turn of the millennium, the ethic of sustainable development is a standard component of policy documents and is formulated in a great many ways.[144] It seems reasonable, then, to agree with the Executive Director of UNEP, Elizabeth Dowdeswell, that the "question of human duty, of our responsibility to others, of morality has been fundamental to the concept of sustainable development." She goes on to give powerful expression to her view of what this morality entails:

> We have to ask ourselves simple but fundamental questions: How should we live? How much is enough? What way of life human beings ought to pursue? . . .
>
> We have to develop the ecological, holistic world view which connects us with our environment and other people and species—both materially and spiritually. Religious traditions emphasize this connection. Our task should be to retrieve these basic symbols and doctrines within each tradition and translate them into clear prescription for public policy and behaviour.[145]

The remainder of this section inquires into the ways in which these fundamental moral and spiritual questions are being pursued in the language of sustainable development.

The Gloss of Global Ethics

The first page of the Brundtland Report makes the enticing claim that "our cultural and spiritual heritages can reinforce our economic interests and survival imperatives."[146] Given this beginning, it is surprising that the Brundtland Report's proclamations actually make little mention of ethics. Only in the final chapter do we learn that "[w]e have tried to show how human survival and well-being could depend on success in elevating sustainable development to a global ethic."[147] If this is indeed what the Commission tried to show, they did so without engaging in anything resembling moral discourse.

Yet, certainly, *Our Common Future* implicitly builds its vision of our common future upon normative foundations. We read that efforts to "combat international poverty, to maintain peace and enhance security worldwide, and to manage the global commons," depend not only on techno-economic solutions but also on "greater willingness and co-operation."[148] And what is the motivating force behind this greater willingness? It is apparently a concern for equity: "Even the narrow notion of physical sustainability implies a concern for social equity between generations, a concern that must logically be extended to equity within each generation."[149] Many if not all of the ethical questions raised by this claim remain unaddressed in *Our Common Future*, including those questions that transcend the commission's strongly anthropocentric perspective and those that ask who, and on what basis, is in a position to decide what marginalized communities and future generations need.

Perhaps in response to the general neglect of ethical questions by the Brundtland Commission, the Earth Summit was charged with the task of producing "a statement of principle as the basis of conduct by nations and peoples in respect of environment and development to ensure the future viability and integrity of the Earth as a hospitable home for human and other forms of life."[150] Testifying to the contested nature of human needs and aspirations, the two years of negotiations that laid the diplomatic foundations for the Earth Summit exposed how much Southern governments saw the North's professed concern about the earth as a subterfuge for their attempts to limit the right of poorer countries to pursue economic development.[151] Comments like the following from a German negotiator give substance to the concerns of the South: "mankind has not a right to exploit nature and natural resources to the extent we [have been] doing . . . [for] the last 100 years. . . . Aesthetics and beauty of life are dependent on a variety of landscapes, of animals, and so on."[152] Given that the bulk of future resource use will occur in the growing populations of poor countries, and given that affluent countries have the wherewithal to become more efficient as a result of the wealth they have derived from the last one hundred years of exploitation, this reference to the rights of "mankind" is clearly politically fraught.

Not surprisingly, the resultant statement of principle, the *Rio Declaration on Environment and Development*, lacks force and clarity. Horse trading between North and South ensured that moral concerns about development and moral concerns about the environment were not skilfully balanced, as the British negotiator Tony Brenton claims, but rather were simply listed sequentially.[153]

Environmental economics only adds to the tentativeness, banality, and general lack of reflection surrounding the moral consideration of human needs in development discourses. Environmental economists effectively seek only minor modifications to the universalizing and utilitarian ethics that inform neoclassical economics. In their *Ethical Foundations of Sustainable Economic Development* (1990), environmental economists Pearce and Turner present sustainable development as a catalyst at the "interface between environmental economics, human ecology, and ethics."[154] Persisting with the highly technocratic tone of this definition, these authors describe "development" as a vector defined by movement toward the following: increases in real income per capita; improvements in health and nutritional status; educational achievement; access to resources; fairer distribution of income; and increases in basic freedoms. Characteristically, these authors fail to acknowledge, first, that the objectives of development are open to incommensurable differences in ethical definition and in moral practice, and, second, that the idea of a global moral community is both ambiguous and contestable.

Techno-Ethical Optimism and Global Civil Society

A universal ethical understanding of social and ecological well-being is inextricably joined to a liberal democratic vision of global civil society in sustainable development discourses. To borrow from Michael Walzer, "global civil society" names the "space of uncoerced human association and also the set of relational networks" that result from the rapidly emerging global interconnection of local economic and technological activity.[155] A crucial ingredient of the techno-ethical optimism fostered by the idea of sustainable development, then, is that it is able to draw upon recent grassroots interest in creating a global mosaic of local civic alliances. But rather than appreciating the defining paradox that "a global civil society is rooted in a highly particularistic Nature and place," this agenda simply seeks to soften the harshness of global technocratic managerialism behind a mask of civility.[156]

We are asked to accept, therefore, that the pursuit of ecoefficiency is coincident with the creation of a global community bound together not simply by technical interconnections but also by an ethical commitment to manage its common home equitably for present human generations and for those to come. It is only by taking up the pursuit of growth, so this techno-ethical argument runs, that a one-world community will overcome poverty, inequity, and ecological crisis, entering into the Golden Age of technology. It is an opportunistic vision well summarised by Frank Popoff, Chairman of the Dow Chemical Company: "If we view sustainable development as an opportunity for growth and not as prohibitive, industry can shape a new social and ethical framework for assessing our relationship with our environment and each other."[157] The message here is that corporate entities are best placed to create the regulatory and civic spaces in which we can care for each other and the world around us.

Ecomodernism typically displays this kind of unrestrained and inarticulate faith in the social progressiveness of modern techno-economic change. It is a faith overlaid with fervent good intentions about the convergence of narrow economic interests with wider moral interests. In *Factor Four*, for instance, three leading energy experts provide a review of the ecoefficiency revolution yet conclude with the following startling claim:

> Productive use of the gifts we borrow from the earth and from each other can gain us more time to search. But like any tool, it can only help us a little towards, and can never substitute for, the renewal of our polity, our ethical principles and our spirituality. The resources that we need most urgently to rediscover and to use more fully and wisely are not in the physical world, but remain hidden within each one of us.[158]

This impassioned, and now all-too-predicable, plea to higher-interests is alarmingly adrift amidst the vast sweep of Promethean solutions this book offers.[159]

The moral possibilities of global community claimed by ecomodernists were displayed in some detail by the Commission on Global Governance—comprised of prominent international bureaucrats, technocrats, and politicians—in its call for "broad acceptance of a global civic ethic to guide action within the global neighbourhood."[160] The Commission's report, *Our Global Neighborhood* (1995), considers that this global civic ethic—founded on no less than the core values of respect for life, liberty, justice and equity, mutual respect, caring, and integrity—will bring with it "new ways of perceiving each other as well as new ways of living." Although global society is increasingly determined by the imperatives of the free market, this report holds out the hope that the core values of global civic virtue will "constrain the competitive and self-serving instincts of individuals and groups" and will "provide a foundation for transforming a global neighbourhood based on economic exchange and improved communications into a universal moral community in which people are bound together by more than proximity, interest, or identity."[161]

The development agency of the UN similarly offers the injunction that members of the global community must not be treated as "instrument[s] of production," but must be seen as equal bearers of moral rights under the principle of universality of life claims: "What must be avoided at all costs is seeing human beings as merely the means of production and material prosperity, regarding the latter to be the end of the causal analysis—a strange inversion of means and ends. . . . [I]t is well to remember Immanuel Kant's injunction 'to treat humanity as an end withal, never as means only."[162] Sustainable development, according to the UNDP, recognises the inherent worth of human life and the equal right of each human individual, including those not yet born, to meet basic needs and enhance their quality of life. Thus, this approach can be distinguished from previous development strategies in that it "regards economic growth as a means and not an end."[163] What this boils down to, however, is the claim that there is no inherent conflict between economic growth and environmental protection because it is not necessary that natural capital *per se* be preserved: "What must be preserved is the overall capacity to produce a similar level of well-being—perhaps even with an entirely different stock of capital."[164]

Contesting Prometheus' Promise

The review of policy, techno-economic rhetoric, and techno-ethical optimism presented in this chapter exposes some of the ways in which environmental concerns are being redefined so as to promote an agenda of ecomodernization. And, "[a]t the centre of eco-modernism," reiterates Richard Welford, "we find the search for eco-efficiency. . . . Eco-efficiency must fit within the growth paradigm and actually, it is subtly designed to reinforce it."[165]

Through changes in economic and scientific theory, applied in new forms of technology, and through the institutional transition to a truly global form of the technological society, sustainable development gives voice to the Promethean promise that the earth can be effectively managed as a device capable of ensuring the indefinite survival and moral well-being of humanity. I shall in chapter 2 set about dispelling the technocratic illusion on which hope draws by displaying the ways in which the language of sustainable development coopts, marginalizes, and oppresses cultural discourses and practices of sustainability not defined by the technocratic agenda of ecomodernism.

CHAPTER TWO

POLITICS

Confusion, Cooptation, and Dissipation

INTRODUCTION

THE EMERGENCE OF sustainable development as an international policy language, a techno-economic agenda, and a global ethic creates the impression that the ideal of sustainability has made possible broad agreement about how the challenge of ecological crisis may best be met. Strengthening this impression, there are many, like Ashok Khosla of the International Union for the Conservation of Nature, prepared to suggest that sustainable development "may well go down as one of the major intellectual breakthroughs of the twentieth century."[1] This idea is a breakthrough, according to Khosla, because it introduces into development discourse a concern with "temporal continuity."

In this chapter, I counter such sanguine assessments with the argument that the appearance of either consensus or intellectual clarity in sustainable development discourses is superficial and deceptive. I accept that the essentially contested nature of the ideal of sustainability invests it with great cultural richness and political promise. Yet this promise has been confused, coopted, and steadily dissipated by the ecomodernist agenda. As we saw in chapter 1, the pursuit of ecoefficiency is commonly represented as politically emancipatory and as coincident with a whole variety of morally progressive aims. It is a promise of coincidence deserving of Sharachchandra Lélé's parody: "Sustainable development is a "metafix" that will unite everybody from the profit-minded industrialist and risk minimizing subsistence farmer to the equity-seeking social worker, the pollution-concerned or wildlife-loving First Worlder, the growth-maximising policy maker, the goal-oriented bureaucrat, and therefore, the vote-counting politician."[2]

Business leaders and environmentalists are apparently united in their pursuit of ecoefficiency. The former because they seek increased profitability, the latter because they hope that this technical agenda carries within it a radical agenda of social equity and environmental ethics. Yet the adoption of sustainable development by powerful economic interests does not indicate that ecological awareness has been smuggled into the core deliberations of the technological society. It indicates the exact opposite—namely, that the interests of the technological society have been smuggled into ecological awareness. Langdon Winner, an astute social critic of technology, reminds us that attempts to adopt the idea of efficiency as the basis for radical moral and political change are perhaps always destined to fail:

> Because the idea of efficiency attracts a wide consensus, it is sometimes used as a conceptual Trojan horse by those who have more challenging political agendas they hope to smuggle in. But victories won in this way are in other respects great losses. For they affirm in our words and in our methodologies that there are certain human ends that no longer dare be spoken in public. Lingering in that stuffy Trojan horse too long, even soldiers of virtue eventually suffocate.[3]

The Promethean agenda not only does not invite fundamental political questions, it shields the technological society from these questions by reinventing ecological crisis as a frontier of vast techno-economic opportunity, which we are urged to enter with profound techno-ethical optimism. The ecological crisis has been reinvented as an urgent imperative for the completion of the emancipatory promise of technology—the promise of freeing its 'masters' from material scarcity and moral ignorance. More than anything else, the combination of the imperative of technological development and the ideal of sustainability, with its consequent linking of efficiency with ecology, has led to the emptying of environmental discourses of their cultural content.

In the section that follows, I explore some of the most important ways in which the rhetorical structure of sustainable development coopts thinking about sustainability. Section II employs the 1992 Earth Summit as a case study of the practical political consequences of this cooptation, particularly for those who exist at the margins of global technological culture. Section III lays some of the political and conceptual groundwork necessary for contesting the agenda of sustainable development as one of many cultural possibilities inspired by the ideal of sustainability.

I TECHNOCRATIC SUSTAINABILITY

One commentator has suggested that sustainable development is a "fashionable phrase that everyone pays homage to but nobody cares to define."[4] In truth, however, people have paid homage to this catch-cry by defining it to suit their own needs, with over seventy definitions in print by 1992.[5] With the term meaning "something different to everyone," remarked Richard Norgaard in 1988, "the quest for sustainable development is off to a cacophonous start."[6]

In the midst of this cacophony, the Brundtland Commission's definition, with its "motherhood" theme of the satisfaction of basic physical and social human needs now and into the future, acts as a kind of common denominator. As Timothy O'Riordan points out, this theme "has staying power because most people want to believe in it. It sounds comforting—human well-being and ecological security in a world of peace, goodwill, and cultural tolerance—not brought to heel by ecological collapse, militaristic anarchy or debilitating greed."[7]

Some, including Johan Holmberg and Richard Sandbrook, claim that the eminently desirable but entirely vague quality of sustainable development discourses "is no real drawback," as long as the debate is held together by the "intuitively powerful concept" of equity.[8] This is a dangerously naive proposition, however, for the universal appeal of the notions of intergenerational and intragenerational equity—and to a lesser extent the notion of interspecies equity—are used by entrenched political interests to obscure real political and moral difference and dissent. We would do better to heed Donald Worster's warning that sustainable development resembles "a broad, easy path where all kinds of folk can walk along together, and they hurry toward it, unaware that it may be going in the wrong direction."[9]

The ecomodernist assumption that there exists some kind of consensus within the human community about its common future gives rise to the expectation that global political consensus is a precondition of any practicable response to ecological problems. This expectation in turn supports the view that sustainability rests finally with the globalized institutions of the technological society, for only within these institutions will humanity be able to develop and act upon such a universal consensus.[10] While political and moral utterance remains at the level of banal abstraction within ecomodernist discourses, the Promethean agenda continues apace almost unnoticed, almost as if technical efficiency were the default position of policy debate. This is not a new phenomenon, of course. In 1973 Ivan Illich observed that it "has become fashionable to say that where science and technology have created problems, it is only more scientific understanding and better technology that can carry us past them."[11] The upshot of this, Illich pointed out, is that policy responses to ecological and development crises are directed toward escalation rather than to genuine review of the technological society.

The policy framework of sustainable development funnels broad countercultural and counterhegemonic concerns about sustainability into a narrow ecotechnocratic agenda. The superficial and deceptive breadth of this framework arises as a consequence of its joining of the legitimately contestable concept of sustainability to the largely impregnable policy agenda of technological progress. Within this framework political and moral aspirations function as abstract goals with no real policy content because the realm of practice seems predetermined by techno-economic imperatives.

Joining Sustainability and Progress

How has the ideal of sustainability come to be joined to the agenda of technological proliferation? To answer this question, we need to return to the historical fact that there have been two distinct waves of postwar environmental concern that have both drawn upon the ideal of sustainability. That is, the ideal of sustainability is ambivalent with regards to the cultural concerns of the first wave and the techno-economic concerns of the second wave.

Like that of the second wave, first wave environmental concern was firmly bound together by an aspiration for "a condition of ecological and economic stability that is sustainable far into the future."[12] However, unlike the second wave, this aspiration was primarily understood in cultural rather than in technical terms. The first wave drew normative questions about development, limits, growth, stability, progress and wealth into political contests about the roots of social unsustainability. The merging of discussion about ecological sustainability with development discourses, which was to give rise to the language of sustainable development, began at the peak of first wave concern, catalyzed in 1968 by two international conferences.[13]

Post war debates about sustainability cannot be explained without reference to the nascent science of ecology and, specifically, without reference to theories of ecological succession and biospheric systems. Theories of ecological succession provided scientific credibility for political arguments emphasising that ecological systems were complex yet fragile, and that ecological stability, once disturbed, is difficult to re-establish. The understanding that life on earth exists within a discrete biospheric envelope, an idea made disturbingly real in the 1960s by the space-born images of earth, emphasized the extreme danger inherent in any degradation of global life-support systems.

In general, first wave scientists like Rachel Carson, Barry Commoner, Paul Ehrlich, and the authors of *Limits to Growth* presented the science of ecology as belonging to a new radical ethic of sustainability. With the rise of ecomodernism, however, the dangers inherent in using scientific arguments to frame political reform were soon apparent. The notion of ecological succession was adopted by the disciplines of resource management, which have a long prewar history of technocratic ideas about sustainability, as I show in chapter 3, to provide scientific credibility for neoclassical arguments that high levels of forestry, fishery, and agricultural productivity can be sustained indefinitely. In a similar fashion, biospheric awareness was skewed away from ecological concerns to make the case for global resource management regimes and planetary technoscience. Biospheric images and concepts have also been vital in second wave ideas like 'the global neighborhood'. These images have been used to suppress political difference and to promote the global market, a fact made evident by the proclivity with which transnational corporations (TNCs) use whole-earth images in their promotional material.[14]

I thus recommend that we recognize as a misapprehension the view that the sustainable development agenda is a vehicle for first wave concerns. This is a misapprehension evident in the way that "sustainability now tends to be used effectively as an abbreviation of sustainable development."[15] It is evident, also, in Ernest Yanarella and Richard Levine's claim that sustainable development derives from "a multitude of individual and group initiatives in sustainable living."[16] Initiatives, they argue, such as those advocated by Fritz Schumacher, Hazel Henderson, John Todd, Barry Commoner, Murray Bookchin, Ynestra King, and Arne Naess. This misapprehension is also shared by at least one in this list, Naess, who has claimed that "the victory of the notion of sustainable development over the post-war notion of 'economic development' and 'economic growth', and the simplistic 'development' is itself the sign of an awakening from ecological slumber and should be greeted with joy and expectation."[17] The sooner the radical ecology movement, especially its most important philosophers like Naess, recognize that this is a hollow victory, and pay closer attention to the subversion of the environmentalist lexicon, the better.

Whilst there is some truth in Tarla Peterson's claim that "Leopold's and Carson's work provides a foundation upon which sustainable development discourse has built," I believe, contrary to Peterson, that we best understand the resultant rhetorical structure as essentially empty. Associating the agenda of sustainable development with the eclipse of developmentalism and with radical grassroots experiments is not only fallacious, it endangers these radical critiques. I suggest, therefore, that we reject Peterson's claim that "[b]oth *sustainability* and *sustainable development* remain contested terms," in favor of the appreciation that whilst sustainability is an essentially contested and culturally rich discursive domain—an idea I develop in concluding this chapter—the language of sustainable development is conceptually incoherent and politically compromised.[18] Attempts to revive the potential of this language as an agent of cultural transformation, including attempts as trenchant and well reasoned as that of Michael Jacob, are valiant but ultimately futile.[19]

The incoherence of this language is wonderfully displayed in George Myerson and Yvonne Rydin's earnest discourse analysis of sustainable development. These authors jumble together terms like *sustain*, *sustainability*, and *sustainable development* and then analyze a jumble of texts from *Our Common Future* and *Blueprint for a Green Economy* to the work of Naess and Henderson. In the midst of the resultant shambles, they claim that "sustainable development represents new ethics, new politics, and new economics."[20] We are told that this language has at its core an ethos seeking to "redefine culture" through an "essential vision" that is both pragmatic (e.g., Brundtland) and spiritual (e.g., Naess).[21] There is danger, clearly, in any form of discourse analysis that is not undergirded by an analysis of the relationships between texts, practice, and politics.

The strong critique of the eurocentricity of development theory provided by Ozay Mehmet in *Westernizing the Third World* (1995) serves to emphasise this point. He takes up the important task of explicating the postwar history of injustice perpetrated in the name of development. Mehmet goes on, however, to undermine the political force of his argument by giving support to the vision of sustainable development set out in *Our Common Future*. He becomes caught within this vision in the mistaken belief that it ensures that "Western capitalist growth is increasingly recognized as the mainspring of unsustainability."[22] Apparently accepting the Brundtland Commission's rhetoric about equity, Mehmet concludes his book by drawing on the "ultimate aim of a sustainable world" to advocate, laudably, a form of social development in which "wealth distribution more closely approximates population distribution."[23] Yet the fact remains that the ultimate aim of a sustainable world is employed within the ecomodernist agenda to vigorously defend a form of biospheric management that maximizes market freedoms and corporate efficiencies.

Ecomodernism has swamped countercultural ecology movements by sheer political force but also by colonizing their discourses. Whereas early environmental debates were characterized by "competing vocabularies," environmentalism at the turn of the millennium is caught up in a farce in which widely divergent views are compressed into virtually identical terminology.[24] This does not mean that radical ecological and radical development critiques have been obliterated. Thankfully, these radical movements continue to evolve in both activist and in philosophical terms. But it does mean that these critiques not only exist on the margins of policy issues, they exist at the margins of environmental issues. The language of sustainable development has profoundly decentered, and thus disoriented, political and moral discourses about ecological crisis.

Confusion in Environmental Discourse

One of the principal indicators of the confusion created in ecological thought by sustainable development discourses is the proliferation of attempts to measure sustainability. These attempts reflect the dominance of narrow econometric concerns in sustainable development debates. Andrew Dobson's delightful retort to these sustainability indicators is worth quoting at length:

> The Scots can heave a sign of relief. They may have come second-to-bottom in this year's Five Nations rugby football championship, but at least, according to the Approximately Environmentally Adjusted Net National Product (AEANNP) method of measuring sustainability, they live sustainable lives. But hang on a minute. According to the Pearce-Atkinson measure (PAM), Scotland was "weakly unsustainable" between 1988 and 1992. . . . Ah, but relief is again at hand: Net Primary Production (NPP) figures suggest that Scotland has a carrying capacity of 5,490,000 people—more than 1990's estimated population of 5,102,400. But, oh dear, what's this? Ecological Footprint and Appropriate Carrying Capacity (EF/ACC) data reveal that, "in Scotland 20% more land is required than available." And things don't look too

> good, either, if Herman Daly and John Cobb's Index of Sustainable Economic Welfare (ISEW) is applied to Scotland. . . . One wonders what would have happened had the Rugby Football Player Assumption (RFPA) been factored in here.[25]

The perspective that is often lost in the scrummage for conclusive numbers is that the "usefulness of sustainability indicators is directly related to the policy context which they are used to address. . . . Using core indicators does not, in itself, provide a basis for devising new polices."[26] More fundamentally, so I will argue in section III, the point is that measurements of sustainability commonly internalize the assumption that this ideal can be understood in unambiguous, objective, and universal terms, yet this assumption is itself ambiguous, normative, and contestable. Let me stress that it is not quantification *per se* that is at issue here.[27] Empirical evidence of ecological unsustainability was crucial to the ideal of ecological sustainability becoming politically important in the first place. Rather, the point is that any attempt at measurement encodes a moral and political perspective that may be legitimately contestable.

Partly as a consequence of attempts to measure sustainability, it is now standard practice to place the range of views on sustainability along a continuum stretching from conventional, technocratic explanations of soft, minimal, shallow, weak or even very weak sustainability to the politically and morally radical extremes of hard, maximal, deep, strong or very strong sustainability.[28] Many, like the environmental economist David Pearce, make the assumption that this continuum is some kind of sliding scale which measures the amounts of sustainability a policy position contains.[29] However, strong sustainability is not in any meaningful sense strong by dint of having a greater quantity of sustainability than weak positions. What these typologies actually attempt to encompass are a range of deeply divergent normative perspectives.

Descriptions of a continuum of contested explanations of sustainability do seek to do justice to the diversity of contestants, as well as to the lack of clear demarcation between many of these views. Nonetheless, the one-dimensionality of these descriptions can only perpetuate the risk that substantive moral and political agendas will be coopted and decentered within technocratic agendas. I suggest that we therefore distinguish two distinct constellations of discourse that overlap in their use of the lexicon of sustainability but that occupy different dimensions of moral and political argument. We can call these discursive constellations *technocratic sustainability* and *cultural sustainability*. This terminology, while still vague, reflects more of the epistemological, moral, and metaphysical scope of contestation about sustainability.[30]

Some critics of ecomodernism, such as David Orr, may be surprised that I do not attempt to contrast technocratic sustainability with the notion of ecological sustainability, rather than with that of cultural sustainability.[31] I have deliberately avoided this contrast, however, for I consider it to be based upon a fundamental category mistake, although I do not wish to deny the value of Orr's important insights. Ideas about sustainability necessarily orient our thinking to ecological reality. But the fact remains that we may understand this reality differently depending on the nature of this orientation: depending, that is, upon the normative and political structures that define our discourses. The discourses of technocratic sustainability entrench and obscure the deeper sources of what is culturally unsustaining in ecological relations typical of the technological society. But this is not to say that the resultant policies of sustainable development will not be able to engineer greater ecological stability and, by implication, greater ecological sustainability of some form. What makes discourses contesting the meanings of cultural sustainability incommensurable with technocratic approaches is that they do not recognize technical sustainability as an end in itself. These contests are predicated upon the recognition that more sustainable forms of social development are guided by political and moral discrimination about what is truly worthy of being sustained in human communities. As I will show in part II, this discrimination does not conflate practical possibilities with technical possibilities.

The Hegemony of Sustainable Development

The conceptual counterpositioning of notions of technocratic sustainability and cultural sustainability is likely to be misleading unless it is founded on an acknowledgment of the way in which these two discursive constellations are also located in the actual political spaces created by the global hegemony of the North. We must acknowledge that technocratic concepts of sustainability exist in the institutional core of the technological society. Substantively cultural questions about sustainability survive at present only at its periphery.

In entering into contests over cultural sustainability, we necessarily enter into counterhegemonic struggles for political legitimacy, and power, and problem definition. The British sociologist Michael Redclift was one of the first to argue that debates over ecological sustainability may be coopted by interests that are united, regardless of their liberal-capitalist, military-capitalist or state-socialist form, in their pursuit of economic growth. Redclift's *Sustainable Development: Exploring the Contradictions* (1987) astutely anticipated the discourses of technocratic sustainability. His *Wasted: Counting the Costs of Global Consumption* (1996) provides an excellent review of the resultant global environmental politics.

In his analysis of the hegemony of sustainable development, Redclift stresses two things. First, contestation about sustainability cannot be divorced from questions about the political economy of global economic systems. We must not fail, demands Redclift, to clearly identify development with the historical patterns of resource exploitation in the South that ensured the colonial hegemony, and continues to ensure in many countries the neocolonial hegemony, of the North. Although he was hopeful in 1987 that the forthcoming

Brundtland Report would pose a radical challenge to conventional development strategies, Redclift was nonetheless aware that a concentration on growth within the language of sustainable development has served to "obscure the fact that resource depletion and unsustainable development are a direct consequence of growth itself."[32] As it turned out, notes Gerald Schmitz, the Brundtland Commission "avoided as much as possible historical-political analyses which might have interfered with its consensus building function around 'technocratized' functions."[33]

Second, Redclift stresses that poor communities are principally occupied with establishing sustainable livelihoods within their local environments. Such communities are only rarely occupied with, predominantly Northern, concerns about the sustainability of "the Environment" *per se*. Global environmental issues such as those surrounding wilderness, population, biodiversity, fisheries, and forests therefore need to be understood in terms of tensions between global and local political economies if they are not to be used as yet another justification for disenfranchising the world's Southern majority.

The failure, deliberate and otherwise, of the North to sufficiently appreciate these two issues meant that much of the environmental concern expressed at the 1992 Earth Summit was used to facilitate the oppression of the "Two-Third's World." In section II I employ this conference as a case study through which to expose the hegemonic forces at play within ecomodernism. These forces are especially evident in the relations between Eurocentric and non-Eurocentric institutions, Northern and Southern governments, global and local interests, and affluent and poor communities.

II GLOBAL POLITICS, NEOCOLONIALISM, AND THE EARTH SUMMIT

The Earth Summit negotiations, which spanned two years and culminated in the heads-of-state conference in Rio de Janeiro, marked the coming of age of global environmental politics and established sustainable development as its pre-eminent discourse.[34] The form of this politics is a pale shadow of the first wave of "global ecopolitics" that Denis Pirages described more than two decades ago as overturning the "international political hierarchy and related system of rules established during the period of industrial expansion."[35] Wolfgang Sachs, an eloquent critic of ecomodernism, suggests that rather than overturning hierarchy, "the UN Conference in Rio inaugurated environmentalism as the highest state of developmentalism."[36] Developmentalism, the Eurocentric ideology behind postwar development strategies, privileges the interests of an elite. However, this elite is not simply comprised of political, business, and military leaders. It is, Sachs rightly points out, "a global middle class of individuals with cars, bank accounts, and career aspirations. It is made up of the majority of the North and small elites in the South and its size roughly equals that eight percent of the world's populations which owns a car."[37]

To be sure, what came into focus at Rio was the fragility of the earth and the scarcity of its resources. But at the same time, argues Sachs, conventional development discourses have transformed environmentalism from a "knowledge of opposition to a knowledge of domination:" ideas such as "'risk', 'ecosystem', 'sustainability' or 'global'—which were once hurled from below to the élites at the top, begin now to bounce back from the commanding heights of society to the citizens at the grassroots."[38]

North/South Tensions

Earth Summit negotiations were shaped by headline-grabbing North/South tensions, the more so as the East/West tensions of the Cold War that had long defined global politics were easing dramatically. Evidence of these North/South tensions can be seen in the failure of many nations to ratify the biodiversity and climate conventions, and in the constriction of negotiations on forests to a nonbinding declaration of principles.[39] Southern governments drew attention to the history of failed development strategies, to the legacy of Southern environmental destruction that was a result of the structural-adjustment (liberalization) policies of the IMF and World Bank in the 1980s, and to the danger of new forms of "ecocolonialism." They gave priority within the emerging global environmental politics to humanity's global responsibility to eradicate poverty.[40] In particular, the South emphasized the North's obligation to dismantle iniquitous structures within the global economy. Conversely, Northern governments gave priority to humanity's global responsibility to be careful stewards of the Earth's resources. The North thus emphasized the South's obligation to curtail population growth, dismantle barriers to trade liberalization, and seek out greener, more ecoefficient forms of production.

As can be seen in any issue of Journals such as *New Internationalist* or *Third World Resurgence*, North/South tensions remain almost entirely unresolved in global environmental politics. The nonbinding statement of principle *Rio Declaration on Environment and Development* was typical of Earth Summit documents in its use of the language of sustainable development to suggest that these contesting positions are in some unspecified way mutually supporting. Consider, for example, the following:

> [Principle 4] In order to achieve sustainable development, environmental protection shall constitute an integral part of the development process and cannot be considered in isolation from it. . .
> [Principle 5] All States and people shall cooperate in the essential task of eradicating poverty as an indispensable requirement for sustainable development.[41]

As to how environmental protection is to be joined to the eradication of poverty, political leaders left it to global technocrats to determine. For, despite significant tensions over the emphasis placed on environment or development issues, sustainable development was universally accepted by political leaders in Rio as requiring sustained economic growth. And of course, the interests of Southern politicians are often closer to those of Northern elites than they are to the poor they putatively represent. According to Nicholas Hildyard, such a consensus simply serves in practice to ensure the continued hegemony of the North:

> Underpinning Ag 21 [*sic*] is the view that the development crisis results from insufficient capital (solution: increase Northern investment in the South); outdated technology (solution: open up the south to Northern technologies); a lack of expertise (solution: bring in Northern-educated managers and experts); and faltering economic growth (solution: push for an economic recovery in the North).[42]

The Earth Summit confirmed that the concept of sustainable development is being formulated so as to represent modern economic development as the only source of solutions to the ecological crisis. "While development as economic growth and commericalization are now being recognized as being at the root of the ecological crisis in the Third World," says Vandana Shiva, "they are paradoxically being offered as a cure for the ecological crisis in the form of 'sustainable development'."[43]

The Limits to Debate

So are we really to accept, asks Worster, that "an eco-technocratic elite" can manage the biosphere so as to "make it yield greater and greater production, until everyone on earth enjoys a princely life, and all that without destroying its capacity for renewability?"[44] Most participants at the nongovernmental *'92 Global Forum* held concurrently with the Earth Summit responded to this proposition with an emphatic "No!" Their protest about the manipulation of the ideal of sustainability by world leaders was loud and strong: "The nongovernmental organizations from all nations . . . do not accept the concept of sustainable development that is used only to produce cleaner technology while maintaining the same patterns of exclusive and unjust social relations for the majority of the Earth's people."[45] Comments like this represented a bold attempt to re-open the cultural contestation about sustainability that the Earth Summit negotiations sought to marginalize.

Many nongovernmental organisations (NGOs), and some Southern governments, emphasized that most of the cultural causes of unsustainability, as distinct from its technical symptoms, were not up for discussion at the Earth Summit. These are root causes of which even a cursory list must include the:

- spread of hyperconsumption in the North
- Eurocentrism in development theories and economic structures
- severe limitations of competitive nation states as actors in world governance

- corrosive cultural impacts of global information systems
- oppressive conduct by TNCs, especially in Southern countries
- drain on economic and ecological resources by the world's military
- historical injustices in the creation of Southern indebtedness
- colonial legacy in the postcolonial corruption and collusion in many Southern governments
- hegemony perpetuating the continued use of fossil fuels
- marginalization of women and indigenous peoples
- degradation and loss of local knowledges, languages, and practices
- domination and devaluation of nonhuman life
- escalation in technologically mediated risks to humans and ecosystems[46]

The neglect of these overlapping issues was not innocent. Most government actors were involved in excluding issues with the potential to call into question the global common ground of sustained industrial development.

The Earth Summit changed "little in the face of flawed democracy," O'Riordan tells us.[47] Governments lobbied hard to strip the agenda of matters that might constrain their domestic economic growth and international competitiveness. The political economy of fossil fuel use, for instance, was kept off the agenda by rich oil-exporting nations, led by Saudi Arabia and joined by the U.S. The Framework Convention for Climate Change was rendered toothless by the U.S. government's opposition to the idea of a carbon tax, refusal to endorse timetables for action, and resistance to setting targets for greenhouse-gas emissions. The U.S. also opposed the imposition of any restrictions on the access of its companies to Southern genetic resources, refusing to sign the biodiversity treaty. This decision was later reversed; however, the U.S. subsequently blocked negotiations over the biosafety protocol called for in this treaty at the 1999 UN Convention on Biological Diversity in Cartagena, Colombia.[48] Atomic energy was kept off the agenda by France. Northern governments traded off with Southern governments to minimize discussion of overconsumption and population.[49]

We should also note Pratap Chatterjee and Matthias Finger's contention that TNCs and their public relations consultants used the Earth Summit to connect "environmental management to economic growth and global trade" and thus "rehabilitate business as a means of environmental protection."[50] As evidence of this, the International Chamber of Commerce and the Business Council for Sustainable Development used the two-year preparatory process to head off attempts to develop the regulatory code for TNCs advocated by the UN Center for Transnational Companies (UNCTC). These attempts were extraordinarily successful. So successful, in fact, that in the lead-up to Rio the

UNCTC was closed down, while the Business Council put forward its own voluntary code of conduct. We cannot miss the significance of the fact that TNCs substantially helped finance the Earth Summit.[51]

Despite the diplomatic complexity of Earth Summit negotiations, some of the issues at stake were basic and fundamental. Rhetoric about intragenerational equity disguised the reality that the South's "single greatest opportunity to embrace sustainable development lies in reversing the net transfer of wealth from poor to rich that has continued since 1985 as interest payments on debt have outstripped ODA [overseas development assistance]."[52] This is a reality in which the North, with about twenty percent of the world's population, has over eighty percent of the world's income, domestic savings, and trade.[53] The only truly equitable responses to intragenerational concerns clearly are massive and direct financial reforms including substantial debtforgiveness (in contrast to the often expedient and tokenistic display of this reform adopted by Northern governments in recent times); democratic control of Bretton-Woods institutions and the Global Environment Facility (GEF); and nonmarket-based mechanisms of technology transfer. Further, these are initiatives which will only be effective if accompanied by concerted political action to support grassroots struggles in overcoming corrupt and authoritarian regimes in the South.

These were basic issues and initiatives about which Northern governments were silent and Southern governments appeared heavily constrained. This silence concealed two crucial reasons for unsustainable development: first, the appalling history of reckless, expensive, and ineffective industrialization by often corrupt Southern governments and their loose-lending Northern financiers, and second, the historical resistance of Northern governments to providing substantial development assistance without attaching strings of strategic self-interest.[54] The failure of political leaders to acknowledge this historical context also obscured the institutional problems facing the UN in implementing *Agenda 21*. Although nation states offered rhetorical support for *Agenda 21*, a few months after the Earth Summit the UN was in "deep financial crisis," and it remains unable to realize even a fraction of the funding required for *Agenda 21*.[55] All were able to witness the depressing irony of the UN's largest debtor, the U.S., advocating tough new global environmental targets while continuing to default on its existing financial commitments.

This irony points up the two premier problems facing the UN and any other institution of global governance in the era of global environmental politics. First, there is the deep incoherence within the UN's organizational structure. To be an effective global manager, the UN must achieve internal policy coordination. But since its inception, the UN's structure has reflected the North/South split, with its distinctly prodevelopment and pro-environment agencies often operating at cross-purposes. These rival agencies must compete with each other for resources, a competition that the UNEP has continually lost to the UNDP. In the implementation of *Agenda 21*, concludes Reg Henry, "the UN should expect agency conflict to be unmanageable."[56] The second problem facing the UN is that it is founded on the discourse of state sovereignty, while regimes of global technocratic management demand much greater emphasis on transnational actors. The discourses of the global economy are in fact deeply incoherent. Consumers are encouraged to support domestic over imported goods. Workers are encouraged to take pride in domestic industries by striving to beat the rest of the world in productive efficiencies. But, simultaneously, the global economy blurs national identities through the loose geographic loyalties of transnational businesses and the shared icons and aspirations of a global consumer class. With the rise of global economic liberalization, Richard Barnet and John Cavanaugh point out, the "power of national governments over the two most critical functions of the nation-state—security and economic development—have eroded, but the myth of national sovereignty is as strong as ever."[57]

Despite rhetoric about national sovereignty, Earth Summit negotiations revealed TNCs to be the "only currently functioning global agents."[58] "Power is shifting," asserts Marian Miller,

> increasingly from nation-states to transnational actors such as TNCs, international financial markets, multilateral banks, and international media groups such as the Cable News Network (CNN). Local and national bureaucracies no longer control information or the important sectors of the economy; these are now controlled by international oligarchies composed of business executives, financiers, and image makers.[59]

These international oligarchies flourished in the 1990s in the political spaces created by the end of the Cold War. Globalization, shorthand for global economic liberalization, is in fact bringing about the deglobalization of world politics. The influence of transnational economic networks—especially Northern multinational blocs such as the G-8 (formerly the G-7) and the Organization for Economic Cooperation and Development (OECD)—is now such that they are constructing a postnational order styled to suit the parochial interests of a decreasing number of ever larger megacorporations (as the recent spate of mergers between giant TNCs testifies).

Neocolonialism

The impotence of the UN in implementing either coherent ecological strategies for social development or a system of equitable global governance means that we must pay close attention to critics from the South for whom global environmental politics raises the specter of two distinct types of neocolonialism.

The first of these, environmental neocolonialism, arises through the professed need for coordinated management of the global commons. This form of neocolonialism has provoked bitter criticism of "elites in the North who want to keep the Amazon Basin as a free holding-ground for genetic diversity, as an air-conditioning plant, as a thermostat of their own madness."[60] The combination of narrating nature as a global whole—as Nature—as external wilderness and as requiring protection, has increased the vulnerability of Northern radical ecology movements to cooptation within the project of global managerialism.[61] For those already oppressed by the hegemony of development, further disempowerment by Northerners voicing the interests of Nature is a real and disturbing prospect.

The second form of neocolonialism, corporate neocolonialism, arises from the stimulus provided to TNCs by global economic liberalization. There is therefore another group of Northern elites who have proven only too willing to, say, help the governments of Brazil, Colombia, Peru, and Bolivia make short-term economic gains from the exploitation of the Amazonian basin. These gains rarely contribute to the sustained alleviation of poverty, commonly siphoning profits back to the North and further widening the gap between rich and poor in the South.[62] As Chakravarthi Raghavan points out, globalization "is really about the expansion of TNC activities to the developing world and on TNC terms."[63] A claim borne out in the OECD's proposed, but justifiably and widely condemned Multilateral Agreement on Investment.[64]

Consider how nation states will be affected unequally by the current deglobalization of political power. Central transnational actors, such as the World Bank, IMF, and WTO remain firmly controlled by Northern economic interests, while TNCs funnel capital back to predominantly Northern stakeholders.[65] The distinction between North and South remains stronger than ever even if their respective national members are becoming harder to distinguish from one another.

Environmental neocolonialism and corporate neocolonialism are mutually reinforcing within the language of sustainable development. Furthermore, they are convergent with the potential for recolonization within some human rights discourses.[66] Transnational business has adopted the language of sustainable development to present itself as the world's premier agent of global environmental care. Recent analyses of the practical environmental credentials of TNCs makes it apparent that their green tinge is more the result of their ability to engage in sophisticated image-making and marketing—greenwashing—than anything else.[67] This state of affairs will not change until wider regulatory and economic structures reverse the present priority of short-term profit maximization over long-term social goals. The Earth Summit provided little cause to think such a reversal is immanent. As Maria Mies puts it:

> The Earth Summit . . . again made clear that solutions to the present worldwide ecological, economic and social problems cannot be expected from the ruling elites of the North or the South. . . . [A] new vision—a new life for present and future generations, and for our fellow creatures on earth—in

> which praxis and theory are respected and preserved can be found only in the survival struggles of grassroots movements.[68]

Unfortunately, the momentum given to the agenda of sustainable development at the Earth Summit also made clear that the survival struggles of grassroots movements are more endangered than ever.

The Cooptation of Grassroots Struggles

Chatterjee and Finger's *The Earth Brokers* (1994) offers a sobering assessment of the way in which "Southern NGOs and NGO representatives through their participation in . . . [the Earth Summit] quite logically became coopted. This added some well needed fresh blood to the old development élite, which had already absorbed the mainstream Northern NGOs."[69] The Earth Summit was founded on an unprecedented involvement of NGOs, but then successful cooptation necessarily depends upon selective inclusion, not exclusion. Predicably, the involvement of NGOs was effectively limited to observing the maneuvers of global elites.[70]

Nonetheless, many NGOs claimed that the focus created by the Earth Summit provided a much-needed opportunity for constructive discussion between Northern conservation groups and grassroots development groups from the South.[71] As Barnet and Cavanaugh observe, grassroots groups are "spinning their own transnational webs to embrace and connect people across the world."[72] Grassroots activism, working at the margins of global information systems, is undoubtably building global civil societies.[73] Yet the Earth Summit made plain that the objectives of those within these transnational civic networks are at odds with ecomodernist hopes for a global civil society that will serve as the human face of global management regimes. Thus, the French environmentalist Alain Lipietz is in my view wrong in claiming that the presence of and cooperation between NGOs ensures that the Earth Summit

> will be seen as a decisive step towards consciousness that humankind is united in facing the catastrophe which its madness has brought to the planet. For ages, ecologists have been actively campaigning for thinking on a global scale, . . . we can congratulate ourselves whole heartedly on the emergence of the desire for an agreement on the world ecological order.[74]

In their advocacy for thinking on a global scale, many Northern ecologists accept uncritically the necessity of action on a global scale—a necessity that imposes Northern technocratic solutions on the South. Chatterjee and Finger can therefore assure us that while some have joined Lipietz in giving themselves wholehearted congratulations, "NGOs are now trapped in a farce by which they have lent support to governments in return for some small concessions on language, and thus legitimized the process of increased industrial development."[75]

One of the crucial deceptions facilitated by the rhetoric of sustainable development at the Earth Summit was the broad consensus that there is no conflict between global and local concerns about sustainability. We need only consider the way in which Secretary-General Maurice Strong sought to sell the upcoming Earth Summit to the grassroots:

> The participation of people at the local level in meeting the challenge of sustainable development has the great merit of providing a mechanism for taking into account local conditions and social issues at every stage of the planning process. . . . [The Earth Summit process] will draw on the traditional knowledge and resource management practices of indigenous people and local communities as contributions to environmentally sound and sustainable development, and intends to integrate them into the programs expected to result from this first 'Earth Summit'.[76]

Strong's comments exemplify the UN's perspective that local participation is "instrumentally valuable to the success of development projects." Gerald Schmitz correctly exposes this approach as treating "NGOs working with the poor . . . as agents of project participation, private sector delivery, and local 'civic culture,' not as agents for wider democratic change."[77] Participation becomes code for what Majid Rahnema describes as "human software" designed to add-on a user-friendly aspect to the hardware of technocratic development.[78] Ecomodernists assume that their understanding of resources, management, and sustainability are conceptually uncontestable—an assumption that denies moral and political difference between the global and the local. And yet, observes Sachs, "rarely has the gulf between observers and observed been greater than between satellite-based forestry and the *seringuiero* in the Brazillian jungle."[79]

Local perspectives cannot be integrated as contributions to global managerialism without rendering them substanceless and without endangering the local economies and technological practices that sustain them. Stripped of their context, the "practices of indigenous people and local communities" are presented to the mechanism of sustainable development as one-dimensional bits of information. The animating vigor of the grassroots is repeatedly cropped and left to wilt by the agenda of ecomodernism. But if nothing else, says Hildyard, the Earth Summit at least "made visible the vested interests that stand in the way of the moral economies that local people are seeking to re-establish in the face of day-to-day degradation of their rivers, lakes, streams, fishing grounds, rangelands, forests, and fields."[80]

Blind Spots in the Global View

The technocratic interests standing in the way of local moral economies have built the language of sustainable development around the lexicon of global awareness, global problems, global responsibility, global ethics, and the like. This lexicon, this global *mythos*, as Gustavo Esteva and Madhu Suri Prakash so aptly characterize it, pushes forward a new notion of community, the "global neighbourhood," as a way of deflecting attention from the iniquitous consequences of global economic integration.[81]

The Brundtland Report begins with the claim that our technological ability to view the earth from space has led us finally to understand that the earth is "a small and fragile ball, dominated not by human activity and edifice but by a pattern of clouds, oceans, greenery, and soils."[82] According to the global *mythos*, we attain a sense of material humility and egalitarian fellow-feeling at the sight of our small and fragile home. In fact, however, the global view is deeply ambiguous. It does not necessarily undergird a culture of humility nor an ethic of care for the earth. It may well produce what Yaakov Garb describes as a "psychological flattening and impoverishment of understanding."[83] Whilst allowing us to view our home as a global unit, the view from space afforded by the technological triumphs of the Cold War also place us outside and distant from this home. It is a short step from this to re-invent ourselves as "observer, manager, and planner of the 'Blue Planet.'"[84]

The popular Northern environmentalist slogan, "Think Globally, Act Locally" masks the fact that the "'*global*' in the dominant discourse is the political space in which a particular dominant local seeks global control and frees itself of local, national, and international restraints. The global," continues Shiva, "does not represent the universal human interest, it represents a particular local and parochial interest which has been globalized through the scope of its reach."[85]

Take Chatterjee and Finger's description of the way that the global view became, at the Earth Summit, a conduit for the argument that humanity is uniting spiritually, as well as technologically, in the face of ecological disaster:

> On the one hand, the idea that individuals are now connected because they all share a common global environmental awareness quite logically leads to some sort of global management. On the other hand, global management is probably in need of some sort of "philosophical" framework that would give it the moral and ethical dimensions necessary for it to be legitimized by the people. Not to mention the fact that many of the global managers are themselves members of the New Age church.

I agree with these authors that this blending of global managerialism with global spiritualism is "a-political, a-sociological, a-cultural, and a-rational," and thus thoroughly problematic.[86] I agree that the "planet-wide extension of economic rationality under the cover of eco-efficiency will . . . further cultural destruction and erosion."[87] In "one world" we are increasingly left with only one form of life, that of optimal efficiency and maximum productivity. Such globalism is inescapably parochial, unitary, and oppressive.

The deceptive ground of the global *mythos* is displayed in Miller's call for "decentralization and the distribution of economic power" in *The Third World in*

Global Environmental Politics (1995).[88] Miller argues that "Third World NGOs have begun to challenge government and corporate practices . . . using sustainable development strategies."[89] And crucial to these strategies is, we read, the realization that "a single biosphere imposes a shared destiny on the Earth's citizens. The challenge then is to increase global awareness of this common destiny."[90] But, in fact, central to the counterhegemonic experiments Miller unwisely calls "sustainable development strategies" is the realization that while some are destined to live in the mansions of the global neighborhood, the majority of those long oppressed by the eras of colonialism and developmentalism will undoubtably swell its ghettos.

III CONTESTING CULTURAL SUSTAINABILITY

"*The struggle for peace is a struggle for sustainable development*—and the struggle for sustainable development is a struggle for true peace and justice."[91] In making this kind of normative claim, Southern environmentalists are attempting to extract sustainable development from the conceptual constellation of ecomodernism and the technocratic sustainability it champions. The danger in this strategy is that in simply using the language of sustainable development, the pluralist and practical objectives of Southern activists are dissipated by the hegemonic interests of the technological society.

Sustainable development has transformed normative questions about sustainability from agents of cultural change into agents of neocolonial hegemony. The lesson we must take from the Earth Summit is that the language of sustainable development must be used only with great wariness, if it is not to be discarded altogether, if culturally substantive ecological concerns are to be recovered in public discourse. The ideal of sustainability provides an important site for contesting latemodern thinking about social progress. Yet, paradoxically, the emergence of powerful policy discourses about sustainable development has only further marginalized moral and political questions about the nature of social development.

But, some are likely to protest, surely this judgment is too harsh. Sustainable development does embody a normative perspective, the perspective of a caring and cooperative global civil society. After all, doesn't the UN also demand equity, peace, security, harmony, global good faith, and universal human rights in the name of sustainable development? The unshakeably sane prose of Sachs brings all this grand talk back to its institutional reality:

> The rhetoric, which ornaments conferences and conventions, ritually calls for a new global ethic, but the reality at the negotiating tables suggests a different logic. There, for the most part, one sees diplomats engaged in the familiar game of accumulating advantages for their countries, eager to out-manoeuvre their opponents, shrewdly tailoring environmental concerns to the interests dictated by their nation's economic position.[92]

Talk of a global ethic of sustainable development not only ignores the prior moral and political commitments of this technocratic background, it builds upon them so that they are further obscured. The overriding moral goal of individual and social life is presented as being maximum sustainable consumption of optimally efficient technologies.

The ethical aspirations surrounding this agenda are mostly sincerely intended. Few of those engaged in sustainable development discourses do not genuinely seek a more just, cooperative global community in which social deprivation and ecological despoliation are the exception, not the rule. Even if it were possible, my purpose has not been to establish claims of corruption or of rapacious self-interest and unfettered greed against those advocating the policies of sustainable development. I have sought to display that the agenda of ecomodernization approaches the ideal of sustainability from within a cultural framework that equates moral development with technological development. This framework denies that our earth is home to a plurality of worlds, that we live not in a universe, but in a "pluriverse."[93] The moral vision it offers, no matter how wellintended, atrophies, leaving only a rhetorical shell that deflects attention from the ways in which the continued development of the technological society depends upon the oppression of nontechnocratic cultures and the instrumentalisation of nature. We are well advised, therefore, to heed Ronald Engel's caution: "Before we accept sustainable development as a new morality, as well as a new economic strategy, we need to know what ecological, social, political, and personal values it serves, and how it reconciles the moral claims of human freedom, equality, and community with our obligations to individual animals and plants, species and ecosystems."[94] Not all aspects of the sustainable development agenda must be jettisoned as a result of this kind of inquiry, however. It is vital that highly technological societies adopt incremental economic and technological reforms, and as urgently as possible. In this way, we may be able to reduce current rates of ecological degradation and resource depletion, thereby increasing the opportunitiy we have for moving toward forms of society more sustaining than our own.

What I think must be rejected is the claim that the agenda of technocratic sustainability is a response to ecological crisis defined in anything other than technical terms. Ecomodernism offers narrow technical "solutions" to the "problems" of unsustainability. The discourses of sustainable development are not legitimately contestable. They do not permit genuinely practical responses to the domination of nature, to the oppression of nontechnocratic cultures, or to the despotic hegemony of Eurocentric elites. The discourses of technocratic sustainability fail abjectly to illuminate the moral relationships of technological culture to other cultures, to the future, to nature, or to its own inhabitants. They fail to shed light on the moral meanings of sustainability. So, if we are not to further entrench the moral and political sources of our social unsustainability, our involvement in the Promethean agenda must remain secondary to and dependent upon our political and moral attempts to contest and undermine the anthropocentric and Eurocentric hegemony of modern and ecomodern forms of development.

On the basis of her book, *Sharing the Earth: The Rhetoric of Sustainable Development* (1997), Peterson would perhaps raise a stronger objection against my diagnosis of terminal malaise in sustainable development discourses. Drawing upon communication theories of symbolic action, she argues that these discourses creatively destabilize development theory by placing an oxymoron at its center: "The coupling of sustainability with development . . . is decidedly incongruous. If we maintain the tension, the concept can provide the comic corrective needed to enable us to lurch into the future without the need to destroy our past. By giving us a terminology for asking old questions in a new way, it enables us to discover new solutions."[95] Peterson was not unaware that at the Earth Summit the "synthetic possibility offered by sustainable development receded into the distance."[96] But she remains hopeful that sustainable development "allows people to select a point of identification between themselves and others from whom they must otherwise remain alienated."[97]

I find Peterson's interesting and sophisticated rhetorical analysis unsatisfying. It lacks a clear appreciation of the actual political consequences—in contrast to the abstract linguistic possibilities—for those outside of her own middle-class, Northern context if they conflate the flexible ambiguity of sustainability theory with the unambiguously ossified, hegemonic practices of technocratic development.[98] The advocates of economic development do not enter the discursive domain of sustainability an equal dialogical partner. Their voice projects loudest and resonates longest.

Local Coevolutionary Development

Redclift guides us toward more meaningful contestation about sustainability in his remark that "many of the social objectives which reflect environmental knowledge, and make the environment a contested domain, especially in the South, are part of a critical perspective which does not fit with conventional wisdom about economic growth and development."[99] Redclift counsels that we contest the Eurocentric and largely urban discourses of the Environment that represent "nature as existing separately from us, as performing functions for us. . . . The question is then," concludes Redclift, "do existing practices leave us confident that we can arrive at other discourses about the environment? Would other discourses lead us to a different understanding of the environment, and what would they imply for our actions?"[100]

The universalizing Eurocentric view of the Environment is anathema to those who exist only locally. The ecological surroundings of rural cultures in the South are typically drawn into social life through a nexus of meanings and practices that encompass capital, labor, technology, and natural resources, not to mention historical, cultural, and religious symbolism. Robert Chambers' influential description of the sustainable livelihood thinking of these communities seems to me nothing less than a description of this nexus.[101] Sustainable livelihood thinking does not seek to internalize the environment within ideas of social development so much as it is incapable of conceiving that systems of ecological regeneration could ever be external to the material, economic, emotional, and spiritual well-being of human communities. The ecological rationalities arising out of sustainable livelihood thinking are not in any modern sense theoretical; they are encoded and sustained in everyday practice.

The message "trickling up from the grassroots" in the South, and from indigenous communities in the North, says Arturo Escobar, is that any contestation about ecological sustainability must involve contestation about economic development:

> Rather than continuing to accept as normal the vision of the Third World as in need of Development, there is an acute need to assert the difference of cultures, the relativity of history, and the plurality of perceptions. For the Third World this means to shake off the meanings imposed on them by the Development discourse, to open up in a more explicit manner the possibility for a different regime of truth and perception within which a new practice of concern and action would be possible.[102]

The grassroots, claim Esteva and Prakash, "are learning to escape from the extravagant idea that political struggle requires, as a premise, a clear conception of the desirable social regime that is the 'ultimate' goal."[103]

Of course, contesting the local moral meanings of sustainability need not be confined to communities at the cultural margins of the technological society. Wendell Berry is a wonderful and important advocate for the resistance of Northern "bioregional" communities, especially farming communities, against global interests. Berry describes for us the deep incompatibility of global thinking and sustainable livelihood thinking:

> Global thinking can only do to the globe what a space satellite does to it: reduce it, make a bauble of it. . . . If you want to see where you are, you will have to get out of your spaceship, out of your car, off your horse, and walk over the ground. . . . The abstractions of sustainability can ruin the world just as surely as the abstractions of industrial economics.[104]

Norgaard, in his important book *Development Betrayed: The End of Progress and a Coevolutionary Revisioning of the Future* (1994), develops this claim further, arguing that sustainability can be contested only within discourses and practices that place the coevolution of social systems and ecological systems as a prerequisite to any vision of technological development.[105] The discourses and practices of coevolutionary development are particularist, rather than universalist, argues Norgaard, because the biosphere is "far too complex for us to perceive and establish the conditions for sustainability" in any universal, global way.[106]

The importance of coevolutionary practices is clear in the example of the Chipko movement of Northern India, as recounted by Shiva.[107] In the political resistance of local women who hugged their trees as the bulldozers of development advanced, it is evident that coevolving forests cannot be understood simply as a utility, a source of fuel, food, and building resources. Forests are one aspect of coevolutionary practices that provide economic security, ecological stability, delineate social (and particularly gendered) roles, house rich historical, agricultural, culinary, and medical knowledges, confer cultural and regional identity, and mediate spiritual orientation to the cosmos. As sustainable development policies redefine this forest as so many feet of timber in need of optimal sustainable harvesting, or even as the earth's lungs in need of protection, these coevolutionary practices are reduced to "bytes" of abstract information in need of "updating" and "reformatting."[108]

"When sustenance is the organizing principle for society's relationship with nature," argues Shiva, "nature exists as a commons."[109] However, when the "pseudo-sustainability of 'sustainable development'" becomes this organizing principle, coevolution is ruptured. Technosystems are seen as more fundamental to human survival than ecological systems. Inevitably, the requirements of the technosphere are seen to be prior to those of the biosphere. Nature becomes the domain of private property rights. Society becomes the domain of capital.

David Goodman and Michael Redclift's *Refashioning Nature: Food, Ecology, and Culture* (1991) exposes the consequences of the dominion of capital, rather than that of sustenance, in the context of food production.[110] These authors show that the South's food crisis is an obligatory outcome of the political economy of Fordist (mass-industrial) agriculture that has concentrated control of world food resources within a handful of Northern TNCs whose defining interest is profit, not human sustenance. The inherently democratic and intergenerational fecundity of a tomato plant, say, has been harnessed within a trap of selective breeding and patent rights. This trap involves making the plant dependent on industrial products—a dependency greatly enhanced by the newly emerging gene technologies—and the majority of the world's farmers dependent on the infertile seed owned by a corporate elite.

It is in terms of coevolutionary practices that we need to interpret Redclift's claim that sustainability "is not so much an invention of the future as a rediscovery of the past."[111] Redclift defends this position against claims of nostalgia. The reconsideration of the past he advocates is not a turning away from the reality of technocratic globalization of human life, nor does it seek to romanticize nonindustrial forms of life. It is a recognition of the right of other forms of life to exist and to adapt according to their own internal histories of sustainable livelihood thinking. It is a rereading of history that can reveal and value the rich diversity of human forms of life, making possible a fresh understanding of the cultural specificity and boundedness of industrial forms of life. The basic need of local people to retain political autonomy over their environments runs counter to the ecomodernist interests of both Southern and Northern elites. These elites attempt to reify sustainable development, conferring authority on expert discourses that describe and measure the features of The Sustainable Economy, The Sustainable City, The Sustainable Biosphere, even The Sustainable Global Ethic. Rather than fostering coevolutionary development, these discourses rupture the commerce between local communities and their more-than-human world, and out of this rupture, "cultural practices that have evolved to sustain both production and the environment are lost."[112]

Normative Questions

Coevolutionary thinking and practice in the South create normative and political spaces within which the inherently moral meanings of cultural sustainability can be contested—spaces that the hegemonic discourses of technocratic sustainability seek to colonize. These spaces exist also within the technological society itself, although they are much less clearly defined than in cross-cultural contexts. The latemodern desire for sustainability is enormously important. Within its elastic embrace, there is room to contest crucial moral questions about technological progress. Questions such as these posed by the ecological economist John Peet: "What are our values? What are our roots? What sustains us? What do we want to pass on to our grandchildren?"[113] And, as Richard Smith emphasises, any "answer to sustainability lies in the commitment of individuals, and by extension of the collective, to a reorganization of everyday life in the direction of a less unsustainable way of living, and this is only possible by the engagement of people in the debate about the normative structures of everyday life."[114] Understood in this way, sustainability is an essentially contested domain of meaning and practice. Moral concern with the ideal of sustainability demands that we pose difficult questions for ourselves, individually and communally, questions whose answers remain open to legitimate contestation and reformulation.

In his influential *The Terms of Political Discourse* (1983), political theorist William Connolly argues that the crucial point to grasp about essentially contested concepts is that they are fundamentally appraisive. They result from the combination of normative perspectives and descriptive statements.[115] If we try to remove from these concepts the moral points of view they embody, we also "subtract as well the rationale for grouping the ingredients of each together within the rubric of one concept."[116] Connolly points out that without an understanding of the moral orientation of contesting positions we are unable to understand the political character of discursive spaces that concepts like 'democracy' and 'equality', and even 'politics' itself, hold open:

> When groups range themselves around essentially contested concepts, politics is the mode in which the contest is normally expressed. Politics involves the clash that emerges when appraisive concepts are shared widely but imperfectly, when mutual understanding and interpretation is possible but in a partial and limited way, when reasoned argument and coercive pressure commingle precariously in the endless process of defining and resolving issues.[117]

The ambivalence of essentially contested concepts, the fact that they resist definition, resolution and closure, is a source of political strength, not of weakness. Their openness makes possible the ongoing mediation of meaningful, legitimate political difference. "The realization that opposing uses [of contested concepts] might not be exclusively self-serving but have defensible reasons in their support," claims Connolly, "could introduce into these contests a measure of tolerance and a receptivity to reconsideration of received views."[118]

It is in this sense of normatively reflective political discourse that I consider the contestability of latemodern ideas about sustainability to demarcate a broad political arena for defining and resolving, for describing and evaluating, the ecological crisis. For "contesting earth's future," as Michael Zimmerman puts it.[119] The contestation that this arena demarcates is not in any fundamental sense new or discrete. It draws within it a cluster of perennially contested terms, such as *democracy*, *equality*, *progress*, *nature*, and *stability* but orients them, descriptively, to ecological problems. John Barry offers a rich expression of the essentially contested character of sustainability discourses and of the need for techno-economic objectives to be founded on this deliberation, rather than the other way around:

> The essential indeterminateness and normative character of the concept of sustainability implies . . . that it needs to be understood as a discursively "created" rather than an authoritatively "given" product. . . . Sustainability is thus more than finding ecologically rational methods of production and consumption; it also involves collective judgement on those patterns. It is not a matter of examining the ecological means to determined ends; ultimately sustainability requires a political-normative judgement on the ends themselves.[120]

Sustainability is at best a wedge in political discourse holding open normative reflection about the transformation of our ecological reality. To contest sustainability is to challenge the idea that there is a single ecological reality, a Nature, that stands outside of particular, embodied social encounters with the living earth; it is to affirm that there are a plurality of contested natures.[121] At best, the ideal of sustainability is contestable, not in the sense of inviting conflict, but in the sense of keeping open "the elemental moral question of what way of life human beings ought to pursue."[122] The resilience of essentially contested concepts arises from their capacity to keep such fundamental questions open, resisting the imposition of unitary, contextless definitions. Paradoxically, this is also the source of their vulnerability. For these fundamental questions may remain open while practical political options have long been closed.

The triumph of ecomodernism has been to hold together a wave of environmental concern that disengages normative ecological questions about our behavior, our everyday practices, from the practical discourses of economic growth, global managerialism, corporate liberalization and technological globalization. Disengaged from the pragmatic, technical answers that determine what ways of life are open for us to pursue, such essentially contested concepts become grist for the mills of intractable, interminable and largely inconsequential debate. They underwrite discourses that are stripped of actual political and moral commitment. The historical possibilities of contesting sustainability have been displaced by an ahistorical agenda for engineering the future. Political possibilities have been largely reduced to managerial regimes for optimising the efficiency of the economy-environment relationship. Moral possibilities have been constricted to those that occur in the global neighborhood.

As the case of sustainable development demonstrates, essentially contested concepts degenerate into confusion and become vulnerable to cooptation within authoritatively given structures precisely when their contestants assume that the aim of the contest is to dispel ambiguity. Such concepts become essentially empty when they are thought to produce one view all share, one definition all agree upon, one victor all obey.[123] Paradoxically, intellectual clarity and truly democratic agendas for social development depend upon contestation about cultural sustainability holding open the political ambiguity of ecological problems and the plurality of practical responses. I thus endorse Orr's claim that, "[u]ntil we see the crisis of sustainability as one with roots that extend from public policies and technology down into our assumptions about science, nature, culture, and human nature, we are not likely to extend our prospects much."[124] I take up the challenge of following these roots in the next chapter by contesting the metaphysical assumptions about technological progress that have shaped the technological society.

CHAPTER THREE

METAPHYSICS

Making Nature Secure

INTRODUCTION

THE IDEAL OF SUSTAINABILITY is of central philosophical significance in latemodernity because it yokes together practical responses to and normative questions about ecological crisis. Yet this ideal is rapidly being either coopted and then eviscerated or marginalized and then ignored by the agenda of ecomodernization. I thus concluded chapter 2 with the claim that any recovery of the animating cultural force of this ideal must treat the language of sustainable development with utmost caution, and it must give ground in environmental discourses to local, and particularly to Southern, coevolutionary agendas for social development and ecological flourishing.

In this chapter the focus of my inquiry shifts from the reality of the political oppressiveness of ecomodernism to investigate the assumptions of nature and culture, ecology and technology, experience and reason that make this oppression seem rational in the technological society. But the ideal of sustainability, as I want to contest it, keeps alive the tension between pragmatic objectives and radical revisioning, between counterhegemonic objectives and countercultural questions, that ecological crisis establishes in our everyday lives. The move from politics to metaphysics is thus not intended as a move away from the grim fact that the degradation of ecosystems and nontechnological cultures translates to the reality of suffering, dislocation, and death; the ultimate reality of mortal beings. On the contrary, we must peel away the metaphysical blinkers that would prevent us from seeing this reality and our own contribution to it more completely.

"Just as Socrates taught us that the good that is our own calls for philosophical criticism to save it from popular misunderstandings, so now does the idea of sustainability," posits Carl Mitcham.[1] What then, in its most basic terms, is the idea of 'sustainability'? I shall argue that it is best understood as the ongoing ability to support, relieve, sustain, or nourish.[2] Unlike the adjective *sustainable*, which is applied as easily to the productivity of a munitions factory as it is to the harnessing of solar energy, the verb *sustaining* holds open the actively normative questions that the idea of sustainability raises. We are required to probe: What truly sustains us? Why? And how do we know? Conversely, we must ask: What are we to sustain above all else? Why? And how may we do so?

In latemodernity, the ideal of sustainability encompasses a discursive domain that celebrates sustaining things (in both senses of this phrase)—what I am calling the domain of cultural sustainability—as well as a discursive domain that celebrates human control over their world—the domain of technocratic sustainability. The center of gravity in both of these domains is the aspiration of human communities to sustain the sources of their sustenance in their coevolving, that is, their cultural and ecological, reality. Yet this aspiration for sustainability functions incommensurably in these two domains. In discourses of cultural sustainability, strategies for sustaining the sources of sustenance are understood first and foremost in political, moral, and spiritual terms. In discourses of technocratic sustainability these strategies are understood first and foremost in descriptive, instrumental terms that allow for little evaluative judgement. In the former, concerns about sustenance are encoded in just political, enriching moral, and invigorating spiritual practices. In the latter, these concerns are encoded in optimally efficient technological structures.

To leave matters here would be to perpetuate a common deception in thinking about sustainability, however. The project of technocratic sustainability is no less a political, moral, and spiritual project than any other; it is undoubtedly a cultural project. But it is essentially unreflective and uncontested. It denies its own origins by presenting the technological pursuit of efficiency as a culturally neutral, instrumental endeavor. Ecomodernist discourses displace questions of cultural sustainability with technical answers, but in so doing they also enact, and in a way that further obscures, a particular cultural set of answers to these questions.

A generation ago Herbert Marcuse observed that, as a "technological universe, advanced industrial society is a *political* universe, the latest stage in the realization of a specific historical *project*."[3] This enduring observation places before us the difficult task we face in contesting an ecomodernist project that seeks to overcome ecological crisis by securing nature more fully to technological purposes. Our task is difficult, for this project redefines social life so that the ultimate task of politics becomes management of the proliferation of technology. The language of sustainable development does not simply represent a superficial political opportunism, nor does the resultant technocratic agenda represent one institutional approach amongst many open to the technological society. This language and this agenda extend the cultural forces that are reducing our social and ecological reality to the dimension of instrumental utility.

Questions about sustainability are therefore best framed in ways that keep open metaphysical questions about the world, that is, the technological universe, of latemodern society.[4] But I do not take this to mean that latemoderns must search for an illusory conceptual vantage point, an Archimedean perspective, outside of this world. It is precisely this kind of search that renders so much philosophical thinking about ecological crisis irrelevant to policy contests. My inquiry is therefore predicated on the perspective of inhabitants searching for internal contradictions, weak points of creative tension, through which they may probe into and beyond the boundaries of their technological world.

In the first section of this chapter I probe the ways in which ecomodernism extends the modern ideal of progress, an extension that supports the capacity of ecomodernism to absorb and grow stronger from apparently radical ideas about biophysical limits, ecological scarcity and global interconnectedness. In section II I draw upon the literature of radical environmental philosophy to show how ecomodernism is shaped by, yet also subtly redefines, the dualistic metaphysics that have impelled a cultural project of mastering nature since Descartes.

I A CULTURAL PROJECT

It is a mistake to see the ideal of sustainability as solely a postwar phenomenon, even though this term was not coined as an abstract noun until the early 1970s.[5] As the recurrent historical theme of local ecological collapse precipitated by unfettered human development testifies, it has been incumbent on all human societies to consider (and prospectively) the ability of their environments to sustain their needs. We should not be surprised, therefore, to discover that sustainable development has long been an important if implicit theme in modernist thinking. Historian Donald Worster, in his excellent essay *The Shaky Ground of Sustainable Development* (1993), traces technocratic concerns about sustainability back to the European roots of this term in the late eighteenth century German concept of 'sustained-yield'. Rather than arising out of ecological concerns, the modern concept of sustainability arose out of economic concerns about the declining productivity—the disappearance, in fact—of Germany's forests. Worster concludes that "'[s]ustainable development' is therefore not a new concept but has been around for at least two centuries; it is a product of the European Enlightenment, is at once progressive and conservative in its impulses, and reflects uncritically the modern faith in human intelligence's ability to manage nature."[6]

As problems of declining productivity increased and spread around the globe with the spread of industrialism, so too resource managers became increasingly aware of the need to conserve the resource base. In this spirit, the North American conservationist Gifford Pinchot claimed in 1901 that "[c]onservation has much to do with the welfare of the average man of today. It proposes to secure a continuous and abundant supply of the necessaries of life, which means a reasonable cost of living and business stability."[7]

Progressive and Conservative Impulses

The Enlightenment traditions of resource management posit that a productive ecosystem, such as a fishery, should be stabilized at the optimum level of productivity that can be sustained indefinitely, thereby maximizing the intergenerational economic benefits that contribute to social progress. Reflecting this, the *Oxford English Dictionary* defines sustainability with reference to the claim that "[s]ustainability in the management of both individual wild species and ecosystems . . . is critical to human welfare."[8]

Modernist thinking asserts that the combination of ecological stability and techno-economic progress confers sustainability on human societies. The principal advance in this thinking encouraged by the language of ecomodernism is that ecological stability can no longer be understood primarily as an inherent capacity of an abundant planet. It must now be understood primarily as a design feature inherent in technosystems themselves. Bill McKibben's well-known assertion that "nature has ended" is to latemoderns what Nietzsche's shattering declaration that "God is dead" was to those entering into the modern age.[9] What is ending are those cultural narratives that presented nature as radically separate from human culture. Narratives of nature as either pure wilderness or as lifeless machinery are both losing their explanatory power. In the narratives of latemodernity it seems that nature cannot now exist independently of the processes of technological evolution. The remarkable idea that humanity must actively save the planet is now central to and inconspicuous within environmental thinking of both ecomodernist and radical bent. The arch-Promethean Julian Simon was regarded with some amusement and much disdain in the 1970s and the early 1980s for his argument that there is no ecological crisis and no limit to growth that human ingenuity cannot overcome. Yet, and with good reason, Simon was able to preface his 1994 essay *Scarcity or Abundance?* thus: "what you read below was a minority viewpoint until sometime in the 1980s, at which point the mainstream scientific opinion shifted almost all the way to the position set forth here."[10]

There is thus historical depth in the claim set forth in the ecomodernist manifesto *Our Common Future* that the concept of sustainable development implies not absolute limits but the "limitations imposed by the present state of technology and social organization on environmental resources and by the ability of the biosphere to absorb the effects of human activities."[11] Thought of in this way, a scarcity of trees or of fresh water becomes not a limitation on the activities of deforestation or irrigation, both of which are essential for sustained development, but, rather, becomes the technological challenge of genetically enhancing trees or of designing large-scale desalination plants.

The failure of many countercultural ecological critics to appreciate that the modern idea of 'progress' incorporates within it a tradition of thinking about stability serves at least as a partial explanation for the rapidity with which their arguments about global ecotastrophe were marginalized in environmental debates during the 1980s. Many early environmentalists assumed that the growing awareness of ecological notions of limits, scarcity and interconnectedness would inevitably discredit modernist arguments about the sustainability of technological progress.[12] William Ophuls, writing on the politics of scarcity in 1977, argued that "nobody really disagrees about the ultimate implications" of ecological scarcity, namely, that "we must learn to live in 'steady-state' or 'spaceship' societies characterized by great frugality in the consumption of resources and by deliberate setting of limits to maintain the balance between man and nature."[13] He boldly declared that "the basic principles of modern industrial civilization are . . . incompatible with ecological scarcity."[14] This claim continues to be worked out in some technical detail in the field of steady-state or ecological economics. Kenneth Boulding, Herman Daly, John Kenneth Galbraith, Nicholas Georgescu-Rogen, Hazel Henderson, and others have pioneered the argument that the neoclassical pursuit of infinite growth in a finite, closed biophysical system is inescapably unsustainable.[15]

The metaphor of *spaceship earth* was an important vehicle in the 1970s for the idea that humanity has to adopt an environmental ethic that acknowledges that we live within a finite system. In the new, twenty-first century, however, this metaphor is used less as an inspiration for Ophuls's ethic of restraint and more as an illustration of the need for global management regimes for updating and refitting this ship. The imperative of scarcity becomes the policy of ecoefficiency, not the policy of human frugality. Fear of scarcity becomes fuel for the engines of progress. Social stability becomes synonymous with the dynamic of progress, for stability is thought to be founded upon the ability of social activity to overcome external limits.

The Dutch philosophers Pieter Tijmes and Reginald Luijf paraphrase this typically modern argument as follows: "if the cause of violence is a shortage of goods, then economic competition and an accompanying explosion of economic growth are the only means to social stability."[16] These authors go on to show how ecological degradation has been construed as just such a form of violence against the interests of the technological society in ecomodernist tracts like *Our Common Future*: "The . . . [Brundtland] Commission in no way bids farewell to economic growth formulated on the basis of the concept of scarcity but simply reformulates the concept of growth as sustainable development. . . . [E]conomic development continues to be defended as a necessary fight against scarcity—indeed, as a fight that must be sustained."[17] The principles of steady-state society enumerated by many in the 1970s are almost identical to the principles of sustainable development articulated by the UN in the 1990s. Nonetheless, while the former bring with them a marginalized agenda of countercultural questions about social stability, the latter bring with them an incommensurable, powerful, revitalised, and deeply political agenda of technological progress.

Few students of Enlightenment history are likely to be surprised at the rise of a technocratic agenda for sustainable development. The appearance of what seems to be a Pheonix rising out of the ashes of industrialism conforms completely to Marcuse's prediction that the technological society will triumph over much of its opposition not because it confronts and wins direct political struggles, but because it "integrates all authentic opposition, absorbs all alternatives."[18] To understand the ability of ecomodernism to so promptly digest and reconstitute claims that ecology must dictate terms to technology, we must first understand how modernist thought countered the premodern preoccupation with stability in the first place. I shall briefly sketch this historical context here, emphasising the themes of limits and scarcity.

Ecological Limits

Fifty years ago, in his study of the idea of stability in the history of European thought, the German philosopher-engineer Frederick Dessauer claimed that the realization that "social conditions are stable or should be stable was not a new discovery, to be credited to any individual or epoch, but part of the living faith of antiquity and the Middle Ages."[19] Within the long occidental history of this living faith Dessauer includes Plato, Jewish moral law, the Christian Bible, Aquinas, and, more recently, Rousseau and Burke. What unites these traditions is not their explanations of social stability, which are diverse, but the methodological fact that they placed the stability of social systems as the organizing principle of social life and as foundational to any notion of social progress. It is, consequently, superficial to characterize these traditions as being opposed to ideas of social progress. Rather, they derived their idea of material progress from their metaphysical certainty that the highest purpose of social life was to maintain an *a priori* order; an order unveiled in natural, social, and divine phenomena. As Dessauer saw, "stability includes a good deal of progress and consists in the ability of a social organism to avoid excessive and one-sided advances as well as the immobility which leads to breakdowns. It is, in other words, the awareness of limits."[20] And the principal premodern limit on material progress in Europe, a limit imposed largely through the classical eschatological metaphysics of early Christian traditions, took the form of belief that transcendental achievements were of greater moral significance than material achievements.

In general terms, the Enlightenment turn in philosophy, which began in earnest with the early seventeenth century world of Descartes and Francis Bacon, viewed the older traditions of stability, with their emphasis on reflecting a cosmic order in social life, as a constriction of individual freedom that stifled the prospects for rational social betterment.[21] Descartes' *Discourse on Method* created space for a challenge to these traditions, although he by no means fully entered this space himself, by establishing epistemology on the maxim that accumulated bodies of knowledge must be dispassionately adjusted to the "plumb-line of reason."[22] Bacon's *New Organon* and *New Atlantis*—the latter a science fiction fantasy that makes for uncannily familiar reading for a latemodern, excepting its pious tone—created space for viewing the future as a realm of infinite possibilities by arguing that humanity can and should bend nature to its creative will using technological force.[23] With good reason, Bacon earns the epithet from Neil Postman of "first man of the technocratic age."[24]

The light that Descartes and Bacon shed on the apparently stagnant world of medieval Europe was that nature is a composite of myriad and lifeless components that functions as does a mechanical clock.[25] Exposed to the glare of unencumbered reason, nature appeared to lack any inherently unifying *telos*. Traditions that narrated nature as sacred, magical, and soulful were revealed to be the ignorance and hubris of superstition or the instruments of religious hegemony. Inexorably, modern scientists such as Newton proceeded to discover the immutable and universal laws that direct the mechanism of evolution, yet that do not embody any rational purpose beyond that of keeping the clock-work universe ticking. Purpose only enters this universe in the form of human self-consciousness.

In the world Descartes and Bacon saw, external limitations are overcome, and thereby progress attained, to the extent that rational knowledge about natural machinery takes over from the inefficient meandering of evolution. A lack of rational development in existing social practices, a lack of material advance, in short, a lack of progress, appeared as the specter of backwardness, idleness, and ultimately moral decay. Yet as we have already seen, notions of progress and stability do not stand over and against each other so much as they inform and shape each other. The Enlightenment idea of 'stability' was derived instrumentally from the antecedent metaphysical conviction that the purpose of social life was to develop the raw stuff of existence into a rational form, a Paradise on Earth. By the end of the seventeenth century C.E., the empiricist Locke had established the new progressivist tradition as a political as well as a scientific project by presenting "stable government in a free society" as a "problem of social engineering" and, therefore, ultimately as a "technical problem with a timeless solution. . . . Since the activities of government are limited and since only government is stabilized," Dessauer concludes from Locke, "sufficient space is left for changes within that sphere of society which remains outside the control of government."[26]

The ideal of progress articulated by Descartes and Bacon remained bound to their Christian cosmology and is quite different from how we understand this ideal today. According to Mitcham, we need to see the rise of the science of progress as arising out of a longer historical transformation of temporal narratives within Christianity:

> But as the second coming of Christ became indefinitely postponed, late classical and medieval Christians began more and more to focus on the practice of spiritual life in the present—a phenomenon that finds expression, for instance, in the rise of Christian monasticism. When the Christian mythology of progress toward transcendent salvation by means of faith was abandoned, it was easily replaced with a scientific ideology of progress toward happiness in this world and satisfaction by means of power and control over nature. . . . Thus was the idea of progress toward some final state that is a renewal of a beginning state replaced by progress toward a final state as perfection of realization of potentiality.[27]

David Noble's important study of the medieval convergence of Christianity and technology, *The Religion of Technology* (1997), develops this argument further. He chronicles the gradual turn from Saint Augustine's fourth-century C.E. description of technology as "but the fragmentary solace allowed us in a [fallen, corporeal] life condemned to misery," to Roger Bacon's rapturous celebration of technology, some nine hundred years later, as nothing less than the divine path of reunification through which "we become one with him [God] and Christ and are gods."[28] In particular, the millenarian doctrines arising in the early centuries of the second Christian millennium, based largely on the prophesies of the Book of Revelation, linked an increase in technological activity with the immanence of the eschaton; indeed, technological proliferation was seen to prepare the way for the reign of Christ on earth. The development of Benedictine monasticism from the eighth to the thirteenth centuries of the Christian era is perhaps the clearest practical expression of the turn to technology as transcendence. Inspired by John Scotus Erigena's claim in the ninth century that the 'mechanical arts' are "man's links with the Divine," the Benedictines overcame the prevailing religious prejudice against the practical, turning their monasteries into nothing less than centres of innovation.[29] It was the monasteries, then, that gave medieval Europe its greatest machine, the mechanical clock, more than three centuries before Descartes and Francis Bacon finally harnessed philosophy to technology. The invention of mechanical time facilitated the observance of the religious offices of the rigidly apportioned Benedictine day. But it did much more than this, of course. It simultaneously launched European culture into the tunnel of linear, chronological time in which progression makes more sense than permanence. Practices centered on the cyclic—diurnal, lunar, and solar—regeneration of nature became less defining of humanity's reproductive mortality. Time, Lewis Mumford observed before most others, "took on the character of an enclosed space: it could be divided, it could be filled up, it could even be expanded by the invention of labor-saving instruments."[30] Progress, the march toward a better future, became a more obvious explanation of mortal life in a clockwork world than cosmologies which assert the infinite regenerative return of the same and, thus, the inherent stability of time itself.

Ecological Scarcity

According to Tijmes and Luijf, it was Hobbes, only a few decades after Descartes and Bacon, who entrenched the Enlightenment celebration of liberty, and the resentment of external restraint this entailed, in a distinctly modern notion of scarcity. This notion has become so central, say Tijmes and Luijf, it is "the organising principle of modern society" and "the origin of a free market economy." Through it we are now mired in the irony that "only modern and affluent societies are convinced about the importance of scarcity as a determinant of social behaviour."[31] We are mired with Keynes in the uncomfortable position that although the love of money is a "disgusting morbidity," it is necessary until scarcity is overcome and we can once again "honour those who can teach us how to pluck the hour and the day virtuously and well."[32] More disturbingly still, our cosmology remains bound to Hobbes's view of the human being as a machine with no inherent ontological relationality to anyone or anything else and whose most noble function is self-preservation in an essentially indifferent reality.[33]

The latemodern fixation with sustained economic growth is the other, often hidden, side of the coin of Hobbesean scarcity and self-preservation. To be sure, many within the Enlightenment tradition, especially during the two-hundred-year transition between the emergence of modern science and the industrial reformation of everyday life, emphasized the need to overcome scarcity whilst remaining uninterested in, even resistant to, the growth of capital. The sources of classical economic theory found in the work of Burke, Malthus, Ricardo, and J. S. Mill thus belong more strongly to the older traditions of stability than the newer tradition of progress.[34] It was Mill, after all, who offered the following vision of the stationary state:

> It is scarcely necessary to remark that a stationary condition of capital and population implies no stationary state of human improvement. There would be as much scope as ever for all kinds of mental culture, and moral and social progress; as much room for improving the Art of Living, and much more likelihood of its being improved, when minds ceased to be engrossed by the art of getting on.[35]

So why, we are compelled to ask, does growth remain an imperative amidst our latemodern prosperity? Why does it remain an imperative amidst prosperity whose extent Mill and perhaps even Keynes could not have dreamed?

The view that economic growth is synonymous with progress did not in fact gain wide currency until the emergence of the Victorian preoccupation with perfecting commercial enterprise.[36] Mitcham suggests that with the acceptance of the need for economic growth in the second half of the nineteenth century came the final abandonment of the residual Christian ideal that social progress meant movement toward a final, perfect condition. Progress became and remains understood as the move away from an imperfect past, drawing upon "the fear of an allegedly unsustainable and inhuman past of scarcity that must at all costs be avoided."[37] Growthmania, to borrow Daly's evocative term, continues to flourish with tumorous abandon because the Art of Living, as modernism understands it, can only be begun once scarcity has been overcome, and scarcity cannot be overcome because the logic of technological progress demands it.[38] Scarcity is the goad that stimulates the productive fervor and consumptive desire necessary to prevent technological society from collapsing in on itself. It draws social activity into a vortex in which material achievements are overwhelmed by an escalation of the human desire for material achievement, explains Hans Achterhuis:

> Before the rise of modern economic society no one suggested that unlimited desire was a natural quality of man. Scarcity arising out of this limitless, triangular desire is, in this general sense, an invention of modernity. . . . Fear largely regulates human behaviour. In order to escape fear, men create by covenant "the great Leviathan," a semi-absolute state, that keeps its subjects in awe and that prevents the permanent scarcity from developing into outright war.[39]

Hobbesean scarcity has little if anything to do with the basic human need for sustenance. As Murray Bookchin pointed out in *Post-Scarcity Anarchism* (1971), scarcity, if it is to mean anything in human terms, "must encompass the social relations and cultural apparatus that foster insecurity in the psyche."[40] But, as Nicholas Rescher also made so clear in his *Unpopular Essays on Technological Progress* (1980), this insecurity and the discontent it fosters is systematically cultivated in the technological society, for it is crucial in maintaining the dominance of a materialist understanding of progress.[41]

Ecological Interconnectedness

Notions of limits and scarcity have been relatively easily absorbed into the ecomodernist agenda. In the language of sustainable development, anxiety about global ecological limits and scarcity has become another source of materialist insecurity, and another imperative for economic growth. In contrast, concerns about global ecological interconnectedness would seem to bear greater promise as a foundation for countercultural concerns about sustainability, because they come as much from deep internal contradictions within the scientific project launched by Descartes and Bacon as from without.

Since the early 1970s, scientists such as David Bohm, Fritjof Capra, Erich Jantsch, James Lovelock, Lynn Margulis, Ilya Prigogine, Rupert Sheldrake, Lewis Thomas, and Gary Zukav have popularized the idea that science is on the cusp of a paradigm shift from metaphysical mechanism to metaphysical holism, the latter idea being variously explained through recourse to a raft of secondary notions including relativity, improbability, nonlocal causality, and chaos.[42] Books like Capra's *The Tao of Physics* (1975) and Zukav's *The Dancing Wu Li Masters* (1979) spread the argument that this new science bears more in common with Eastern traditions and with the organicist traditions of premodern Europe than it does with the history of modern science itself.[43] "What we need then," claimed Capra in his later book, *The Turning Point* (1982), "is a new 'paradigm'—a new vision of reality; a fundamental change in our thoughts, perceptions, and values. The beginnings of this change, of the shift from the mechanistic to the holistic conception of reality, are already visible in all fields and are likely to dominate the present decade."[44] In particular, says Capra, the new paradigm will displace our existing "hard" technology, for it is "fragmented, rather than holistic, bent on manipulation and control rather than cooperation, self-assertive rather than integrative, and suitable for centralized management rather than regional application by individuals and small groups. As a result, this technology has become profoundly antiecological, antisocial, unhealthy, and inhuman."[45] Almost two decades on, Capra's new vision of reality, his view that we are shifting from a materially progressive to a spiritually progressive age, has about it an air of unreality. Unlike Marcuse's earlier analysis, Capra's vision provides few resources for understanding the rise of the Promethean project of ecomodernism because, unlike Marcuse, Capra saw the implosion of technological modernity as inevitable and immanent.

The ongoing paradigm shift that is displacing Newtonian assumptions from the foundations of science displays all of the confusion that we would expect, having read Thomas Kuhn's characterization of the movement from "normal" to "revolutionary" science.[46] A baffling array of holisms now compete for paradigmatic status. More baffling still is the question of what the social and ecological consequences of this conceptual fluidity might be. In particular, what does holism mean for those who wish to contest the agenda of ecomodernism? Certainly the language of holism is crucial to the way in which sustainability is being articulated within the radical ecology movement, especially within the disciplines of ecological economics and ecological technology.[47] And evidence of the "rising culture," of the New Age and New Paradigm movements, in which Capra and others place so much hope, remains. But are latemodern experiments in spiritual, psychological, medical, ecological, and agricultural holism, important though they are, being crafted together into social structures, into a new culture, capable of resisting modernist narratives of technological progress?[48] In responding to this question, we must note that in addition to becoming a mantra in the radical ecology movement, the language of holism can be found in everything from sex manuals to corporate management strategies, from sports psychology to natural medicines, from programs for personal development to computer modelling, from investment planning to fruit fly control. And as we saw in chapter 1, the language of sustainable development has sought to absorb the challenge of holism in the form of ecoefficiency initiatives such as cradle-to-grave product life-cycle management, industrial ecology networks, planetary science, and futures management. More broadly, it has absorbed this challenge in the idea of a managerial ethic of global stewardship. We find that even conventional developments in neoclassical economics and global technological trajectories, such as information systems, gene manipulation, and space exploration, are repackaged by ecomodernists as embodying the new systems paradigm. In effect, the idea of holism has been integrated within ecomodernism by reconstituting it as the challenge of globalism, a challenge demanding the solutions of technological holism. With the destiny of the earth linked to the progress of technology in ecomodernist narratives, ecosystems are being brought under the management of technosystems. Ecological scarcity is being met by the attempt to subsume the ecosphere within a self-sustaining technosphere.[49]

To assert that the language of holism has been trivialized and diffracted is not to deny the importance of countering the mechanistic structure of modernist thinking. But clearly we must be careful to explicate the political vision underlying any particular use of the term *holism*. We need to observe that, despite its commitment to holism, ecomodernism remains faithful to the view that the earth is a neutral instrument (an instrument for human survival, no less), albeit an instrument more amenable to systemic rather than mechanistic metaphors. Within the innovations of technological holism, nature is no longer thought of as a collection of parts akin to the cogs of a clock, but as parts akin to the electronic information suspended within the interconnectedness of the Internet and defined, not by atomistic autonomy, but by networks of interconnection that allow for their appearance, but not their location, at any point. In fact, to paraphrase McKibben, within ecomodernism nature can no longer be thought of at all; every vantage point in the latemodern web of life affords us only a view of technology.[50]

The New Age of Ecomodernism

Technocratic sustainability and the technological holism on which it rests will unavoidably precipitate further ecological degradation. It is already reinforcing the structures of global injustice. Nevertheless, it may well be able to sustain itself much longer than is commonly assumed in radical ecological thought. I suggest that it is not difficult—especially after enduring the pulsing techno-ethical optimism of now commonplace documents like the World Resource Institute's *Transforming Technology*—to envisage that sustainable development of the ecosphere will produce lush technologically managed eco-enclaves: something like the Biosphere 2 dome in the Arizona desert writ large.[51] The oft-repeated Holy Bendiction of biosphericists is their invocation of "the ability to change an entire planet to produce environments to support life."[52] Inside these enclaves, provided environmental engineers can overcome the substantial and expensive problems that led to the abandonment of Biosphere 2, the cycling of gases, liquids, and solids, as well as the functioning of biological components (species), will be engineered for maximum stability, amenity, and aesthetic appeal. In addition to blue roses in every vase, we can picture industrial ecology estates hidden behind plantations of hypergrowth trees, vast concrete dykes on the post-ice cap shores, buildings fitted with air-purification systems, electric cars whizzing by, and, everywhere, verdant lawns fed on an endless supply of desalinated water. We can envision vast laboratories for hydroponic production of vegetables and production of meat from artefanimals designed to be without sensory function. The whole hive of activity can be expected to be powered by a distant desert (ex-tropical forest perhaps) littered with uncountable numbers of photovoltaic panels; a network prudently backed up by a stratospheric network of wind turbines. And looking up to the heavens, which are clear except for the traverse of laser beams scrubbing the sky of pollutants, we see capsules of sulphur dioxide launched into the upper atmosphere to reflect excessive ultra-violet radiation.

It was the burden of chapter 2 to show how the majority of humanity, let alone the majority of other living beings, can only be further disadvantaged by the environmental and corporate forms of neocolonialism that biospheric instrumentation brings. The avowed holism of the global view falls prey to "'the globalitarian temptation' of a techno-natural selection of human beings."[53] The already dismal prospects of the South are likely to be vastly worse if a network of elite eco-enclaves, managed by transnational business, absorbs the majority of the world's natural, technological, and financial resources while exporting large amounts of waste and magnifying technologically mediated risks of all sorts. And what of those destined to be among ecomodernism's chosen few? Even if "technocratic caretakers could be mandated to set limits on growth in every dimension," I agree with Ivan Illich that a beneficiary of ecomodernism "would live in a plastic bubble that would protect his survival and make it increasingly worthless."[54]

It is consequently of concern to me that ecomodernism has drawn unexpected support from the convergence many radicals ecologists see between the new agenda of technological holism and the new expressions of cosmological holism. Many holists have taken up the theme popularized by technoutopians like R. Buckminster Fuller that technology embodies a principle of cosmic evolution, and have ended up in an uncomfortable embrace with the World Business Council for Sustainable Development.[55] They have found themselves in what Guy Beney so lusciously depicts as a "multi-ambiguous collusion" arising between awareness of "global emergencies," the new scientific paradigm of "Geopóiesis" and a techno-scientistic revival of "Mother-Earth" imagery.[56]

Gaia and Ecomodernism

The Gaia Hypothesis of the late 1970s, put forward by scientists James Lovelock and Lynn Margulis, posits that the earth's development and functioning can only be understood as arising out of the maturation of a totally integrated, interacting, and self-sustaining organic whole: an organism. The Gaian argument, despite its suggestive title and adoption by many as a nonanthropocentric environmental ethic, provides ample scope for showing how apparently radical holistic concerns may be absorbed within the agenda of ecomodernisation. Holism leads as easily to technocratic control of the planet as it does to animist respect for Gaia. Consider two particularly unwelcome political directions in which Lovelock himself develops this argument. First, Lovelock speaks of "the people plague" which inflicts Gaia like a tumor, an unhelpful phrase at best, and one that persists still in environmentalist literature.[57] He decries the "heresy of humanism" and warns that "humanist concerns about the poor in the inner cities of the Third World . . . divert the mind from our gross and excessive domination of the natural world."[58] The upshot is a twist on dire warnings about ecotastrophe, a twist we can call the revenge of Gaia. This is, so it seems, an apolitical revenge in which rampant human populations will be expunged to make way for other experiments on Gaia's path to cosmic maturation—"her" path of sustainable development. The mass starvation and suffering of the majority of the world's population becomes not a consequence of exploitation by the North but a consequence of Gaian immunology. In rejecting this argument absolutely, I agree with Beney that "to applaud the new paradigm, one must . . . feel assured of being on the right side of 'geobiotic' history!"[59]

The second unwelcome feature of Lovelock's account is his justification of a scientific platform for planetary control and sustainable management on the basis that humanity, as the agent of Gaia, is fulfilling its cosmic destiny. Lovelock, an expert in gas chromatography who has worked on projects such as the NASA space program, champions geophysiology, a holistic technoscience seen by some as the key to the further evolution of the Gaian immune system.[60]

The combination of the trend toward planetary medicine with the globalitarian overtones of Gaia's revenge brings the disturbing result that affluent technological societies, represented as the fittest in the battle of technological evolution, emerge as those destined to become the nervous system of Gaia. Between the lines of Lovelock's argument I see the extremely problematic claim that all other forms of society are being pared away by Gaia to make room for "her" Promethean leap into awakening. The transcendentalist Peter Russel reflects the spirit of Lovelock's hope that modern technological development represents nothing less than the evolution of Gaia's consiousness itself: "As the communication links within humanity increase, we will eventually reach a time when the billions of information exchanges shuttling through the networks at any one time create similar patterns of coherence in the global brain as are found in the human brain. Gaia would then awaken and become her equivalent of conscious."[61]

The Growth of Ecomodernism

I have sought to show that the agenda of ecomodernism is capable of resisting and more importantly of absorbing and reconstituting cultural discourses about sustainability arising out of concerns about ecological scarcity, biophysical limits and biospheric interconnectedness. Understanding the cultural narratives that inform ecomodernism, we can better understand what was at stake in Bookchin's open letter to the ecological movement in 1980, written just as ecomodernism was beginning to take shape:

> The hoopla about a new 'Earth Day' or future 'Sun Days' or 'Wind Days', like the pious rhetoric of fast-talking solar contractors and patent hungry 'ecological' inventors, conceal the all-important fact that solar energy, wind power, organic agriculture, holistic health, and 'voluntary simplicity' will alter very little in our grotesque imbalance with nature if they leave the patriarchal family, the multinational corporation, the centralized bureaucratic and political structure, the property system, and the prevailing technocratic rationality untouched.[62]

Bookchin warns that, without explicit political and moral practices within which it is expressed, anxiety about global ecotastrophe, population bombs, and the exhaustion of resources will be coopted. It will be redefined so as to legitimize the revitalization of economic growth, the technological domination of nature, and the political oppression of nontechnological societies. While an ecologically sane culture that has moved beyond the reach of patriarchy, global capital, and technocratic organization will inevitably adopt organic forms of agriculture and renewable forms of energy, Bookchin reminds us that this logic does not necessarily work in reverse. Organic produce can be niche-marketed to global elites—the glossy brochure linking new "organoveg," say, to better facial skin-tone and higher IQ in your children, is easy to visualise—as readily as can the latest handgun.

II THE METAPHYSICS OF MASTERY

Bookchin was one of the first scholars, along with Arne Naess, Rosemary Radford Ruether, and others, to attempt to articulate a metaphysical, epistemological, and moral basis—an ecophilosophy—for the radical ecology movement. Radical ecophilosophy is now a well-established, burgeoning subdiscipline.[63] Although it has until recently been occupied with internal disputes, frequently eschewing policy debates for the finer points of cosmology, this literature provides a promising domain within which to contest the project of ecomodernisation because it directs us to the metaphysical sources of the Enlightenment project of technological progress. [64] In what follows, we shall see how the dualistic structure of these sources continues to impel the technocratic will to dominate nature. We shall see that this will is impelled both by the fear that we live in a purposeless, indifferent reality and by the hope that unrelenting dominion over nature is the best way to invest security and meaning in human life.

The ecophilosophical mode of inquiry offers many perspectives from which to explore what is culturally unsustaining in and about the technological society. However, radical metaphysical critiques often produce a sweeping condemnation of Enlightenment philosophy, as well as the call for a deep shift in consciousness, without adequately characterizing the practical political possibilities open to us in our technological world. Indeed, in challenging the abstractly theoretical and universalizing character of modern thought, many critics offer equally abstract, universalizing alternatives. When ecophilosophers as wise as Michael Zimmerman are "encouraged by movements that encourage sustainable development," and those as historically attentive as Carolyn Merchant suggest that the "SD movement . . . is consistent with many of the goals of ecofeminists," we must accept that the danger of being coopted within the project of ecomodernism is very real.[65]

Through the diverse landscape of radical ecophilosophy interweave three primary paths—social (or political) ecology, deep ecology and feminist ecology, (or ecofeminism)—and a much larger number of overlapping secondary paths—Gaian ethics, bioregionalism, neopaganism, postmodern ecology, ecosocialism, ecopsychology, ecotheology, and modern interpretations of Buddhist, Taoist and indigenous philosophies. What makes all of these diverse approaches radical with respect to ecomodernism is that they place an ontological rehabilitation, if not repudiation, of modernist traditions antecedent to any more sustainable ethical understanding of our relationship with nature. These approaches thus deepen our understanding of the ways that ecological concerns about limits, scarcity, and interconnectedness are now caught, and are being misshapen, between colliding cultural objectives. They are caught between the modernist, technocratic project that seeks to master nature, increasingly rendering it the mere stuff of domination (to echo Marcuse), and a loose coalition of ontological traditions that seek to recognize the ways culture emerges out of, and is embedded within, the meaningful and more-than-human history of the living earth.

A strong postwar literature (in English) on the mastery of nature includes Marcuse's *One Dimensional Man* (1962), Lewis Mumford's *The Myth of the Machine* (Vol. 1 1967, Vol. 2 1970), William Leiss's *The Domination of Nature*

(1972), Carolyn Merchant's *The Death of Nature* (1980), Morris Berman's *The Reenchantment of the World* (1982), Bookchin's *The Ecology of Freedom* (1982), Vandana Shiva's *Staying Alive* (1989), Val Plumwood's *Feminism and the Mastery of Nature* (1993), and Noble's *The Religion of Technology* (1997).[66] In what follows I shall consider Plumwood's book in some detail as it is valuable to the present inquiry in four ways. First, her concentration on metaphysical dualism allows her to link the oppression of nature with forms of oppression based on gender, class, and race. This linkage more accurately reflects the oppressive reality of latemodern development strategies than do many solely ecological critiques. The breadth of this critique also constructively undermines the artificial divides many radical ecophilosophers erect amongst themselves, a fact I consider a distinct advantage as social ecology, deep ecology, and feminist ecology (and the many variations on these themes) are all vital constituents of a subtle, nonuniversalising and richly plural critique of ecomodernism. Second, Plumwood's account evaluates a number of earlier critiques of the mastery of nature in the process providing us with a clear description of the political dangers that reside within the recent surge of holistic thinking. Third, as many other latemodern forms of social inquiry also offer critiques of dualism, Plumwood's work helps us position radical ecophilosophy within the wider latemodern reaction against Enlightenment philosophy. Lastly, her account clarifies the ways in which instrumentalist epistemologies emerge from dualistic metaphysics, thereby establishing a firm point of departure for my inquiry into the philosophical character of latemodern technology in part II of this book.

Dualism and Mastery

Plumwood provides a description of the connections between the dualistic structure of modernist thought—understood as "the process by which contrasting concepts . . . are formed by domination and subordination and constructed as oppositional and exclusive"—and the cultural roots of oppression based on race, class, gender, and species.[67] In her view, forms of "oppression from both the present and the past have left their traces in western culture as a network of dualisms, and the logical structure of dualism forms a major basis for the connection between forms of oppression."[68] Plumwood's articulation of the inherently political structure of dualism provides a countervailing argument to the common confusion that would equate dualism with dichotomy or mere distinction.[69] The essence of dualistic thought is not the attempt to draw distinctions, even radical distinctions, but is the attempt to establish the hierarchical supremacy of a superior (the master subject) over an inferior (the master's subject). Dualism constructs "central cultural concepts and identities so as to make equality and mutuality literally unthinkable."[70]

In this century, philosophical debates about dualism have centered primarily on the crumbling foundations of the Cartesian philosophy of mind.[71] Phenomenologists and existentialists, with their shared emphasis on the mind's embodiment, have blazed the way for many who seek to recover lived experience from the deanimating rigor, the *rigor mortis*, of Cartesian objectivity.[72] Latemodern pragmatists such as Richard Rorty similarly question the epistemological traditions that present the universe as a radically separate reality that can only be known to the extent that it is reflected in "our own Glassy essence."[73] Feminist theorists have sought to expose dualism based on gender, and their strong orientation to *praxis* makes it clear that even though dualism has been relegated to many a philosopher's dustbin, it remains a powerful political construction in the actual oppression of women, nontechnological societies, and nature.[74]

Much of this wider literature locates the origins of metaphysical dualism in Descartes' philosophy of mind.[75] Plumwood proposes, however, that of the three steps necessary before dualism could take hold in Western thought, the first two were taken via Plato's transcendentalism and only the last via Descartes' rationalism. In her account, it was Plato who, first, developed an ontology of the human that had mind at its center, with the mind in turn being centered in reason, and second, conceived of mind and reasoning as oppositional to nature. From this position, Descartes took the final step of imposing an absolute ontological divide between humanity and animals, thus representing nature as both mindless and meaningless without human interpretation and re-creation.[76] In his infamous call for humanity to become the "masters and possessors of nature," he established culture as the master of nature.[77] Aligning the body with the feminine and coding both as "natural," he also offered a rational defense of the Christian politics of patriarchy.

Nature's Master

One of the many strengths of Plumwood's approach is that she does not try to make too much of particular dualistic pairs. She contends instead that particular dualist concepts "form a web or network. One passes easily over into the other, linked to it by well-travelled pathways of conventional or philosophical assumption."[78] What weaves this web together is the "multiple, complex cultural identity of the master formed in the context of class, race[,] species and gender domination."[79] This master subject "resists the recognition of dependence, but continues to conceptually order his world in terms of a male (and truly human) sphere of free activity taking place against a female (and natural) background of necessity."[80] In the hands of the master identity, dualistic thought becomes a tool for a conceptual colonization of otherness. Plumwood describes the logical structure of this colonization as having five sequential stages.

1. Dualism foregrounds the reality of the master subject while invalidating and backgrounding the reality of the subordinate subject. The "master's view is set up as universal, and it is part of the mechanism of backgrounding that it

never occurs to him that there might be other perspectives from which he is background."[81]

2. As already noted, dualism requires not merely distinctiveness of the dominant from the subordinate but their complete hyperseparation. This is achieved through a process of radical exclusion by which the dualistic pair cannot be seen as in any way continuous.

3. The feature of incorporation builds upon radical exclusion to redefine the subordinate by negative contrast to the master subject. The master subject becomes the source of cultural meaning while the slave, as background, is pushed toward the cultural periphery.

4. Lacking inherent meaning, the subordinate subject becomes instrumental means in service of the master's ends. The logical structure of dualism supports the master's claim that the slave is better off as a slave than in any other type of relationship.[82]

5. Finally, homogenization of the inferior conceptual quality and of the politically subjugated subject results as a consequence of the slave being reduced to a lack, becoming an undifferentiated, unimportant, and decentred other set against the master self.

This description of the logic underlying dualistic arguments is of great benefit to us, for it enables an understanding of how the modernist project of mastery is sufficiently flexible to allow the "boundaries of the concept of nature [to] shift to encompass changing social circumstances and opportunities for colonization and mastery."[83] In ecomodernism, then, the background of nature is now increasingly defined in terms of the global, homogenizing crisis of ecological scarcity. In turn, this crisis is construed as being the result of the insubordination of humanity's slave, nature—insubordination that threatens the very liberty of its master—in its basic task of sustaining the conditions of human life. This is to be dealt with by stripping the subordinates of technological culture—chiefly nature, non-European traditions, and women—of any autonomy that may remain after the ravages of developmentalism. In the discourses of technocratic sustainability, nature is no longer deemed responsible for cleaning the air and water, regulating the weather, directing genetic evolution, and absorbing waste. Nature will now tackle these tasks under the whip of global environmental managers, and then only in the absence of more ecoefficient and increasingly cybernetic substitutes.

Holism and Relationality

The account of dualism set out in *Feminism and the Mastery of Nature* dispels some of the confusion surrounding the current, on-going paradigm shift from mechanistic to holistic theories. Plumwood unmasks the fact that many of the new holists have responded to the longstanding dominance of Newtonian atomism with an extreme dissolution and denial of difference that attempts to coalesce all life within a single metaphysical principle, an overarching principle of the Whole. She identifies three forms of this argument. First, there is the scientistic rationalism of sociobiologists who suggest that all beings are equal, yet equally mindless and equally self-seeking; little more than automatons, they are physically and emotionally engineered by genetic programming according to the dictates of the overarching principle of evolution. Second, there is the principle of Gaian geopoiesis, a global principle of self-organization, that trumps the interests of individuals and species. Third, there are those ontological therapies, which she associates (too strongly, in my view) chiefly with deep ecology, that conflate dualism and difference to assert the oneness, the monism, of reality.[84]

I agree with Plumwood that many attempts to explain individual life in terms of an overarching Whole construct their holistic metaphysics in a way that accords to the Whole the hegemonic status of the master subject to which individual beings are enslaved. These extreme, reductive holisms remain as dualistic as the extreme forms of atomism they would seek to supplant. They reduce life to a cosmic principle in which "beings do not escape thinghood and move to the status of being in their own right. They do so only through participation with the center."[85] Thus, where Cartesian mechanism strips the body and nature of *telos*, reductive holism strips individuals of teleology, agency, and intentionality, investing these only in the Whole. Recalling us to my earlier comments about the risk of many apparently antimodernist holists being sucked up within the ecomodernist agenda of technological holism, Plumwood points out that "it is a fallacy to assume that negating atomism is negating mechanism."[86] It is a fallacy because a dualistic understanding of holism still conceives of a subordinate subject that is mere instrumental resource, a mechanism, for the master subject. Whether the earth is conceived of as totally lacking a unifying *telos* or as embodying the *telos* of a Whole, it can still be understood as instrumentation. In the technological holism nourished within discourses of sustainable development, human and natural agency is subordinated to the principle of stability encoded in the technosphere, that is, the technological Whole. The center through which ecomodernist identity is formed is that of technological efficiency, which is to say that beings attain thinghood only to the extent that they become efficient producers. Ecomodernism is headed toward the immanent and final stage in the technological mastery of nature; what Plumwood calls the stage of "devouring the other." This stage bodes "the final destruction of all resistance that the earth as other has to offer, as biotechnology and other mastering technologies repopulate the world with assimilated artefacted life and the master science strives to harness all global energy-flows to the Rational Economy."[87] Yet, although this claim represents an excellent summary of the emerging project of ecomodernism, I find Plumwood's response to this project too abstract and lacking a noninstrumentalist explanation of technology.

RELATIONALITY

Plumwood's response to the project of mastery is to identify nondualistic frameworks "which recognize dependency on the earth as sustaining other as the central fact of human life." These are frameworks of what I have been calling cultural sustainability. She continues:

> In the sphere of human society, the best examples of such mutually sustaining relationships are found in care, friendship, and love. Increasingly the project of expelling the master from human culture and the project of recognising and changing the colonising politics of western relations in other nations converge, and increasingly too both these projects converge with the project of survival.[88]

Her description of care, friendship, and love as sustaining correlational virtues has obvious appeal, but it also lacks clarity. Not least because the project of ecomodernism is first and foremost a project of survival. The ecomodernist concern for the survival of technologically abundant forms of life feeds an agenda of technological holism that extends the master's colonizing power into the structure of the ecospheric web itself.

Plumwood, in naming the ecomodernist project unintelligent, even irrational, suggests that the master is unable to realize sustainability because "he" systematically destroys "in biospheric nature a unique, nontradeable, irreplaceable other on which all life on the planet depends. . . . [T]he master rationality is unable to grasp its peril."[89] She thus admits of only two possible endings to the project of the technocratic domination of nature. The first is an ecotastrophic finale in which the master rationality destroys the "sustaining other," resulting in the death of all life (including that of the masters). The second, quasi-utopian finale is one in which the master subject's own latent concern for survival in the face of immanent destruction leads to the abandonment of the project of mastery through domination. However, as I have already discussed, we must also consider the possibility that this project may not have a foreseeable end and may yet prove at least partially successful in establishing deeply iniquitous but relatively sustainable (that is, stable) technocratic structures capable of supporting a global ecocracy for some time yet.

Plumwood's description of nondualistic frameworks for sustaining the other draws heavily upon feminist accounts of the moral status of the relational self.[90] Seeking to counter both extreme atomism and extreme dualism, Plumwood proposes a nonreductionist basis for thinking about holism that she names the intentionality criterion:

> On the intentionality criterion, mindlike qualities are spread throughout nature and are necessary to its understanding, but there is a high level of differentiation between different sorts of mindlike qualities and between different sorts of beings which have them. Intentionality is an umbrella under which shelter more specific criteria of mind such as sentience, choice, consciousness, and goal-directedness (teleology).[91]

In a colorful illustration of how differentiated entities are embedded within relations of continuity, Plumwood points out that "we can be delighted that our local bandicoot colony is thriving without ourselves acquiring a taste for beetles."[92] In contrast to the absolute ontological isolation of the self described in modernist narratives, articulations of the multifaceted and interactive expressions of intentionality within self-developing living systems offer us an account of the self-in-relationship that "can not only explain how instrumentalism can be avoided but also provides an appropriate foundation for an account of the ecological self, the self in noninstrumental relationship to nature."[93]

Plumwood's ontology of moral relationality is undeniably helpful in displaying the severe constrictions imposed on modern moral thinking by dualism. She provides a convincing account of how the modern, hyperseparated self, shaped by self/other dualism, presents obvious problems for ethics because the world that is radically separate from this self becomes mere instrumental means. "The domain of ethics," Plumwood reminds us, "is the domain of those who have not been instrumentalized, whose needs and agency must be considered."[94] Yet the logic of dualism ensures that this ethical domain appears to contain only a collection of hyperseparated selves who can only relate to one another according to instrumental procedures that maximize their individual autonomy. To argue that this flawed ontology has tragically deformed modern traditions of thinking about moral experience and technology, in particular, is the burden I take up in part II.

While I believe that Plumwood's diagnosis of our latemodern moral malaise is headed in the right direction, the therapeutic dimension of her argument remains largely unexplored in *Feminism and the Mastery of Nature*. Her account of relational selfhood emphasizes the significance of practices of care that are capable of nurturing virtues such as respect, gratitude, sensitivity, reverence, and friendship. She contends that under the reign of dualism there has been a "loss of the particular practices of care through which commitment to particular places is expressed and fostered."[95] Plumwood does not clearly define these "particular practices of care," however, nor does she explain how and why they are lost, nor how they are to be differentiated from practices of mastery. Neither does she show how this loss is linked to dualistic thought. Finally, she only hints at the connections between the ecological self's "commitment to particular places" and the social character of the virtues of selves-in-relationship.

The abstractness of Plumwood's constructive argument about practice is particularly strong with respect to technology, the dominant determinant of our everyday experience of relationality. Plumwood's articulation of the importance of care, friendship, and love in cultivating an ethic of relationality needs to be balanced with a recognition of the problematic status of these virtues in the technological society. If anything, the mythology of romantic love, communities of friendship based on personal preference, and the care lavished on cars, lawns, and other personal possessions has been heightened in latemodernity. These relations commonly serve to soften the isolation experienced by hyperseparated selves, but they do not necessarily undermine the cultural carelessness, disregard, and disdain shown for anything that exists outside of the procedures of technocratic management. Care, friendship and love have become instruments for maintaining the comfortable unfreedoms, to paraphrase Marcuse again, afforded by private consumption. They have become, in many ways, emotional tools for ensuring the passive loyalty of the subjects of technological culture to the reign of the master rationality. We need only consider how the technological culture obscures the master project of technomedicine—with its instrumentalism of the ideas, intuitions, and practices of health—behind the caring, warmly human, yet subordinated profession of nursing. We might consider also how the physical alienation of suburban dwellings—each adrift in their own ocean of lawn and each resisting the presuburban essence of its place—is obscured beneath the traffic of telephonic and on-line friendship, beneath the loving attention given to vibrant televized celebrities and beneath the careful way in which we take care of that which we possess.[96]

Dualism, Instrumentalism, and Technology

Plumwood's account provides clear explanations of the instrumentalization of human beings, nonhuman beings, and other natural things. We can better appreciate how the viewing of non-European racial groups instrumentally led European colonialists to either employ these people as slaves or to eradicate them as unwanted dross in the building of civilization out of the *terra nullius* of the New World. Similarly, the domestication, selective breeding, and direct genetic modification of animals become rational projects aimed at making animals, such as the newly cloned sheep, more efficient instruments for meeting the needs of the technological culture. Equally, nonliving, natural things are defined and valued in accordance with their most useful instrumental functions. So, for instance, when a river is understood first and foremost as a store of kinetic energy, it becomes rational, indeed necessary, to develop this resource so as to harness its energy to human purposes.

Part of Plumwood's response to the instrumentalization of our world is to call on us to reflect on our relationships based on love, friendship, and care because they reveal the enrichment that comes through relationships in which we do not instrumentalize the other. It is Plumwood's hope, as it is my own, that we will extend this experience of enrichment beyond our human lovers, beyond our immediate genetic progenitors and descendants, beyond those who share our forms of life, to encompass the worlds of people, beings, and things that are currently enslaved by the master rationality. But to understand the sources of our enrichment we need to recognize that there is a fourth category of thing that is commonly instrumentalized yet little discussed in radical ecophilosophy. This fourth category of things are the products of technological activity.

Dualistic metaphysics assert that an artefactual object can have no value-in-itself because it is designed and created for the sole purpose of serving an instrumental function. Indeed, it is because artefacts are the products of human ingenuity and action, in contrast to the other categories of thing, that the instrumentalism of artefacts has seemed to many simply a commonsense view. Yet the recognition that artefacts have an instrumental dimension arising from their design and application is not the same as the instrumentalist claim that technologies have only an instrumental dimension and thus exist as neutral servants of human purposes. Drawing on Plumwood's description of the logical structure of the web of dualistic thought, we can see that technology appears in modernist thought as background to the foreground of scientific knowledge and is thus defined through incorporation as applied science. Defined in this way, as a void needing to be filled with the meanings of science, technology appears as a unified, homogenized phenomenon that is radically excluded from the human mind and thus excluded from the human essence. Technology is seen as an expression of and a servant to this essence, not as constitutive of it. As an external application of mind, technology is simply instrumental means to the ends established in pure reason. The stuff of nature, which seems void of purpose, gains meaning to the extent that it is transformed into artefacts that are neutral conduits of the rational, scientific purposes of the master subject. Framed in this way, it is easy to see how issues of morality and ontology have long been radically excluded from issues of technology in the technological culture. In the next chapter I draw upon recent literature in philosophy of technology to resist this exclusion. I shall prepare the ground here, however, by emphasizing the importance of a noninstrumentalist account of technology to our understanding of the project of mastery.

Dualism in Practice

What is required, if we are to draw metaphysical questions about nature toward our everyday experience, is an extrapolation of Plumwood's analysis of dualism to the preeminent fact of our lives; our technological practices. We need to scrutinize the widely neglected cluster of dualistic explanations—principally of theory/practice, human/technology, science/technology, and artifice/nature—that are interwoven within our contemporary understanding of technology. This is a cluster absent from Plumwood's account, a fact that prompts the observation that dualism so defines the structure of our technological world that even its philosophical critics often implicitly perpetuate the theory/practice dualism that so pervades the Academy.[97]

It is often assumed that dualistic thought can be overthrown by an intellectual commitment to rejecting the "concept" of dualism. Conversely, it is often assumed that nondualistic cosmologies establish the starting point from which nondualistic forms of life will be established. Both of these approaches fail to appreciate that the web of dualism is woven as much throughout our technological practices as it is throughout our theories.[98] In every detail of our latemodern technological lives, we experience nature as an instrumentalized and homogenized background against the colorful foreground glamor of our technological lives. In our cars we are hyperseparated, not just by ethical dualism, but also by sheet metal. Behind our computer screens we are indeed disembodied minds which exist in no-place. Flying into our ecotourist resorts, we do read nature as a collection of colors, textures, noises, and shapes, organized, and re-organised by developers for the sensual pleasure and relaxation of the earth's masters. Dualism, instrumentalism, mechanism, atomism, anthropocentrism, androcentrism, Eurocentrism, and a host of other isms are most completely understood as descriptions of our embodied, technological forms of life. They are descriptions of our lived technocentrism. It is in understanding how we live these unsustainable isms that we may understand how to begin living beyond their reach.

Just as dualism grew and developed though centuries in everyday life, so too its recession will necessarily encompass changes to everyday life over many generations. Our forms of thinking and our forms of practice have coevolved under the guidance of dualism. Dualism has most commonly been discussed as if it were a 400-year-old, or even a 2,400-year-old, theoretical *Gestalt*.[99] But if there was any kind of *Gestalt* heralding the modern era, it was that performed by Galileo when he turned the telescope on the heavens. As Hannah Arendt stressed in her classic description of the birth of our modern World Alienation, "not ideas but events change the world . . . and the author of the decisive event of the modern age is Galileo rather than Descartes."[100]

Even with our focus on events, it is hard to find a single decisive event, including Galileo's uplifting of the merchant's telescope from the harbor to the heavens, that stands outside of wider historical processes of technological transformation.[101] Lynn White, Jr., for one, has been an articulate opponent of the view that the millennium before European Enlightenment was simply a dark or middle period between the illumination of Greco-Roman and Modern civilizations. White points to the significant technological innovation, much of it a result of global diffusion—at which point we might do well to note that the Chinese developed a remarkably precise mechanical clock 1,000 years before the Benedictines—that was transforming the lived world of Europe (and coevolving with the emerging eschatological interest in technology) in the centuries before pure mind's enlightenment.[102] The world of Descartes and Bacon, with its heavy ploughs, stirrups, waterwheels, windmills, mechanical clocks, printing presses, giant organs, navigational and cartographic instruments, maps, paintings with three-dimensional perspective, mirrors, eyeglasses, microscopes, and telescopes revealed a natural world that could be known and that simultaneously paled beside the ordered cerebral world of progressive urban life.[103]

The gradual diffusion of glass mirrors into every home of Western society, for instance, may have been of profound philosophical significance. The mirror provides the instrumental function of allowing humans to view themselves as a disembodied image. This was a function previously served only by occasional glimpses reflected from water, from a liquid eye, from polished metal, or via the pictorial and verbal representations of others. It is interesting to speculate how growing up in a world where mirrors are commonplace might also encourage a heightened sense of personal ego; greater faith in the possibilities of the detached objective observer; increased identification with individual difference rather than communal similarity; greater concern with the physical transformations of aging and, thus, with mortality itself; stronger emphasis on the reality conveyed by imagery of all kinds; and, ultimately, apparent confirmation of the atomistic reality of discrete and unique physical selves. Looking with a cultural eye at our mirrors, we are encouraged to ask: was the narcissistic possibility of the lake writ large in European culture via image technologies such as the mirror? Is it a coincidence that Western economic and political systems are based upon an understanding of atomistic individuals in competition with one another? Is it a coincidence that science is premised upon the possibilities of detached, objective observation? Is it a coincidence that the act of material consumption is founded on a desire to express our unique individuality, our personal creativity, our idiosyncratic style? Did the infant Descartes stare into a disembodied reflected image of himself long before he disembodied his mind with his insight, *cogito ergo sum?*[104]

Certainly no one artefact such as the mirror can really be considered in isolation from wider patterns of technological innovation in any meaningful analysis of cultural change. But the point of this brief and admittedly superficial historical digression is simply that we must acknowledge the ways in which the possibilities for articulating our ecological relationality are at one and the same time the possibilities for experiencing our ecological relationality. Sustaining practices that nurture the relational virtues of love, friendship, and care are at least as much a matter for careful discrimination about the technology choices available to us as they are for careful discrimination about how we think. The important question seems to me to be this: How can we begin to live more sustaining, less alienating forms of life? How can we, that is, sustain those relationships that are the source of our sustenance?

Radical ecophilosophers identify the cultural roots of our unsustainability as metaphysical and moral rather than as technical. This is a profoundly important undertaking. Yet many radical metaphysical critiques often relegate the issue of technology to secondary status in their search for greater philosophical clarity. They often assume that the necessary reform of technology will emerge as a consequence of a prior and more fundamental philosophical reformation. In this regard, it is significant that much radical ecophilosophy draws its main therapeutic inspiration from notions and places of wild or pristine nature—areas largely devoid of human artifice. So inspired, many ecophilosophers disregard the ecological possibilities of more overtly social environments.

In its poetry and politics, radical ecology emphasizes the value of immersing ourselves in the nontechnological purity of nature. "The natural environment is that which affords inspiration to the ecologically minded," Freya Mathews asserts in *The Soul of Things* (1996), an essay that other radical ecologists would do well to heed, "whereas it is artificial environments that leave them feeling flat or otherwise dispirited."[105] For radical ecologists, the world of artifice typically has about it the air of impurity, of perfection tarnished by technological endeavor. The danger here is of course that the Enlightenment dualism of technology and nature will simply be inverted in ecological thought. The danger is that nature will be elevated over technology, thus perpetuating the logic of hyperseparation in our thinking about our ecological and social reality. What will be missed in any such construction is the fact that every explanation of nature, be it deep or shallow, is at the same time a politically laden explanation of human culture.[106] Rather than drawing ever deeper lines in the sand between artificial and natural materiality, then, I think we should see the dangers of any dualistic construction of our social and ecological reality, whether this construction be one that accords supremacy to technology or to ecology. What is needed is an understanding of human practice that acknowledges that human embeddedness in their world is at once technological and ecological.

My aim in this chapter has been to unearth the cultural project of mastering nature enough so as to expose its broad outline. I have argued that the agenda of ecomodernization not only furthers this project, it advances greatly the prospects for its culmination. In chapter 2 we saw that sustainable development translates into a politics of colonization. In this chapter, we have seen that the logic of colonization runs through this cultural project back to its origins in metaphysical dualism. It is vital to appreciate that the push toward ecoefficient mastery is demanded by the logical structures that are established in our minds and in our practices. But, as we have seen, attempts to counter the totalizing project of technological mastery often produce equally all-encompassing critiques that, while they may make great sense as an alternative theoretical construct, are able to make little impact at the level of our everyday lives. What is needed, and what I attempt in part II, is an analysis of technology that holds together an ontological awareness that we are engaged in the hegemonic project of building an unsustaining world with a practical awareness that our actual experience of technology is always far too rich and ambiguous to be fully enframed by this world. I shall argue that the ideal of sustainability cannot be abstracted from our lived philosophical commitments and that technology lies at the heart of these commitments.

II

The Technological World

Here is the test of wisdom,

Wisdom is not finally tested in schools,

Wisdom cannot be pass'd from one having it to another not having it,

Widsom is of the soul, it is not susceptible of proof, is its own proof,

Applies to all stages and objects and qualities and is content,

Is the certainty of the reality and immortality of things, and the excellence of things;

Something there is in the float of the sight of things that provokes it out of the soul.

—Walt Whitman

CHAPTER FOUR

BUILDING A DEFORMED WORLD

INTRODUCTION

THE CONTEST OVER the ideal of sustainability is at once a contest over the future of technology. But what is it in or about technology that is to be questioned? The lack of substantive cultural responses to this question has smoothed the way for the Promethean project of ecomodernism. "At every turn the prospects for sustainability hinge on the resolution of problems and dilemmas posed by that double-edged sword of unfettered human ingenuity." But, continues educationalist David Orr, at the "point where we choose to confront the effects of science and technology, we will discover no adequate philosophy of technology to light our path."[1] Philosophers César Cuello Nieto and Paul Durbin emphasize the corollary: "Sustainability is one of the most important issues in the history of humankind—indeed, in the history of life on Planet Earth—so as philosophy of technology comes into its own, it clearly *must* address this issue."[2]

As sustainable development discourses display, ecomodernism simply rehearses the inquiry that has established technology as the apotheosis of late-modern life, that is: can technology be made more efficient? This form of inquiry establishes technology as a province in which engineers reign with the authority of high priests, making unquestionable the political and moral fact that everyday life is being refashioned so as to maximise the production and consumption of technologies.

I have to this point used the expression *the technological society* in a purely descriptive way. Our society is certainly technological if the pervasiveness of technological objects and techniques is any guide. But we should now note that our society is also technological in a much more fundamental sense. *Technological society* names a peculiar political and moral condition in which the greatest common good is understood as the greatest possible productivity of technosystems. David Strong's aphorism is apt: "Our culture's vision of the good life is the goods life."[3]

Given the primacy of technology in modern life, one might expect, suggests Langdon Winner, "that the philosophy of technology would be a topic widely discussed."[4] But technology has been most commonly regarded as lacking cultural substance, as being the neutral instrument of human purposes: a proposition underwritten by the dualistic logic of instrumentalism. Thus, the "most accurate observation to be made about the philosophy of technology is that there really isn't one," Winner lamented in 1986.[5] Fifteen years on, this observation is, thankfully, now less accurate.[6] And one of the first goals of the new philosophers of technology has been to expose and coordinate the extensive yet previously ignored, marginalized, or misinterpreted resources for thinking about technology that exist in the philosophical canon.[7] Nonetheless, the philosophy of technology is nascent, struggling to be recognized as a credible academic pursuit and dwarfed still by the philosophy of science, the oldest and largest subdiscipline of philosophy.[8] As evidence of this struggle, the philosophy of technology has to date played virtually no direct role in debates about sustainability.[9] This is unfortunate, not only because it eases the passage of ecomodernism, but also because it allows radical ecological critics to reject this project either on the abstract basis that it represents flawed theory or on the blindly reactionary basis that modern technology is somehow inherently bad. Consequently, both technocratic and cultural discourses about sustainability have substantially failed to investigate how technologies, woven as they are within the fabric of our practices, *express, shape and perpetuate* our philosophical commitments.

The failure to take philosophical account of technology contributes to the perception that the proliferation of postmodern theories, particularly the critiques of grand universalizing narratives associated with poststructuralism, reflects the rise of nonmodern ways of life. It has "become commonplace to claim that modernity fragments," notes Anthony Giddens, leading to the "emergence of a novel phase of social development beyond modernity—a postmodern era."[10] My concern with the neocolonial hegemony of ecomodernism should display why I am uneasy with this now familiar claim. Like Sean Cubitt, "I cannot believe that the triumphant western imperium is about to crumble because of an epistemological event. . . . I prefer to understand what we are living through as an acceleration of modernity, not a break from it."[11] And like Giddens, I think that the "unifying features of modern institutions are just as central to modernity . . . as the disaggregating ones. The 'emptying' of time and space set in motion processes that established a single 'world' where none existed previously."[12]

While a great deal of our contemporary technology, especially information technology, is postindustrial, this does not thereby establish that it is in any meaningful sense out of the reach of the grand technological practices of modernism that largely define our experience as corporeal, temporal, and earthly beings. Our practices undoubtably display a malleable, even ephemeral, surface that gives rise to an appearance of unprecedented choice and diversity. One of my aims in this chapter, however, is to chip away enough of this surface to reveal that the technological substructure of these practices is the embodiment of modernist narratives. Indeed, with the continued purging of traditional, premodern remnants and illusions, a purge that is by no means complete, our entry into the twenty-first century is perhaps best viewed as being more fully, unashamedly modern than anything that has gone before.[13]

As I observed in concluding part I, the challenge posed by the ideal of sustainability is that of holding together an appreciation of technology as a cultural project of mastery, through which we are building an unsustaining world, with an appreciation of technology as the forms of life in which the unquenchable ambiguity and richness of our human condition is played out. In part II, my approach to the normative questions held open by the ideal of sustainability takes the form of an inquiry into the nature of moral life in the technological society. I inquire into the fact of our search for orientation toward questions of what is good in our world. And as Hans Jonas has remarked, the focal fact of modern life is technology.[14] I shall argue that this focal fact is twofold: what we call technology encompasses the phenomenon of world-building and the experience of correlational practice.[15] This and the next chapter are primarily concerned with the technological world, while chapters 6 and 7 seek to recover within moral discourse an awareness of our everyday correlational experience in this world.

In this chapter, I develop the claim that we are blindly building a deformed world. Our latemodern world is unsustaining because it is deformed. And the blindness of our technological agency is the cause of our world's deformation. Yet our blindness is not irreversible; it has not been inflicted upon us. Following Winner, I contend that it is the result of a dangerous cultural sleepwalking in our understanding and experience of technology.[16] Understanding technology as world-building is first of all a matter of recovering our experience of technology, that is, our practices, from the philosophical strictures of instrumentalism. This act of recovery, and the movement toward philosophically rich explanations of technology that it entails, I undertake in section I before describing the phenomenon of world-building in section II. Then, in section III, studying modern technology as forms of life rather than as tools, I describe the actual dislocation of means and ends in the technological world that so deforms our practices and the technological world we are building.

I RECLAIMING TECHNOLOGY AS PHILOSOPHICAL TEXT

Any philosophical inquiry into technology should produce the observation that we need to distinguish between first order and second order explanations of technology. First order, or instrumental, explanations produce descriptions of artefacts and the making and use of artefacts. In Carl Mitcham's helpful modal terminology, these are definitions of "technology as object"—e.g., tools, machines, utensils, apparatus, utilities, clothes, structures, and automata—and "technology as activity"—e.g., invention, design, manufacture, maintenance, and craft.[17] Second order, or substantive, explanations move on to consider implications arising from the making, presence, and use of artefacts in human experience. These explanations uncover the ways in which technology is constitutive of our experience and thereby has substantive social character in its own right. Mitcham consequently adds to object and activity the substantive modes of "technology as knowledge" and "technology as volition." By *knowledge*, Mitcham refers to phenomena such as unconscious sensorimotor awareness, rules of thumb, descriptive laws, and technological theories. *Volition*, although defying simple definition, is best located with ideas of human need and human will.[18] We can quibble here with the labels Mitcham chooses, and I will later closely align knowledge and volition with practice and ontology respectively, but his modal explanation of technology is insightful and sufficiently open to encompass many different substantive explanations of technology.[19]

Instrumentalist Explanations of Technology

First order, instrumental definitions are partial rather than false. *Instrumentalism* names the fallacy of taking this partial truth to be the whole truth about technology. The Promethean discourses of modernism and now ecomodernism are founded upon this fallacy. So too are many cultural critiques of these agendas. Our everyday use of the term *technology* executes this fallacy of externalization by referring simply to the external objects that surround us and are apparently external to the human condition.[20] The formal, dictionary definition of technology executes this fallacy in a different and far less obvious way by declaring technology to be the actualization of factual (that is, artefactual), scientific knowledge. Technology thus becomes, not objects, but "the scientific study of the practical or industrial arts."[21] In effect, technology is represented as having no identity outside that of science. Technologies become the external, static objectifications of the culturally neutral knowledge provided by science. Once again technological objects stand divorced from the essence of human knowing and experience.

The theoretical opacity of technology is juxtaposed with the increasingly pervasive, known presence of technological objects in our lives. For most of us, the word *technology* brings to mind images of lap-top computers, compact disks, FA-18s, satellites, operating theatres, miniature hearing aids, combine harvesters, and cloned sheep. If pressed, we may add to these images some of the inconspicuous yet indispensable artefacts that reside in the dim corners of our lives. Things such as copper pipe, steel saucepans, sticky tape, latex condoms, and foam mattresses, not to mention cheese and beer. What the idea of 'technology' rarely elicits from latemoderns is an appreciation that technology has become the defining characteristic of our social and ecological relationality.

Here we should note that Mitcham's description of technology as knowledge should not be constrained to our knowledge of artefacts. Knowledge, and the discourse in which it exists, is itself a crucial component of the functioning of technosystems. Technique is in effect a relational, in contrast to a material, artefact. Take latemodern money, a technosystem that manifests as objects such as coins, stock exchanges, and automatic teller machines, but that also manifests in the centralized, bureaucratic codification of social and ecological relations. The fact that the collapse of the Russian ruble triggered global economic meltdown, and hence increased my mortgage repayments to an Australian bank, is principally a function of the fact that fiscal knowledge about global social relations is one of the artefacts of the technosystems of money. And, of course, as information is itself becoming the preeminent product of technosystems, technology is creating a novel and potent mode of technology as knowledge which Albert Borgmann, in *Holding on to Reality* (1999), astutely calls information-as-reality. In this mode, the idea and experience of reality themselves become artefactual, bringing an almost limitless knowledge of virtual realities, yet simultaneously occluding accumulated cultural knowledges about our actual, embodied reality.[22]

Technology "has come to mean everything and anything; it therefore threatens to mean nothing," Winner declared in *Autonomous Technology* (1977).[23] What then is the nature of contemporary public discourse about technology? As risk minimizers, prudent consumers, political agents, knowledge maximizers, and leisure seekers, our public discourses implicitly and, less frequently, explicitly answer this question in the form of the hope that technology, like nature, is at the loyal service of its human masters. Technological societies function on the assumption that we will employ the most efficient, the least risky, and the most convenient of our technological servants vigorously and optimistically, accepting that there is little sense in any antitechnological position. We may reject one artefact in favor of another that is more (eco)efficient or safer (with safety itself increasingly seen as a subspecies of (eco)efficiency) according to the dictates of technical calculus. But we are not commonly encouraged to openly question the normative implications of technological change. Why question something that is neutral, transparent, rational, and thoroughly in our power?

Our discussions about technology therefore return by default to technical concerns. Many of us develop a degree of technical literacy simply keeping abreast of the latest innovations that propose to minimize hazards in our lives—innovations such as electrical circuit breakers, water filters, antilock brakes, and acellular vaccines. Yet it is rare indeed that this literacy extends to normative reflection about cultural notions like 'risk'.[24] How often does public debate extend to a consideration of the ways in which risk generation and risk amelioration emerge out of the same technological dynamics, for instance? We are aware that cars are perfected to make possible greater speed of transit, while air bags are perfected to minimize damage in high-speed collisions. But what are we to make of this observation? When we learn that airbags themselves pose new risks, such as the decapitation of young passengers, the resources we bring to bear on our reflections are mostly technical. All we can assert with conviction is the need for better air bags. When we learn that radiation from Australia's mobile phones—and there are now more than six million in a human population only recently grown to nineteen million—might be suppressing immune function, damaging memory, and inducing brain tumors in their users, the best we can do is hope that a radiation shield will soon be on the market: it seems inconceivable that we would hang up our phones.[25]

With Eyes Shut Tight

In his 1937 *Road to Wigin Pier*, George Orwell observed that "[p]eople invent new machines and improve existing ones almost unconsciously, rather as a somnambulist will go walking in his sleep."[26] Winner, echoing Orwell's prophetic voice, notes that "the interesting puzzle in our times is that we so willingly sleepwalk through the process of reconstituting the conditions of human existence."[27] Our technological somnambulism is clearly due in part to the uninteresting drone of instrumentalist descriptions of our world. But Orwell raised the question of whether our sleep is not also induced by some inherently soporific feature of modern technology:

> The machine has got to be accepted, but it is probably better to accept it rather as one accepts a drug—that is, grudgingly and suspiciously. Like a drug, the machine is useful, dangerous, and habit-forming. The oftener one surrenders to it, the tighter its grip becomes. You only have to look about you at this moment to realise with what sinister speed the machine is getting us into its power.[28]

This sixty-five year old warning is sobering. Could it be that we are in the power of technology?

It was popular amidst the postwar boom of the 1950s and 1960s to answer this question in the affirmative, giving rise to a spectrum of determinist accounts of technology ranging from the deeply pessimistic, akin to Orwell's view, to the deeply optimistic.[29] Optimistic determinists, exemplified by the founder of cybernetics Norbert Weiner, the engineer-poet R. Buckminster Fuller, and the science fiction writer Arthur C. Clarke, argued that technological evolution is as inevitable, natural, and as inherently good as ecological evolution.[30] On the pessimistic side, many cultural critics—Jacques Ellul, Siegfried Giedion, Friedrich Juenger, Ivan Illich, Herbert Marcuse, and Lewis Mumford foremost amongst them—argued that technology was becoming an autonomous and totalitarian force.[31] These scholars were disparagingly dismissed by many as reactionary neo-Luddites.[32] In this regard, it is worth noting Kirkpatrick Sale's argument that, contrary to the much-mocked caricature drawn by progressives, the social movements of Luddism have, since their beginnings in England's mid-counties in the early 1800s, been more concerned to denounce narrow political agendas masquerading as technological progress than they have been to denounce technology *per se*.[33]

The superficial nature of much debate about whether or not technology is out of control is nowhere clearer than in the fact that the thinker most commonly credited, or discredited, for popularizing negative technological determinism, Ellul, did not in fact assert that technology had moved beyond the reach of human agency. Although I cannot hope to do justice to this controversial proposition here, I recommend David Lovekin's *Technique, Discourse, and Consciousness* (1991), which does. Lovekin argues that Ellul's dialectical analysis presents the apparent reality of technological autonomy to be an illusion cast by the soporific logic, what Ellul called the *lethotechny,* of modern thought about technological progress.[34] The target of Ellul's polemic was not the demon of modern technology. His targets were the technological somnambulists, the frantically productive human agents who persist in the illusion of instrumentalism while building a deformed world in which humanity can only ever encounter itself as the agent of technology.[35]

Given the superficial if not downright spurious nature of much debate about technological determinism, it is encouraging that social constructivist literatures on technology have grown strong over the last few decades.[36] These literatures display for us many of the ways in which technologies are cultural products embodying particular, especially historical, social interests and perspectives. Feminist theorists, in particular, have taken up the task of showing how modern technological change does not encode the essentially apolitical interests of a transhuman, evolutionary force, but rather encodes the all-too-political interests of patriarchal and other elites.[37] The latest of Arnold Pacey's important historical works on technology, *Meaning in Technology* (1999), has added another string to the constructivist bow by presenting the case that we must explore and give voice to the meanings the making and use of artefacts have for us, not just in our social relations, but in our actual tactile, sensory, and aesthetic experience.[38]

Winner has described the welcome attempt in the sociology, history, and philosophy of technology to "redeem the technological prospect from both the facile optimism of liberal, enlightened thought and the pessimism of cultural critics." However, in light of this trend toward more nuanced and subtle understandings of technology, it is ironic that "in the world at large it appears that the experience of being swept up by unstoppable processes of technology-centered change is, in fact, stronger than it has ever been."[39] Winner points out that, as useful as many social constructivist methodologies are, the pervasive phenomenon of technological somnambulism, that is, the character of technological culture itself, is largely beyond both their interest and their scope. He presents—in a way analogous to my critique in part I of the facile optimism of ecomodernism—the fields of computers and telecommunications as examples *par excellence* of the latemodern experience and discourse of technological autonomy.[40]

Instrumentalism and Determinism: Strange Siblings

The question that Winner's observation about the pervasive experience of external, transcendent technological forces prompts is this—How has the appearance of technological autonomy taken hold in a polity whose predominant explanations of technology are instrumentalist? Surely the idea that technology is neutral is antithetical to the experience of technology as an unstoppable force? These questions return us to the fallacy of externalization, for determinist and instrumentalist explanations of technology share the *a priori* logical structure of dualism. Both theorize technology as hyperseparate from our human essence.[41] As paradoxical as it may seem, latemoderns typically display both instrumentalist and determinist traits in their attitudes to technology. Instrumentalism represents technologies as being our external servants. Technological determinism inverts this hegemonic construction to postulate that technology is hyperseparate from us, not as our faithful servant but as our governing master. The master of the deterministic trope is all-knowing and benevolent or tyrannical and vicious or disinterested and unpredictable, depending on your proclivity for and autobiographical sources of hope, despair, or ambivalence about the technological society.[42] The unifying characteristic of the varying formulations of technological determinism is that they portray humanity as the instrument of technology.

It is the peculiar combination of instrumentalism and determinism that brings about our strange, epic sleepwalk in the embrace of technological efficiency. Despite our inability to articulate the nontechnical aspects of technological change, we are not asleep to the substantive character of technology. We are familiar with the confusing and conflicting feelings that have no doubt been evoked by the proliferation of technology since the first axe made both kindling and revenge more readily available. But as dualistic thinking provides little purchase on the cultural phenomenon of technology, and as instrumentalism demands that we consider only discrete objects, we latemoderns typically crystalize our ambivalence about technological change around the clarity of either fear or adoration for particular artefacts. As if we were feeling our way with our hands, our technophobic and technophilic reactions help orient us as we unconsciously build a thoroughly technological world.[43] Neither positive nor negative forms of determinism are likely to be widely accepted as comprehensive explanations of the technological forces that govern in our lives, for they clash too directly with our instrumentalist hopes and thus would make us uncomfortably aware of their incoherence. Our interpretation of our experience is thus commonly one of ambivalent or even schizoid determinism.[44]

Viewing our world with our mind's eye, we may smile in our dreams about good technology and sweat in our nightmares about bad technology. Mostly, we simply shuffle forward, wondering where technology is taking us, strangely convinced that wherever it is, we will still be in charge of the map. We grope about in the dark trying to decide if this or that object, newly imposed upon our lives, is good or bad.

Despite the rhetoric of rabid ecomodernists about the coming Golden Age of Technology, public discourses reflect in equal measure fear that technology has created ecological crises and hope that it may provide ecological salvation. Many of us reel in fear and disgust from nuclear reactors, strip mines, and genetically modified organisms. And who cannot be anxious at reports that synthetic chemicals are causing nonhuman and human testicles alike to shrivel? Yet most of us would welcome the security of crawling into a marvellously strong yet light-weight tent as a blizzard sets in over our isolated camp. Many activists know the wonderful sense of power the Internet affords in obtaining and disseminating subversive information, and in holding together networks of marginalized voices. And who cannot but marvel at the solar-powered desalination plant that turns the sea into drinking water? Apparently affirming that we can see through technology, our technophilic and technophobic reactions sustain the illusion that we can yet sift good technologies from bad technologies, finally realising the Enlightenment promise of mastery.[45]

We are aware that our lives are being changed. Yet Winner reminds us in *Technology Today* (1997) that we are extraordinarily inarticulate about the nature of this change:

> We realise that the technologies that surround us affect matters we deeply care about—the satisfactions of working life, the character of family ties, the safety and friendliness of local communities, the quality of our interactions with schools, clinics, banks, the media, and other institutions. But finding ways to deliberate, organize, or act on these institutions is not part of our education or our competence.[46]

The cause of our blind, pliant embrace of technology is the misunderstanding that technology stands outside of the essential ambiguity and ambivalence of our human condition. Technology is neither neutral nor an unstoppable force—good, bad, or indifferent. Technology in all its manifestations is as ambivalent, as unpredictable, as honorable, and as depraved as are human agents themselves. Technology is not the neutral vehicle of human agency, it is the essence of human agency.

The task of waking up from our technological sleepwalk, thereby reclaiming technology as a crucial space of moral and political judgment and choice, is the task of recognizing technology as the practices through which we come to know ourselves, each other, and our shared world. We know, for instance, that many latemoderns now orient their lives away from the oppressive judgments they find others make of them in RL (real life) and toward the liberated selfhood of VR (virtual reality). "As of mid-1998," reports Margaret Wertheim, "there are one hundred million people accessing the Internet on a regular basis. . . . In just over a quarter century, this space has sprung into being from nothing, making it surely the fastest-growing 'territory' in history."[47] And as Sherry Turkle has demonstrated, those young people who have grown up with the Internet provide us with one of our presently most revealing views of technology as a mode of self and world understanding.[48] Closing our eyes to views of this sort, in the fallacious conviction that technology stands outside of our selves, we walk through an apparently foreign landscape that we are able to understand only to the extent that we are able to project technophilic and technophobic reactions onto the objects within it. We deny the reality that as the builders of this landscape we encounter ourselves everywhere and anywhere. With the logic of an infant who closes their eyes so you cannot see them, technological sleepwalkers hide only from themselves.

Toward Philosophically Rich Explanations of Technology

This is not to say that focussing our philosophical gaze on the phenomenon of technology will cause it to resolve into a unequivocal, unambiguous entity. On the contrary, I do not think that we should seek to provide unambiguous, discrete definitions of the equivocal phenomenon of technology. In the spirit of amazement championed by George Kateb, I accept that "[n]o particular story about modern technology could be adequate, no interpretation sufficient."[49] Philosophically rich explanations of technology are better directed toward clarifying and illuminating the focal fact of our existence rather than toward the counterproductive pursuit of defining it. They are best directed toward showing us how technology is constitutive of rather than external to our humanity. These explanations will need to have a broad public and transdisciplinary character if they are to counter the introversion of professional philosophy that has contributed so greatly to the neglect of the focal fact of our lives.

The proposition that the atomistic self of latemodernity is defined more by its self-reflexive character, by which it establishes its own conditions for self-understanding, than by pre-given sources of cultural meaning, has become well known, especially through the work of Giddens.[50] But this reflexivity commonly stops at the boundaries of the hyperseparated self. As Borgmann has long argued, latemodernity is remarkably unconscious of the public character encoded in its material reality:

> Concrete, everyday life is always and, it seems, rightly taken for granted. It is the common and obvious foreground of our lives that is understood by everyone. Therefore it is almost systematically and universally skipped in philo-

> sophical and social analysis. But if the determining pattern of our lives resides and sustains itself primarily in the inconspicuous setting of our daily surroundings and activities, then the decisive force of our time inevitably escapes scrutiny and criticism.[51]

The phenomenon of technology, more than anything else in our latemodern world, declares the need for public resources of philosophical reflection. In my view, meaningful responses to this declaration begin with the acknowledgment that our collectively shared technological practices constitute a philosophical text far richer than any penned at a desk.

The first thing this technological text teaches us is that our celebrated late-modern freedoms take place against the background of technological necessity. That this truism is too little understood is borne out by Borgmann's remark in *Theory, Practice, Reality* (1995) that academia has a "hard time discerning the essential uniformity of contemporary culture; its halls resound with the cacophony of interminable moral debates and the din of the culture wars."[52] It seems beneath our lofty interests to comment on the fact that we can only attend that workshop on, say, Transpersonal Psychology and the River Craft of the Hmong of the Mekong Delta, if we first catch a plane or jump in a car. This background necessity is typically presented as the irreducible, biological, and thus acultural, necessity of having to travel; or to eat, drink, excrete, rest, or seek shelter. But the reality remains that artefacts such as planes and cars are representatives of political and moral commitments that we sustain by the mere fact of our use of them. We sustain them regardless of whether or not their production provides meaningful work, contributes to just economic exchange, makes wise use of resources, or creates minimal ecological harm. As I have argued elsewhere, simply reaching for a cup of coffee in a moment of conviviality at a sociological conference on globalization is an event of momentous, yet typically unnoticed, paradox, irony, delusion, and danger.[53]

The dominance of instrumentalist philosophies of technology has left us largely unable to see—and thus largely unable to speak of—the ways that technological skill is the medium in which we must establish life-orienting relationships with people and places, beings and things, past and future, known and unknown. What we have lost sight of in our technological activity is the fact of *world-building*. For, like Winner, I think that the question technology ought to prompt in any meaningful public discourse is this: "How can we limit modern technology to match our best sense of who we are and the kind of world we would like to build?"[54] Like Winner, I think "new worlds are being made."[55]

II THE DEFORMATION OF EVERYDAY PRACTICE

In its most abstract sense, the term *world-building* introduces the unfamiliar idea that our technological digestion and reconstitution of material reality is constitutive of our understanding of this reality. I believe it can be argued that technology, like language, is not simply a manifestation of human knowing, it is also a precondition for it. Through technology we build domains of correlationship, worlds, in which daily life becomes intelligible and through which humanity becomes at least partially oriented to the unintelligible vastness of existence. These are matters of ontology that I will take up in chapter 5 by engaging with Martin Heidegger's postwar explanation of modern technology. But the notion of 'world-building' leads us gradually away from the concrete familiarity of our world. It leads us toward obscure—or, rather, profoundly obscured—matters of ontology, but it begins from what is most familiar to us, and this is, I think, its particular strength as an explanation of technology.

World-Building

To understand world-building we can do no better than consult Winner's essay *Do Artifacts Have Politics?* (1986):

> Consciously or unconsciously, deliberately or inadvertently, societies choose structures for technologies that influence how people are going to work, communicate, travel, consume, and so forth over a very long time. In the processes by which structuring decisions are made, different people are situated differently and possess unequal degrees of power as well as unequal levels of awareness. . . . Because choices tend to become strongly fixed in material equipment, economic investment, and social habit, the original flexibility vanishes for all practical purposes once the initial commitments are made. In that sense technological innovations are similar to legislative acts or political foundings that establish a framework for public order that will endure over many generations.[56]

Perhaps the most obvious illustration of Winner's point is the automobile. Understanding the automobile as the representative of a wider world-structure, we can understand how the emergence of automobiles is at one and the same time the emergence of automobile-dependent places. The center of gravity of pre-automobile cities has necessarily been dispersed by the centrifugal force of automobile-systems; it could not be any other way. The majority of latemodern cities are structured to facilitate the movement of road-based vehicles and only indirectly to facilitate the movement of people. Structured in this way, cities sprawl, necessitating vast technological infrastructure. The juxtaposition of private and public space, of work and home life, of built and natural form, of neighbourhood and region, of urban and rural communities has been thoroughly patterned by the practices of automobile transport.

We should note two things here. First, technological change is at once a change in the ways in which we live, and this in turn represents a change in the ways in which we relate to each other and to the things around us. Technological innovation is a form of political and moral innovation. Every technological change is a form of social experiment; it is impossible to foresee all of its consequences. "In the technical realm we repeatedly enter into a series of social contracts," says Winner, "the terms of which are revealed only after the signing."[57] Second, we should note that technological changes in human life, once encoded in technological structures, become inconspicuous and largely independent of conscious, explicit policy. In fact, our material technological contracts frequently become an unquestioned determinant of policy. Clothed in practice, artefacts create an inconspicuous background upon which the drama of self-conscious agency unfolds. Because our daily, habitual patterns of living come to rest upon this background, and because any change to this background may therefore involve tremendous social upheaval, these structures are usually only open to fundamental change over extended, often intergenerational, periods of time. This is not to say that technological society has become stagnant, however. This would be a preposterous claim; latemodernity is after all characterized by a remarkable flux of artefacts. My argument, then, is that much of this flux flows and eddies over the bedrock of our fundamental cultural commitments.

Taking account of technological change as social experiment and technological structures as unquestioned background and drawing from Ludwig Wittgenstein's theory of language games in *Philosophical Investigations*, Winner recommends that we come to see technologies as forms of life.[58] Seeing technology in this way, we are reminded always that we are engaged in building a social world that predetermines to a large degree the ways of living that are possible. To resist socially dominant technologies is to resist socially dominant forms of life and thus to put into jeopardy the sources of our social identity. Pullitzer-winning author Richard Ford has no illusions about what is at stake: "I don't have e-mail. I'm not on the Internet. I don't have a cell-phone or call waiting or even a beeper. And I'm not proud of it, since my fear, I guess, is that if someone can't find me using any or all of these means, they will conclude that, for technical reasons, I don't exist any more."[59]

But does Winner's analysis mean that we must see a nail as a form of life? And a bowl? What about a sock? This line of questioning appears absurd if we accept the fallacious assumption that instrumentalism has made pervasive, namely, that the primary fact of an artefact is its material discreteness, its apparently independent, atomistic existence. Yet, just as is the case with living beings, artefacts emerge out of complex systems of interrelationship. The analogy to be drawn between ecosystems and technosystems is strong.[60] Technological artefacts are representatives of wider social patterns. Any account of world-building must accord primacy to patterns rather than objects, that is, to technosystems, rather than to technological artefacts. Further, the boundary between object and knowledge is inevitably blurred in an account of world-building. So Rayon socks, imported into Australia from a Thai factory staffed by Hmong peasants, sold in plastic wrapping at my local air-conditioned shopping centre, are a revealing representative of latemodern politics. These socks are products of ethical frameworks as much as of factories. Despite its benign appearance, our jug of tap water, no less than synthetic socks, is a product of forms of life entirely different in kind to a vegetable gourd dipped in a stream or a bucket drawn from a village well. In technological societies, practices for clothing our bodies and collecting water belong within material structures very different in technical design and in cultural character from those of other worlds. These material structures have moral and political character. In latemodernity, Eurocentric hegemony and anthropocentric arrogance are under overt challenge, but they are also increasingly being encoded in the functioning of technosystems. The resultant "split between what we think and what we do is profound," observes Wendell Berry. Because we make decisions about technologies as if they were neutral tools, rather than representatives of moral and political commitments, it is to be expected that our world "must inevitably produce asthmatic executives whose industries pollute the air and vice presidents of pesticide corporations whose children are dying of cancer."[61]

Mumford's enigmatic *Technics and Civilization* (1934) and his bleak, two volume *The Myth of the Machine* (1967–70) are of enduring significance, despite his strong determinism, because he was acutely aware of the need to understand technological change in terms of the aggregation of artefacts into social worlds. The methodology Mumford sought to introduce into the history of technology was similar to that with which Thomas Kuhn so successfully decentered positivist explanations of scientific progress. Rather than describing the technological society as the product of a long march of undifferentiated technical progress from primitive to sophisticated technologies, *Technics and Civilization* provides an account of three distinct worlds, or paradigms, each founded on a distinct complex of energy-harnessing technosystems built in the European transition from the largely rural life of the Middle Ages to the early twentieth century. The creation of these new paradigms of practice was an essentially social process requiring "a reorientation of wishes, habits, ideas, goals."[62]

Unfortunately, the attempts of Mumford and others to bring culturally substantive awareness to the history of technology have been less successful than those associated with Kuhn's convergent project in the philosophy of science. Narrowed by instrumentalism, our understanding of the history of technology remains wedded to the belief that the technological society represents the present apex of a unidirectional and essentially acultural process of technological evolution.[63] Certainly the science and technology of latemodern society affords vastly more predictive control over material events. But, attending to the process of world-building, the paradigm of practice established through modern technology cannot be meaningfully described as more politically and morally advanced than any other possible paradigm on the basis of technical design alone. Indeed, as this expansionist paradigm is founded on the fallacy of externalization, we can say that latemoderns are likely to be remarkably ignorant of the political and moral character of their world.

World-Building as Cultural Trajectory

What exemplars such as the automobile make abundantly clear is that while latemodern life is very receptive to the further evolution of our existing global technosystems, it is, concomitantly, very resistant to changes which would seek to diminish or displace these systems. It is important not to confuse the internal development of a technological complex with the process of substantive cultural change. So, despite profound, even frenetic innovation in automobile-systems over the last twenty-five years, the fundamental character of the automobile as a form of life has changed little. What has changed, as Ivan Illich foresaw it would in *Tools for Conviviality* (1973), is that this form of life, with its characteristic elevation of speed as a universal and unequivocal good, has steadily become the reference point for the majority of our everyday practices.[64] The principle of individual autonomy through private consumption that the automobile embodies has been ever more fully realized, but the principle itself remains static. With the rise of the information age, the recent appearance of the "e-car", that turns the inertia of a traffic jam into a bonus opportunity for pursuing our Internet identities, should surprise no one.[65]

That the automobile embodies an organizing principle of latemodernity is made very clear in the tremendous difficulty with which technological societies are responding to the deleterious unintended consequences of automobile-systems such as global warming, not to mention the host of other automobile-dependent social problems.[66] The very structure of latemodern economies depends upon the presence and proliferation of automobile-systems. The automobile is the embodiment of a specific regime of social power relations.[67] The least disruptive, and thus the apparently most feasible, responses to unintended consequences become those that tamper as little as possible with this form of life. Feasible responses include: unleaded fuel; more efficient combustion engines; smaller, lighter, quieter, safer, and more recyclable automobiles; and alternatively powered automobiles, electric, solar, and the like. Apparently less feasible alternatives, such as communal forms of motorised transport and nonmotorised forms of transport—based on high-density, decentralised, and small-scale urban form—pose radical and unpalatable challenges to the constitutive pattern of technological culture.

The direction of cultural change stimulated by unintended consequences in the technological society helps build a world in which technological innovation is increasingly necessary to maintain, improve, and remedy existing technosystems. This is a constraint encapsulated in Winner's use of the term *reverse adaptation* to indicate "the adjustment of human ends to match the character of the available means."[68] The logic of reverse adaptation is evident in the way that political and moral concerns about ecological crisis have largely collapsed into the attempt to internalize ecoefficient artefacts within latemodern forms of life. The unintended consequences of global atmospheric change demand responses of global scale. This gives rise to the irony that these problems only reinforce the global political order which the exploitation of fossil fuels was important in establishing in the first place. Sadly, awareness of this irony is faint because, as Ulrich Beck has shown us, the world of technological society refutes our everyday experience of technology's ecological and social failures.[69] Our common sense appraisal that something is wrong here lacks conviction and dissipates in the face of the expert discourses of technocracy.

Consider the way that technological societies race blithely toward the end of the age of oil. Unlike the enhanced greenhouse effect, there are few if any grounds for arguing that the depletion and ultimate exhaustion of an obviously finite and nonrenewable resource could be regarded as an 'unforeseen' consequence. Global production of oil is likely to have peaked around the year 2000. As early as 2005 and certainly by 2010 the extraction of remaining oil reserves will rapidly become more expensive—this point will occur shortly before output will decline in real terms—as the remaining reserves that are yet to be tapped are the most inaccessible, the smallest and of the poorest quality.[70] Yet, remarkably, latemodern society continues to expend vast sums on freeways, politicians intent on re-election dangle the lure of cheaper petrol prices, businesses invest in new fleets of trucks, and consumers wander showrooms assessing the latest fashion in headlights.

As the discourses of sustainable development attest, ecomodernism has been able to deflect attention from the history of our past practices to the Promethean task of building a better future in which unintended consequences can be foreseen and accommodated. Such a future, we are told, only unfettered technological progress can provide. Yet, contrary to the dreams of biosphericists, the phenomenon of unintended consequences in technological development is inescapable, for our practices are interdependent and interactive within wider social and ecological worlds.[71] Thus, unintended consequences do not just result from design faults, they are an inherent feature of technological change. This is a point not entirely lost on latemoderns, as the popularity of the film *The Gods Must be Crazy* shows.[72] This comic story tells of the social destabilization—the world wobbling—that resulted from the entry of a Coke bottle into the world of the Kalahari bushmen. In the end, the bushmen threw this disruptive gift off the end of their world, back into the laps of the crazy Gods who had proffered it. Despite our ability to laugh at the way in which a Coke bottle cultivated envy and violence in primitive society, however, we remain unreflective about the technological sources of our own moral destabilization. The philosophy of instrumentalism stifles most of the resistance that may develop against technological progress, even when technical advance produces paradoxical and often catastrophic results. Normative reflection over the uses of technology is routinely reduced to the equation: autonomous user + neutral artefact. Therefore, if research to develop lightweight prostheses should produce an artefact of unprecedented hardness, a latemodern equivalent of the glass bottle, say, its uses will undoubtably include ever-more efficiently murderous weapons, for military and drug cartel alike; ever-bigger ships, for the transport of live cattle and oil; and ever-higher office buildings, for salving the urge for heavenly penetration, an urge that so afflicts the titans of capitalism.

Just as our latemodern societies may be called technological to name their vision of the good life, so too our world may be called technological to name the character of the practical relationships that define our forms of life. In this world, it is easy to see how many take refuge in the nihilism or utopianism of technological determinism. But human autonomy has not been nor can it ever be vanquished by technology, although without a doubt technological change is one of the most important determinants of latemodern life. The interplay between unintended consequences and reverse adaptation establishes the dynamics of world-building and creates a path of least resistance—a trajectory of technological evolution—along which technological change will flow unless the broader shape and character of our world and its forms of life are brought into question.[73] Stephen Hill's description of the apparently potent force of technological trajectories is helpful here:

> [T]he apparent social *power* of a technological trajectory lies in the embodiment (or 'objectification') of cultural meanings *within* technology, and within the *alignments* that come into being between present cultural meanings and the objectifications of the past that we confront in contemporary technological artefacts. Thus the social negotiations that go on in the development of particular technologies are themselves shaped according to a broader culture-technology frame.[74]

The culture-technology frame, or paradigm, according to which our practices have been structured, gives our latemodern world its characteristic deformity, namely that, within it, technology begets technology out of the reach of normative deliberation.

III THE TECHNOLOGICAL WORLD

Borgmann's 1984 book, *Technology and the Character of Contemporary Life*, remains one of the most illuminating and empowering accounts of the patterned coherence of the technological world. His explanation of modern technology, which emerges largely out of his ambivalent reaction to Heidegger's account, draws our attention beneath the vast novelty and dazzling diversity of latemodern artefacts and toward the peculiar, deformed practices they make possible.

Borgmann's observation that technology is "the characteristic way in which we today take up with the world" is at one and the same time the observation that technology has become the defining characteristic of our world of practice.[75]

What gives the technological world its essential uniformity? Borgmann's answer to this question is, in short, "the device paradigm." This notion has the great merit of holding together in our thinking the latemodern world and latemodern practices. It holds together paradigmatic social systems and human experience of devices, the whole and its parts, without reducing one to the other. Borgmann thus develops his account of the latemodern world in the sure knowledge that "technology is most consequential in the inconspicuous dailiness of life."[76] The focal fact of the technological world is that the pattern to which our practices conform is determined by the pervasiveness of devices.

The Modern Promise of Technology

The essence of a device is that it seeks to make manifest the instrumental promise of technology, a promise formulated paradigmatically by Descartes and Bacon and that I showed in chapter 3 to be as strong as ever, if in new guise, within the project of ecomodernism. The device functions in our experience as external means to the ends of the master. This "availability of mere means is itself a remarkable and consequential fact. Historically, it is just in modern technology that such devices become available."[77] The device is the actualization of the promise that Reason is the path of liberation and freedom. It promises to liberate humanity from physical hardship, scarcity, and danger and to present before us, instantaneously, ubiquitously, safely, and easily the things for which once we had to strive, and strive for with great difficulty and at no small risk to life and limb. Outside of concerns about whether this promise is capable of being delivered—concerns typical of the first wave of environmental concern—one can see why such an apparently desirable and ennobling undertaking has been so little investigated. Nonetheless we can, and Borgmann counsels that we must, ask whether this promise is "worth keeping, whether the promise is not altogether misconceived, too vaguely given at first, and harmfully disoriented where technology is most advanced." The question of whether this promise is worth keeping returns us to "the peculiar way in which the promise of technology guides and veils the shaping of the modern world."[78] It returns us to the fact that we seem unable to see the deformation of our social and ecological world that is resulting from our technological agency. The defining fact of a device is its ability to dissociate ends and means in our experience. It brings the ends—commodities—to the foreground of our experience, whilst ensuring that the means—technosystems—recede from view. The more these means withdraw from our immediate bodily experience, the more readily does the commodity become available. The device thus permits little or no insight into or engagement with its machinery. Its promise is to make commodities available as "mere end, unencumbered by means."[79]

By way of illustration, Borgmann describes how the device paradigm selects the central heating system over the hearth because this device produces the commodity of warmth more readily, efficiently, and uniformly. In the development of large-scale technosystems for providing heat, everyday practices have changed from the time-consuming and complex management of a wood fire in favor of the flicking of a switch. The promise of relieving us of meaningless and debilitating drudgery would seem to be fulfilled in such a change. Certainly, in this change warmth is more assured. The hardship of kneeling in the predawn chill—heavy with child, perhaps, eyes smarting in the smoke of ill-seasoned timber—is banished forever. Yet also irrevocably banished is the family's careful, skilled, and cooperative maintenance of fickle yet life-giving fire. The practices of the hearth are replaced by an activity that requires little forethought, skill, or cooperation. It is not that these attributes are banished altogether. Rather, they are relocated and encoded in the sophisticated technosystems and expert managers that are called into play by the flick of a switch. But in this new context, they are defined entirely according to the instrumental calculus of efficiency: they bear no testament to our relation to tradition, place, home, or community.

The fulfilment of the promise of liberation in everyday practice is problematic because we are simultaneously liberated from what burdens us and from what we care about. "The promise leads to the irony of technology," says Borgmann, "when liberation by way of disburdenment yields to disengagement, enrichment by way of diversion is overtaken by distraction, and conquest makes way first to domination and then to loneliness."[80] In other words, the irony of technology is that the pure freedom afforded by pure ends, the freedom that devices afford, is a form of freedom that simultaneously undermines what makes the ideal of freedom valuable and commanding in human life. "[T]echnology *fails where it succeeds most* at procuring happiness," observes Strong.[81] Devices undermine our relationships to those things, places, and people we want to be free to be able to cherish. The social and ecological reality around us is thus de-formed (in contrast to being in-formed) by the medium of devices within which our practices are suspended into an aggregation of sensory instrumentalities. On this point, the conclusion of Frank Dexter's interview with "Satan" is enlightening:

> [Satan] . . . Having a digitised copy of the world, the original can be deleted. Freed from personal identity, with its dependence on stored memory, which wastes time and space, sensory experiences can be supplied when needed, made up as one goes along, so to speak. Without extraneous distractions, or means of return, the mind will be able to inhabit a reality entirely of its own making. . . .
>
> [Frank Dexter] *And this abolition of the world is the knowledge you bring to us?*
>
> [Satan] Everyone shall live in their own cybernetic sensorium, where reality is whatever can be experienced. . . . The ultimate meaning of enlightenment is not to create contentment, but to subvert all the dumb certainties of existence,

> to make everything problematic—to make everything into a problem—in short, to make life intolerable, and everyday life impossible. Intellectuals are committed to this in theory; every advance in knowledge deepens the mystery and confusion of existence, as you know. Technology is this enlightenment in practice.[82]

The Deformation of Practice

The dissociation of means and ends through devices represents nothing less than a deformation of practice and, I will argue in chapter 6, nothing less than the disorientation of our moral experience. Through the creation and use of devices, we are building a world that would "disburden and disengage us from the care of things."[83] Our practices bifurcate into "mindless labor and distracting consumption," drawing us into an increasingly careless stance toward those things that sustain us.[84] Practices cease to be centered around *world-revealing* things, instead becoming centered on objects that produce what we want without our attention, aid, or skill, and thus without our joy. Devices function so as to obscure their ecological and social relations. Only in so doing can they liberate us from the cumbersome burdens of corporeal reality. "In a device," comments Borgmann, "the relatedness of the world is replaced by a machinery, but the machinery is concealed, and the commodities, which are made available by a device, are enjoyed without the encumbrance of or the engagement with a context."[85]

We live in a hyperreal foreground of commodities of unprecedented flexibility, variety, and distracting vivacity.[86] But the more choice and freedom we have, the less consequential the exercise of our choice and freedom seems. Innovative theorists, such as the postmodern feminist Donna Haraway, who celebrate the subversive possibilities of the technological transformation of human identity—the rise of a posthumanist cyborg identity—also typically play down the significance of this bifurcation of means and ends. Modern technology certainly permits great playfulness in the foreground of choice: playfulness with gender, sexuality, body, and community in particular. I agree with Haraway that there is much to celebrate here. Yet I would add the caution that engaged in such play, we still occupy much the same place as consumers within the underlying technosystems that provide us with our basic sustenance.[87]

Although I shall not investigate the detail of Borgmann's proposal for reform of technology until chapter 7, we need to be clear that his response to the technological deformation of practice does not call for us to arm ourselves with sledge-hammers and bitter resentment, nor does it call for us to throw up our hands in sullen resignation. He calls upon us "to make the world truly and finally ours again" by orienting our lives in focal practices through which our social and ecological reality can once again speak to us.[88] He calls upon us to recognize, participate in, and bear witness to practices in which ends and means interpenetrate and coevolve. These are necessarily practices though which we encounter ourselves as carers of significant places, things, beings, and people. Further, he not only claims that orientation to a focal reality is possible within a world dominated by devices, he makes the much more powerful assertion that, provided they do not define our ultimate ends, devices may illuminate, heighten, and facilitate our opportunities for focal encounters with the things that truly sustain us.[89] Inspired by Borgmann's example, although wary of the fact that I find his political contentment with modernity too comfortable,[90] I elaborate on the ontological, moral and epistemological reduction of practice from correlational engagement to consumption in the technological world in chapters 5, 6, and 7 respectively, while the recovery of correlational engagement in practical discourses about sustainability is the theme of part III.

Borgmann reveals that our world-building becomes invisible not simply because dualism of means and ends blinds our thinking, but because it blinds our experience in everyday practices. Technological somnambulism bespeaks our literal inability to see what truly matters in our lives. We sleepwalk even as we build the earth itself into a device for ensuring human survival, every step of the way banishing focal things to the margins of our lives, because our device-laden practices obscure from us the fact of our world-building. The pregnant promise of technology miscarries. While technological progress is able to serve as external means toward the technical end of mastering nature, it is unable to serve as external means to the ultimate moral end of human well-being in a welcoming, sustaining world.

In my view, the way forward to building more sustaining worlds does not lead us to a confrontation with hostility, or even indifference, to the moral ideal of sustainability. We face something altogether different. We face the peculiar powerlessness that afflicts citizens of the technological society as they seek to question and transform the technological structures that define everyday life. More than this, we face the transformation of our moral reaction against the unsustaining features of our world—a transformation that I have argued takes the form of an ecomodernist project—into an enthusiastic revitalization of these very technological structures. What we have to confront, then, is nothing less than our dwindling cultural capacity to see our practices as an expression of our moral condition. Our lives are governed by the sovereignty of technological proliferation because, increasingly, we cannot see any other ways of living. We find it ever more difficult to accept that there are any other worlds, any other forms of practice, that could demand our allegiance, and we participate like sleepwalkers in the process of building a thoroughly technological universe.

CHAPTER FIVE

REVEALING AN INHOSPITABLE REALITY

INTRODUCTION

I HAVE BEEN ARGUING that technology names not simply a collection of instrumentalities but also, and more fundamentally, the inherently human capacity of world-building. And I have described the latemodern world we are building as deformed and consequently unsustainable because it encodes in our everyday practices a profound dualism of means and ends. In chapter 6, I explore the way in which this dualism moves systems of social and ecological oppression out of our immediate experience and thereby out of what we take to be our domain of ethical responsibility. But before considering the moral implications of building a thoroughly technological world, I want first to take the analysis of world-building further by proposing that the material deformation of our world is simultaneously a deformation of our understanding of reality.

My proposition is this: through technological activity we not only build embodied domains of correlationship, worlds, out of the vastness of existence, we simultaneously establish domains of intelligibility within which existence can be narrated in terms that give meaning to human life. Our understanding of what reality is, of where undeniable truths are and are not to be sought, is mediated always, although not solely, through our technological engagement with the things around us. Technology is not only world-building, it is simultaneously world-revealing and must therefore be read as ontological text.

In this chapter, I offer such a reading and argue that the technological world is deformed most fundamentally because it blocks the view of human agents of their own essential relatedness to the things and others around them. I describe how this world reveals our ecological and social reality to be an arbitrary aggregation of instrumental objects and atomistic individuals, while its builders are revealed to themselves as producers of stability and order in a perilous, inhospitable reality. A reality perilous because it is utterly purposeless and thus utterly indifferent to human aspirations. In this world, latemoderns suffer and labor under the illusion that, to be real, they must harness the meaningless materiality around them to their purposes. The sustained production of security inevitably becomes the ultimate purpose of human life in the technological world.

I engage here with several of Martin Heidegger's later essays as a way of holding modern technology in ontological focus.[1] As will become evident, I find Heidegger's vastly ambitious critique of modern technology both fascinating and troubling. He was one of the first philosophers to lift technology out of its "subjectivistic and merely instrumental interpretations," making of it "a primary philosophical question."[2] He warned that instrumentalism "makes us utterly blind to the essence of technology," and he saw before most others the accelerating pace of our technological sleepwalking that is strengthening the impression that human freedom is being lost to technological determination. In response to both instrumentalism and determinism, he argued that technology in its essence is "by no means anything technological;" it is the disclosive human encounter with reality.[3]

Heidegger scholarship has become such a productive, rarefied, and insular industry of late that it seems incumbent on me to emphasize that my interest in this philosopher is solely in moving the present inquiry toward central and profound philosophical questions without being deflected from an overall movement towards transdisciplinary contestation about the ideal of sustainability. I do not claim that Heidegger is the only source of insight on technology's ontological depth, nor that his insight is without flaw, nor that he is indispensable to a rich philosophy of sustainability. Further, I am aware that the full complexity of Heidegger's questioning of technology cannot be dissociated from his wider philosophical concerns or from his lamentable politics. I shall touch on these matters as they pertain to my attempt to explain the nature of our technological world. However, I engage with Heidegger with little interest in developing further the voluminous debates that center on his work.

The first section of this chapter sets out Heidegger's argument that modern technology is the culmination of an Occidental history of domination that began with the classical origins of philosophy itself. I then introduce, in section II, Heidegger's description of the worlds that modern technology displaced, as well as the worlds that may lie beyond it, while resisting, in section III, his totalizing and morally disorienting claim that the only genuine *praxis* left open to us in the technological world is that of meditative, poetic thinking.

I HEIDEGGER ON TECHNOLOGICAL ONTOLOGY

During his long professional life Heidegger was principally occupied with the ontological difference between things and the originating ground of things: the difference between beings and "being." In illuminating our relationship with the originating ground of things, Heidegger argued that the essence of human existence is an openness to the mystery that lies behind material reality. In what follows, I provide a sketch of Heidegger's postwar characterization of our technological world. In attempting even a cursory description of Heidegger's thought, however, I am obliged to acknowledge the host of semantic obstacles that his philosophy of technology places before us. Of particular note is his idiosyncratic, precise, and controversial use of the classical Greek terms *logos*, *techne*, and *poiesis*.[4] I shall preserve Heidegger's specific use of these terms in capitalized form as this conveys something of the magnitude he claims for them and as I engage in debates that make use of these terms in a more general way in following chapters.

Technology, Language, and World: A Thumbnail Sketch

For Heidegger, humanity stands essentially in the ontological gap between beings and being. Humanity participates in the manifestation of the twofold unity of things and their transcendent yet immanent thingness. William Lovitt, a noted translator and interpreter of Heidegger's philosophy of technology, puts it thus: humanity stands as "the *openness* to which and in which Being presences and is known."[5] The openness of human existence (*Dasien*) resides in the essence of language, *Logos*, and in the essence of technology, *Techne*. Both language and technology belong, in their essence, to genuine thinking and both hold open the ontological spaces in which things may reveal themselves. Through *Logos* and *Techne* things point beyond themselves, disclosing the twofold unity of existence.

The spaces held open by *Logos* and *Techne* constitute temporal and spatial worlds. It is for Heidegger "impossible to separate man and world," claims Joan Stambaugh. The disclosure of being takes place within a historical, located world of communication and practice. Things manifest in a world of practice and point beyond themselves to engage humanity in the mystery of things beyond their world and beyond time and space. As Stambaugh explains it, the essence of humanity is bound inescapably to the disclosive possibilities of a particular historical world:

> [Heidegger's] statement that world is an existential of *Dasien* thus means that it belongs to the nature of *Dasien* to be in a world which is disclosed to it. World is not the sum total of things in nature. Rather, world is that within which we encounter things and it is what gives them their connectedness. The character of this "within which" is more basic to the phenomenon of world than the things encountered in it. The "within which" makes the coherence of our experience possible.[6]

A world is not an anthropocentric construct, nor is it a pregiven materiality. A world is a space in which relationships are established between humanity and beings in such a way so as to disclose the essential relationship between beings and being, and through this, between humanity and being. Without the dialogical and technological engagement of humanity in the things around them, there can be no human insight into the nature of things.

Understanding human existence as the worldly disclosure of the unity of origination and appearance, Heidegger does not equate truth with facticity or with the reliable presence of things. Confounding the instrumentalist tenor of modern thought, Heidegger understands truth as unconcealment. But, Lovitt points out, "[u]nconcealment is simultaneously concealment. Unconcealment, truth, is never nakedly present to be immediately known."[7] The disclosure of the truth of things takes place in the limited, historical, built, worldly spaces held open by language and technology. *Logos* and *Techne* cooperate with being in, and are coresponsible with being for, the revelation of truth. Truth does not manifest as universal certainty but as fleeting glimpses of the essence of things.[8]

Our Unsustaining Un-World

In terms of my interest in what is unsustaining in and about our latemodern world, we have arrived at the core of Heidegger's account of human worldhood. Heidegger claims that the technological world is deformed, for it fails to hold open ontological spaces within which the truth of things can be seen. The world we are building occludes our view of the twofold nature of existence altogether. It is, Heidegger tells us, so deformed as to be an un-world in which un-truth reigns.[9]

In the technological world things disclose only their materiality. The originating ground of this world appears to be human productivity. Language and technology cease to engage human practitioners in the revelation of the originating mystery of things. Language and technology only facilitate thinking to the extent that thinking serves the goal of stamping human authority on the indifferent objects around us. Technology reveals only objects. Language serves only the articulation of facticity.[10] Modern technology encloses our experience in such a way as to imprison things within human purposes.[11]

The central assertion of Heidegger's philosophy of technology, expressed most fully in his now widely known 1954 essay *The Question Concerning Technology*, is that in the technological world technology remains "a way of revealing," but has become a concealing revealing.[12] Its essence is to conceal itself: to conceal the ground of truth and thereby to conceal the essence of human agency as that which holds open the spaces in which being can make itself known. In this deformed world, the ultimate purpose of human experience appears to be to wrest the truth from things. Humanity's presence amongst things is reduced from artful to technocratic action. The challenging of things is obvious in the action of a bulldozer felling an ancient rainforest or in the perplexing presence of genetically engineered cows. Yet Heidegger's claim is that the truthlessness of modern technology is not a function of bulldozers, enhanced cows, or artefacts of any kind. The root of modern technology's truthlessness is that humanity no longer occupies its essential role, its home, as that which thinks the truth of things.

Plato and Technology

In his later essays Heidegger discerns a decisive ontological break in Occidental history marked by the thought of Plato. Prior to this break, some four centuries before the Christian era, Occidental experience was characterised by the disclosive illumination of *Logos* and *Techne*. Humans dwelt amongst things, letting them "be". Things spoke eloquently and gained expression in the song and craft, in the stories and buildings of everyday human worlds. After this break, however, begins a long historical transformation in which language is sundered from *Logos* and shrinks into logical deduction. Technology is sundered from *Techne*, becoming instrumental calculation. The central purpose of human life was inexorably transformed from an openness to the mystery of eloquent things into the now familiar assertion of mastery over indifferent objects.

The essence of modern technology arises with the turn toward what Heidegger calls "metaphysical" thinking and what I shall call technological ontology, inaugurated in Plato's philosophy.[13] Technological ontology arises with Plato's relocation of truth from unconcealment to enduring presence. Heidegger's explanation is this: "As unhiddeness, truth is still a basic feature of beings themselves. But as correctness of 'looking', truth becomes the label of the human attitude towards beings."[14] In this new, thrusting attitude toward reality, Occidental humanity no longer participates as a witness to the truth disclosed by things. Truth is thought to lie hidden behind a deceptive facade of materiality. Human agents begin to search for truth, and for the security this is thought to bring, convinced that they permit themselves to be deceived at their peril. Beginning with Plato, philosophers searched behind this facade with the penetrating glare of reason. After Bacon, this search proceeded with scientists and engineers employing technology to tear this facade down altogether. And in the latemodern age, humanity ceases to search for truth in the world around them at all, concluding that the only truth in the universe, the only reality of which we can be certain, is that which lies buried within each atomistic human being. The search for truth becomes the search for self-expression in an indifferent universe. Instrumentalism becomes the most powerfully convincing explanation of technology; information becomes the most powerfully convincing explanation of language. In fact, technology and language are, in latemodernity, collapsing each into the other and both into the hyperseparated self.

The Eclipse of *Poiesis*

Heidegger characterizes technological ontology primarily through contrast with his account of the pre-Platonic understanding of *Techne*. He asks that we note two things about this ancient term:

> One is that *techné* is the name not only for the activities and skills of the craftsman, but also for the arts of the mind and the fine arts. *Techné* belongs to bringing-forth, to *poiésis*; it is something poietic.
>
> The other point we should observe with regard to *techné* is even more important. From earliest times until Plato the word *techné* is linked with the word *epistémé*. Both words are names for knowing in the widest sense.[15]

Poiesis is most commonly translated as that productive activity that belongs to technical thinking.[16] Heidegger, however, intends something entirely different in his use of these terms. He understands *Techne* as the skilful human attention upon being which belongs to *Poiesis*, that is, to disclosive activity.[17] *Techne* is the wisdom that arises from human cooperation in the manifestation of the truth of things. What burst forth directly from nature or was brought forth through human skill in the everyday appearance of things in the Homeric world was the mystery of their origination. Through disclosive activity, humanity saw—that is, humanity thought—the truth about things. *Techne*, artful knowing, belonged to a nexus of causality in which humanity and being were coresponsible for letting things be what they most truly are.[18]

What is it, then, that Heidegger sees in Plato's philosophy to justify his claim that Plato stands at the beginning of a historical transformation that ends with the eventual eclipse of *Poiesis* by modern technological production? He sees the beginning of the transformation of human thinking from a celebration of the engaging appearance of things into a technique, a technological ontology, for securing knowledge.[19] After Plato, truth is not to be observed in the appearance of things, but is to be produced by those that think. The philosopher sought to "grasp and consider reality, to discover whatever might be permanent within it, so as to know what it truly was," says Lovitt. In seeking the indissoluble kernel of knowledge hidden within things, the philosopher simultaneously sought to master things. And this, claims Lovitt in his reading of Heidegger, is "the real origin of the modern technological age."[20]

Productivism

Val Plumwood's historical account of the emergence of dualism—that dualism begins its appearance as the defining logical structure of Occidental thought with Plato and achieves consummation with Descartes—is relevant here, for dualism provides the footing on which technological ontology was built.[21] After Plato, pure mind searches for certainty by turning away from an apparently indifferent materiality. Knowing and acting, philosophy and everyday life became increasingly separate spheres. By the time of the world of Descartes, the maturation of technological ontology is completed with the assertion that human self-consciousness discloses the originating ground of all that is real and true.[22] The principal achievement of the Enlightenment was, in fact, to make explicit the epistemological implications of technological ontology. After Descartes, knowledge about the world was founded on a logical structure that presupposed the framework of technological ontology. Schematically, this metaphysical framework can be represented as having four sequential layers: (1) dualism, (2) anthropocentrism, (3) subjectivism, and (4) productivism. That is to say, knowledge of the world was structured on a framework that revealed (1) humanity, whose essence is pure mind, to be radically separate from its world; (2) mind to be the sole source of meaningfulness and dignity, the sole subject, in a world of objects; (3) objects to be raw material for the self-expression of minds, and (4) the ultimate end of human life to be the technological transformation of its world so as to produce security.

With the ontological cleaving of self and world and the separation of a world into subjects and objects that this makes possible, humans' experience of their world, indeed the earth itself, is thought to be predicated upon being seen and represented by human observers.[23] As Heidegger demonstrates in his study of Leibniz, the originating ground of the modern world appears to be the production of reasons for every thing. Thus, "every thing counts as existing when and only when it has been securely established as a calculable object for cognition."[24] Every thing in the modern world suffers under the resulting holding "sway of the fundamental principle of rendering sufficient reasons."[25] And inevitably humanity, too, suffers. As the master of things, humanity is unable to be at ease, to be at home, amongst things. Humanity becomes alien to an inhospitable earth. Human agents restlessly transform the earth, paradoxically destroying the possibilities for the homecoming and rest for which they long. With the ultimate end of a human life understood as the imperative to produce meaning in a perilous world, we witness the final stage in the eclipse of *Poiesis*: the building of a technological world in the likeness of technological ontology.

From Technological Ontology to the Technological World

The rise of technological ontology within the philosophical traditions of ancient Greece, Rome, the Middle East, and Christendom reaches its apotheosis in the promise of technology articulated by Descartes and Bacon. We must not forget, however, that this transformation of thinking belongs, finally, not to philosophy but to the actual historical transformation of Occidental forms of life over 2,000 years. The worlds of *Techne* and *Logos* faded slowly as practices and ideas changed to reflect the growing sense that the purpose of human life was to be found by taking control of things; wresting from nature "her" secrets, laying "her" mystery bare before the glare of Reason, and thereby ameliorating the risks placed on human life by an inhospitable reality. Occidental worlds in which humanity was related to the mystery of existence by eloquent things gradually disintegrated. They were finally eclipsed by a new totality, a new world, with the establishment of the global paradigm of devices in latemodernity.[26]

The philosophical production of knowledge has given way to the technological production of realities in which instrumentality is the dominant form of human engagement with their world.[27] Thus, declares Heidegger, "metaphysics is now for the first time beginning its unconditional rule in beings themselves, and rules as being in the form of what is real, in the form of objects, lacking a conception of truth."[28] The technological world is the actualization of the thinker's will to security in an unwelcoming world. Yet in seeking to overcome the perils of existence in this world, latemoderns are now barred by the objects around them from encountering their own essence, and thus from understanding what is truly good in their experience. With the closure of the ontological "gap between what there is and how it comes to be like that," technological development takes on the appearance of a cosmic principle, and world-building appears to move beyond the scope of human agency. The world, notes Joanna Hodge, shifts from a living, unfolding disclosure of humanity's ambiguous possibilities into an unquestionable given:

> There is then a shift in the kinds of historical events which can take place and in the kinds of historical narratives which can be constructed. Both become detached from human living, since technical relations and their consequences appear given, not invented. Thus the world as technical relations, far from seeming like a projection of human activity, appears as a given matter of fact.[29]

Humanity: That Which Orders

In the technological world, ontological inquiry ceases; human agents sleepwalk through the process of building a thoroughly technological world. Philosophy is reduced to making factual descriptions of the functioning of our technological world. The morality of anthropocentrism becomes merely a description of the lawful, logical working of things; a more-than-human reality is reasoned as less-than-human. Human agents experience themselves as the source of all meaning and value and are only capable of building a world in which what appears most commandingly real is the instrumentalizing human.

In the project of modernization, nature must be organized as raw material for the technological production of human well-being. "Given the fact that the primary goal of the subject is greater and greater power," notes Michael Zimmerman, "we can conclude that the subject values security and certainty above all else. Mankind as subject can be secure only if the entire cosmos can be subjugated to the unending quest for power."[30] The subject "must organize the planet for the sake of security."[31] In the emerging project of ecomodernization, this imperative has, if anything, gained in strength and urgency. The ideal of sustainability has been compressed into the technocratic idea of security, giving rise to an ecomodernist agenda for maximizing the production of sustainable development. In this agenda, ecological relations are subsumed within technical relations that have as their ultimate purpose human survival. Yet as the technological world has become increasingly self-sufficient, directed by its own imperatives, humanity too has become increasingly instrumentalized as raw material. Humanity, argues Heidegger, is set upon by technology, that is, is coresponsible for setting upon itself, as that resource which orders: "What threatens man in his very nature is the view that technological production puts the world in order while in fact this ordering is precisely what levels every *ordo*, every rank, down to the uniformity of production, and thus from the outset destroys the realm from which any rank and recognition could possibly rise."[32] Human lives are integrated within technosystems as instruments for maximizing production and consumption. Every viewpoint in the technological world reveals only an endless horizon of instrumental objects: computers, frozen peas, embryos, forests, neighbors, and Mars are all caught up in the mobilization of the earth for human security.

The much talked about postindustrial possibilities of technology do not represent the collapse of this world. They represent its culmination. The increasingly flexible, delocalized and "user-friendly" nature of the human "interface" with technosystems—systems that engage, more than anything else, in the production of ephemera—testify to the actual anthropomorphism of the technological world.[33] The culmination of technological truthlessness is, paradoxically, the overcoming of the cleft between subject and object as the world is absorbed within the subjectivity of the human self. The full actualization of technological ontology demands that world and cyborg become one, giving birth to a cyborld, a hyperreality that begins and ends with the human will for security.

II HEIDEGGER ON THE WORLD BEYOND TECHNOLOGY

Heidegger's account of modernity seems unrelentingly bleak. Yet it is wrong to understand him as promoting a fatalistic determinism about technological development. Technology, Heidegger tells us, is "no blind destiny in the sense of a completely ordained fate."[34] Rhetorically posing the question of whether the reign of the essence of technology means that "man, for better or worse, is helplessly delivered over to technology," he responds: "No, it means the direct opposite; and not only that, because it means something different."[35] Technology is, in its essence, profoundly ambiguous.

The Turn Away from Truth

Heidegger's account stands absolutely outside of conventional accounts of autonomy and determination.[36] We need to be clear, therefore, that he is not suggesting that the truthlessness of the modern age can be traced back to a mistake made by a long-dead philosopher. According to Heidegger, Plato's turn to the self-securing of certainty was the thinkers response to the withdrawal of being from humanity.[37] The distancing of being from humanity necessarily brought with it the spectre of overwhelming emptiness and insecurity in human experience. The conditions of human worldhood shifted, drawing humanity into an ever more willful, assertive stance toward things. In thinking and in practice humanity began to grasp at rather than to receive reality. For Heidegger, the human will to mastery arose in response to the epochal turn of being away from disclosive human practices.

Without an appreciation of Heidegger's peculiar account of the destining of being—his claim that the history of being plays itself out in human history as an unfolding of epochal destinings—we will be unable to understand the source of his hope that our modern age of truthlessness will one day end. Nonetheless, in the present limited discussion I leave the intractable debates about the credentials of Heidegger's transcendental view of history to others.[38]

In Heidegger's postwar thought, the destining of being enframes the ontological conditions of human history. But this destining is no unequivocally preordained destiny, and it cannot obliterate the possibility of human freedom. Human freedom remains the "supreme necessity" that can never be absolutely destroyed.[39] The world of technology reveals reality by concealing the truth of its origination. Yet this world still holds open the possibility that humanity will come to dwell in the proximity of truth once again. But much more than this, the essence of technology holds open the possibility that humanity will dwell more fully in truthfulness than ever before.[40] Technology seduces us away from our essential nature as those who dwell amongst things, yet it thereby makes the unalienable possibility of our dwelling only more precious.

A Re-Turn of Truth?

The turning toward a new epoch of truthful disclosure in the history of being cannot be achieved through the human rejection of technology, asserts Heidegger. This possibility is not open to human solutions at all. At his most elusive, Heidegger confronts us with the claim that "the turning" to a new age, that "special moment that sends . . . [being] into another destining," is a possibility that lies with being itself.[41] If we ask, Where is freedom to be sought? we are already headed on the wrong path. The idea that humanity can search out truth and secure its freedom is technological ontology in its essence. Heidegger tells us that the recovery of human freedom rests with nothing less than a recovery of our authentic place, our home, as those who can see the appearance or dis-appearance of truth. This is the essence of the now infamous claim in his 1966 *Der Spiegel* interview that only a god can save us: "I think the only possibility of salvation left to us is to prepare readiness, through thinking and poetry, for the appearance of the god or for the absence of the god during the decline; so that we do not, simply put, die meaningless deaths, but that when we decline, we decline in the face of the absent god."[42] "At best," he continues, "we can awaken a readiness to wait."[43]

WAITING

Heidegger calls for us to "be the one who waits."[44] The obvious response to this is, of course, What are we to wait for? Yet Heidegger cautions that it is not a waiting for anything that is needed. Waiting is, in Leslie Thiele's interpretation, "an attendance upon the reawakening of our capacity for fundamental questioning, nothing more or less."[45] We wait without any guarantee that the epoch of technology will end. But waiting is not, stresses Heidegger, a passive resignation to forces beyond our control. Being and humanity are coresponsible for the possibility of a new epoch within which things can welcome humanity, calling them home. Although the turning toward truth lies with the history of being, the possibility that truth may once again be manifest in a world lies with the capacity of human agents to once again dwell in their essence.[46]

How might humanity dwell in its essence? In 1955 Heidegger gave his clearest response to this question in an address to a general audience in his hometown. Humanity dwells through modes of thinking that are meditative rather than calculative in nature. Through meditative thinking, says Heidegger, we see technology in its essence. Our comportment to our world expresses "releasement toward things" and "openness to the mystery."[47] Releasement names the freeing relationship we enter into with modern technology when we see clearly the truthlessness of our world, and when we see the futility of the human struggle to produce truth. We are released from our imprisonment as that which orders so that we can be that which thinks. We dwell poetically, says Heidegger "in the proximity of the essence of things," and this means, in the technological world, that we dwell in the proximity of the mysterious absence of truth.[48] Released from the claim of technological ontology, we attend to things artfully. Although by *art* Heidegger does not mean the human aspiration to be artistic, nor does he mean the aesthetic in human experience. Both art—poetic bringing-forth—and technology—productive challenging-forth—are fundamental modes of revelation.[49]

ARTEFACTS, OLD AND NEW

Heidegger typically explained the difference between these two modes of ontological disclosure by distinguishing between what he called old and new technologies. So, for instance, we read in *The Question Concerning Technology* that the "old windmill" lets the wind be. The "windmill does not unlock energy from the air currents in order to store it."[50] The artful windmill belongs to the "older handwork technology" and is "completely different" from modern technology.[51] In contrast, the technology of fossil-fuel extraction and combustion challenges the earth to reveal itself "as a coal mining district" and "the soil as a mineral deposit."[52] Heidegger's argument here is that the old windmill and a new coal-fired turbine, or the old bridge and a new dam, to take another of his examples, are not just technically different, they are essentially different. They reveal different realities. They belong to different worlds.

Although it is always something of a relief to find Heidegger offering concrete examples, they are frequently more confusing than clarifying. In particular, his stress on old, engaging things is deeply problematic for at least two reasons. First, artefacts such as windmills are recent innovations if Plato and Descartes are taken as historical reference points.[53] Windmills were part of a "new" generation of artefacts that emerged in the Middle Ages, shifting the locus of power production from human, oxen, and horse muscles, to wind, water, and finally, coal. This trajectory of power-harnessing technologies was reinforced by technologies of social organization, such as the mechanical clock and the printing press, that made possible the shift from rural subsistence to urban affluence. And Heidegger himself adopted a position in his early writings that undermines his later characterization of artefacts like windmills as engaging. In *Being and Time* (1927) he argued that even with the first wooden waterwheel the river is experienced predominantly as a power source, and "when this happens, the nature which 'stirs and strives', which assails us and enthrals us as landscape, remains hidden."[54] From this we could conclude that all that is different between the old waterwheel and the new hydroelectric dam is a matter of technical scale, not of essential orientation toward things.

Second, the distinction Heidegger makes between old poetic things and new technological things is disconcertingly autobiographical. Take his description of Van Gogh's painting of a worn pair of boots in the 1950 essay *On the Origin of a Work of Art.* Heidegger claims that the artist's skill, his *Techne*, is claimed poetically by these boots so as to bring-forth the "surrounding menace of death" in a way that does not attempt, as does modern technological production, to keep death at bay.[55] The artist draws upon the ever-evolving technologies of paint, turpentine, brush, palette, easel, and canvass to illuminate human life from beyond its own sphere of production—to bring forth our mortal trembling in the face of death. In the case of Van Gogh's boots, the artist drew upon the revolutionary techniques of three-dimensional perspective that originated only in the fourteenth century C.E. Some have argued that these techniques prepared the ground for Descartes' later dualism of subject and object, and the split between observer and observed that this entailed.[56] Yet Heidegger waxes poetic on the everyday life of the female peasant in rural pre-industrial Europe to whom he considered these boots to belong. These well-worn boots are not instruments. They are, for Heidegger, things that speak of "the accumulated tenacity of her slow trudge through the far-spreading and ever-uniform furrows of the field swept by a raw wind."[57] In fact, however, these boots belonged to Van Gogh himself, and were painted during his years in Paris.[58] Rather than the "dampness and richness of the soil" or the "loneliness of the fieldpath as evening falls," these shoes tell of the cobbles of Paris.[59] They tell of a city that in the 1880s was one of the world's great megalopolises and in the tempestuous throes of industrialization. Regardless of such factual inaccuracy we might say that Heidegger's point, which is about the ways in which art sets things "thinging," remains. Yet it seems reasonable to speculate that had Heidegger known that these shoes had been trudged back and forth down dingy coal-blackened streets, rather than the lonely fieldpath, he wouldn't have been moved to reflect upon the story they tell. More disturbing for his readers is the fact that he may have been tempted to claim that these shoes were mute technologized objects with no story to tell.

Heidegger's life began in 1889 and spanned the final eclipse of premodern forms of life in Europe. He was thus provided with a front-row view of the concatenation of modern devices into a new world. And I am not suggesting that Heidegger is in any sense wrong to draw upon his own early experience of rural, peasant life in Swabia, with its windmills and wooden bridges, to help him make sense of this new world. On the contrary, I think that the orientation to space, place, history, and time established within the forms of life dominant in our early years exert a powerful influence over our life-long reactions to technological change.[60] Heidegger no doubt resonated to this celebration by Friedrich Hölderlin, whom he considered the archetypal poet, of the world that reared him:

> I was reared by the euphony
> of the rustling Copse
> And learned to love
> Amid the flowers.
>
> I grew up in the arms of the gods.[61]

Plautus's lament for the time of his own boyhood, only two centuries after Plato, evokes a similar longing for the euphonic simplicity of youth:

> The gods confound the man who first found out
> How to distinguish hours! Confound him, too,
> Who in this place set up a sun-dial,
> To cut and hack my days so wretchedly
> Into small portions. When I was a boy,
> My belly was my sun-dial; one more sure,
> Truer, and more exact than any of them.[62]

Many of us have shared with Plautus resentment at having our time rigidly apportioned. Yet, from our location within the frenetic pace of latemodern life, structured as it is around digitized time, the sun-dial, with its attuning of human affairs to the diurnal rhythm of heavenly motion, appears the antithesis of technologized life. Many of us have also shared with Hölderlin the euphony of a childhood filled with enchanted things. In my own youth, however, the sacred place was not a rustling copse but a polluted creek that flowed through a gully wedged between a busy road and a scrap metal yard.

Although Heidegger's examples vividly recount the rupturing of rural forms of life through industrialization, they do not by that very fact justify his all-encompassing claim that there is a radical ontological discontinuity between windmills and hydroelectric dams. The question we must ask is this: to what extent is Heidegger's perception of the unauthenticity of modernity simply an account of his dislocating experience of straddling two worlds? To what extent is such an account, indeed, such an experience, a deep insight into the ontological deformation of the new world? And the most the crucial question, perhaps, for those of us who have known nothing other than the disburdenment of technology, those born into the technological world, is whether we are yet at home, and at rest, in this, our world.[63]

By turning to artefacts such as windmills and bridges to illustrate his description of the essence of modern technology, Heidegger exposes the vacillation in his claim that the essence of such technologies is nothing technological. Even as he points to artefacts that bear testament to the essence of technology, Heidegger is not drawing our attention to external objects but to human practices. The essence of modern technology is the transformation of the ontological character of everyday practices.[64] This is a transformation that has been taking place for many, many centuries via technological artefacts, and not just in the Occident, of course.[65] Heidegger points to the hydroelectric dam as an example of the enclosure of our world by technology, but it is the form of life sustained by the technologies of large-scale electricity that he seeks to expose.

Forms of Life, Old and New

The old windmill is an artefact that cannot be dissociated from the decentralized, small-scale life of rural, pre-industrial subsistence. This was a life largely without clocks, of making and repairing shoes, of going to bed with the sun, rising with the roosters, and waiting patiently for the wind to blow. In all their everyday technological practices, the old peasants responded to the dictates, the times of abundance and lack, of a largely nonartefactual world. For Heidegger, the old bridge built by these people is not a utility for crossing a river, it is a focal thing which draws the river and the local people, the landscape and the village, each into the other. The building, use, and maintenance of the old bridge is embedded in communal forms of life by which humans truly belonged in the place of their birth and in the sure knowledge of their coming deaths.[66]

Acknowledging the place of the old bridge or windmill in relating people to their world, we can be confident about how Heidegger would explain the latemodern, ecoefficient windfarm. With its regularly spaced wind turbines atop massive concrete pillars marching toward the horizon in the service of a distant grid, the windfarm is no different in kind than the hydroelectric dam.[67] The old windmill, waterwheel, or bridge and the new wind or water turbines or freeways are embedded within totally different forms of life.

While Heidegger's descriptions of old forms of life are percolated through his nostalgia, his discussions of the new are unashamedly reactionary. We have, for instance, his claim that "we do not yet hear, we whose hearing and seeing are perishing though radio and film under the rule of technology."[68] Made during the infancy of latemodern communications technologies, this statement is all the more alarming in the twenty-first century. After all, such technologies now entirely dominate our forms of life. Our lives would be unthinkable without them. Yet Heidegger declares that our life with them is rendered ever more unthinking. The technology of communication flourishes, he asserts, because the all-pervading principle of providing sufficient reasons—the epistemology arising from technological ontology—dominates cognition in the form of information. The instrumentalist epistemology of information and the modern technology of communication are united in our technological forms of life so as to place "all objects and stuffs in a form for humans that suffices to securely establish human domination over the whole earth and even over what lies beyond this planet."[69] Even the apparently benign and nowadays wholly inconspicuous telephone constitutes everyday practices which forcefully reveal language not as the primal dimension of human understanding, but as another instrumentality at the service of humanity. On our mobile phones, especially, we hold conservations with strangers while defecating or while taking a solitary walk in the "wilderness;" we invite people into our innermost worlds because we do not feel their presence.

One of Heidegger's clearest descriptions of technological domination as a form of life can be found in the opening paragraph of *The Thing* (1954):

> All distances in time and space are shrinking. Man now reaches overnight, by plane, places which formerly took weeks and months of travel. He now receives instant information, by radio, of events which he formerly learned about only years later, if at all. The germination and growth of plants, which remained hidden throughout the seasons, is now exhibited publicly in a minute, on film. Distant sites of the most ancient cultures are shown on film as if they stood this very moment amidst today's street traffic. . . . The peak of this abolition of every possibility of remoteness is reached by television, which will soon pervade and dominate the whole machinery of communication.[70]

And how right he was proven to be as television marched into the heart(h) of our lives. Heidegger discloses here how we are building a world that reveals time and space to be essentially parameters of technological systems. As technological ontology is actualized in our forms of life, our awareness of the shrinking of space and time in our modern experience fades. It takes on the form of a cosmic principle. The awareness that fundamentally different human experiences of time and space are possible becomes increasingly more abstract.[71]

Take our experience of time. If we have any sense at all of the pace of Swabian peasant life within which Heidegger first lived, it is most likely the ironic consequence of television documentaries. The latemodern experience of time is increasingly defined by the temporal compression typical of the experience of car-dependent cities and of computer-dependent communication. The seconds drag heavily on our impatient and often aggressive hands during our restraint by a red light, or as we fume at the tedious lope of last years word processor. A traffic intersection, through its tarmac-laden, noisy, dirty, and potentially dangerous character, is inherently hostile to lingering. The once commanding rhythms created by the motion of sun, earth, moon, and within our bodies, are now thoroughly replaced by the rhythms created by electronic signals. The technological manipulation and control of time as a linear calculable quantity—a quantity that can be recorded as well as saved, filled, killed and lost—is now our predominant temporal experience.

Our predominant spatial experience is likewise one of compression. Through the technosystems of transport, it is increasingly the case that our point of departure and our destination are compressed together in our temporal experience, no matter what actual distance separates them. Departure and destination are no longer separated by an attentive journeying through ecological and human worlds, but by a kind of void experience. In our cars, we focus our attention fixedly and advisedly on the technosystems of road transport, although, our bodies on automatic pilot, we perhaps let our minds wander along the airwaves of quadraphonic sound, killing time until we arrive. As aeroplane passengers, we relieve the tedium of being transported across a continent between breakfast and lunch by watching the latest movies. Topography, climate, animals, and other travellers are encountered within these technosystems predominantly as momentary scenery or as factors in calculations of risk, comfort, and performance.

Claiming that aeroplane-mediated travel is a novel experience in human history is to state the obvious. The question, of course, is this: what are we to make, if anything, of this novelty? It is around this question that Heidegger's postwar explanation of technology is most revealing. It was his achievement to show us how modern technological change is altering not just the material conditions but also the ontological character of our experience. The technological compression of space and time constitute novel points of reference in forms of life around which we are oriented to questions of what a world, and what our place in this world, could be. For Heidegger, the world revealed in modern technology is essentially unworldly, essentially deformed, because it renders humanity omniscient but homeless. As our technospheric power to reshape and order the earth grows, our understanding of our relatedness to each other and to earthly things withers away.

Increasingly homeless, our thinking pervaded by instrumentalism, we are blind to the disparity between our lived philosophical commitments and our conceptual philosophical commitments. Although there are now vastly more philosophical texts published than at any other point of history, we are ever less cognizant of the philosophical import of our simplest actions. It is now an unremarkable fact that the profession of philosophy proceeds via increasingly frequent conferences at which philosophers descend on a place from all points of the globe. As Heidegger observed, in the early days of air travel, "one finds the Nietzschean doctrine of the Will to Power detestable and [nevertheless] cheerfully flies over the Norwegian fjord. Perhaps one gives an address, laden with the history of spirit, against 'nihilism' and flies around in the airplane, uses the auto and razor-blade, and finds the Will to Power detestable. Why is this grandiose mendacity possible?"[72] It is possible simply because we routinely fail to question the ways in which technologically mediated experiences are of profound significance in shaping our world, and through it, in shaping our view of the earth itself. Our unworld bifurcates into a glamorous but increasingly inconsequential foreground of choice and a background of dull machinery through which we are related to our wider social and ecological reality. Ever less are we aware of the paradoxical dynamic of conceptual plurality and experiential homogenization at work in our lives. The vast plurality of our consumer options, which includes a plethora of fashionable theories, rests on the homogenization of our forms of life within a technospheric blender. Bioregional anarchists, theocentric revisionists, as well as aboriginal elders and Tantric sex therapists, sit side-by-side with solar engineers and UN development workers just down the aisle from intellectual property lawyers and transnational executives (who are, admittedly, embraced in the comforts of bigger chairs and bottomless glasses of champagne) on the aeroplanes of the technological society: each, these days, liberated from the other, as much as from the earth beneath them, by the marvel of autonomous multimedia in-house entertainment systems repeated in every seat.

Can we take it that Heidegger is suggesting that we turn off our radios, televisions, and computers, then? Is he suggesting that we use our cars and aeroplanes less often, so that our minds can once again open toward the poetic? Does being released from technology necessitate changes to our everyday practice? Can we assume that a 24-hour poetry channel on cable television, with Hölderlin as one of its features, would only continue the slide away from essential thinking? What are we to make of Heidegger's suggestion that "[h]ere and now and in little things . . . we may foster the saving power in its increase"?[73] Should we concur with Thiele's claim that Heidegger would be an "advocate of what is called 'alternative', 'soft', or 'appropriate' technology?"[74] Can we, to state the matter plainly, take any advice from Heidegger about our innumerable everyday technology choices?

Technology: Essence and Artefact

It seems clear that Heidegger himself dismissed this line of questioning.[75] He claimed, as late as 1966, that he had "never spoken *against* technology or against the so-called demonic in technology."[76] He berates those who would misunderstand his work as technophobic or deterministic. The supreme danger in our technological world is, for him, not posed by any object, not even by a nuclear bomb, but by the withering away of disclosive thinking. He insists that "there can be no question of a resistance to or a condemnation of technology; rather, I am concerned with understanding the *essence* of technology and the technological world."[77] I find Heidegger's sharp distinction between artefacts and their ontological essence troubling, however. For a start, as the foregoing examples demonstrate, Heidegger himself does not apply this distinction consistently. Despite his protestations to the contrary, he did speak out against many modern technological devices, bringing to light the ways in which they hinder our dwelling. He thus bears some responsibility in the misunderstanding of his position as antitechnological.[78] More importantly, however, in this distinction we encounter Heidegger's disconcerting displacement of questions about political and moral agency by questions of thinking in its highest, most genuine sense. In what remains of this chapter, I shall argue that his distinction between essence and artefact in technology results in an unhelpfully esoteric and apolitical description of the genuine activity within which dwelling is still possible.

Heidegger often claimed to think what was left unthought by the authors of philosophical texts. Hodge argues that Heidegger's own texts should, in turn, be subjected to this kind of scrutiny. Following her lead, I seek to show that Heidegger's texts implicitly offer some support for the view that possibilities for a less instrumental relationship with our world rest with truly practical, rational forms of moral judgment about technological choices as much as, if not more than, they rest with the elevated gaze of meditative, otherwordly rationality.[79] As I shall argue in chapter 7, possibilities for nourishing moral carefulness toward our world arise from an unalienable dimension of human practice. The fact that Heidegger himself shied away from this conclusion is tied, I shall contend, to his infamous and crippling encounter with National Socialism in the 1930s.

III THE QUESTION OF GENUINE ACTIVITY

Heidegger says that the question that can free us from technology is this: "*How must we think?*"[80] To ask "What shall we do?" is, apparently, to be prey to instrumentalism. It is to seek solutions, rather than to understand ever more deeply our plight and the saving power that grows within it. According to Hubert Dreyfus, an influential interpreter of Heidegger and one who defends much that I find disturbing in Heidegger's account, the supreme danger Heidegger describes is one beyond all practical concerns. But, because "Heidegger has not always been clear about what distinguishes his approach from a romantic reaction to the domination of nature," and because his approach confounds our instrumentalist epistemology, Dreyfus contends that "we are tempted to translate it into conventional platitudes. Thus, Heidegger's ontological concerns are mistakenly assimilated to ecologically minded worries about the devastation of nature."[81] Dreyfus emphatically rebukes those who would get caught up in everyday problems:

> Heidegger's concern is the human distress caused by the technological understanding of being, rather than the destruction caused by specific technologies. Consequently, he distinguishes the current problems caused by technology—ecological destruction, nuclear danger, consumerism, and so on—from the destruction that would result should technology solve all of our problems.[82]

What I find unacceptable here is the absolute disjunction Dreyfus identifies in Heidegger's account between our ontological distress in adopting a technological understanding of being and our embodied distress at the degradation of our ecological and social relationships. By defining our distress at the destruction of nature as "worry" and further labelling this worry a "conventional platitude," we see clearly the lurking danger of intellectual elitism within the philosopher's elevated gaze on the essential issues. The line between opposing our calculative orientation to thinking and disregarding embodied human suffering and exploitation is fine indeed. I shall come to Heidegger's transgression of this line shortly.

Dreyfus's conclusion that concerns about the devastation of nature verge on platitudinous is drawn all too easily from within the comfort of secure professorial life amidst technological affluence. Just as the ancient Greek thinkers were insulated from practical everyday matters by the normalized tyranny of slavery, so too are many modern thinkers insulated from the submerged tyranny perpetrated through modern technosystems.[83] Take, for instance, Dreyfus's assessment of the role of technology in contemporary Japanese life: "The television set and the household gods share the same shelf—the styrofoam cup coexists with the porcelain teacup. We thus see that the Japanese, at least, can enjoy technology without taking over the technological understanding of being."[84] In a more recent paper, he takes this point further, claiming that in the Japanese coexistence of television and household gods, "it becomes clear that the technological understanding of being can be dissociated from technological devices." We learn that "pretechnological practices" are sufficiently robust so as to sustain a culture in which modern technologies remain but in which technological ontology is decentred.[85]

I cannot accept this proposition. Technological ontology is so rapidly becoming ubiquitous in most societies precisely because of modern technology's surface of plurality, adaptability, and cultural flexibility. We live in the midst of the hyperreal illusion of having gained television sets while not having lost our gods as, all the while, our world is ever more flattened towards the one-dimension of technological commodification. No doubt many Japanese visitors to Caucasian technological society are as impressed by exotic medieval cathedrals looming alongside the gleaming arches of McDonalds, as was Dreyfus amidst the exotica of Japanese living rooms. Yet behind these surfaces, the technosystems of hegemonic oppression thrive. In turning on their light switches to better see their altar to divinity and TV, Japanese citizens activate, as surely as do those of France, the technosystems of nuclear energy. The shelf itself is likely, as it is in Australia, to be an unacknowledged memorial to the ancient rainforests of Borneo. In eating their traditional *sushi*, this culture now stimulates the technosystems of drift-netting, chemical-intensive aquaculture, and genetic engineering. In importing rice, they encourage cash crops rather than self-sufficiency in Africa. In prudently saving to purchase a new Buddhist statue, they contribute, through their gigantic banks, to the technosystems of finance that power global capitalism.

In the face of the neo-Heideggerian appeal to higher concerns, I think it important to emphasize that the destruction of nature is for vast numbers of people not an abstract worry but an immediate and direct threat to livelihoods, cultural practices, and routinely to human and nonhuman life itself. The suggestion that deep thinking lies in the turn away from the technological world toward the study of ancient Greek philosophy or meditation on eighteenth-century poetry seems pretentious and politically fraught.[86]

We are cautioned by Heidegger not to rush headlong into action aimed at solving an evident but, he assures us, nonetheless inessential problem such as the destruction of a river valley through the construction of a hydroelectric dam. Heidegger insists that in our urgent hurry we will miss the real threat, which is not to the valley or even its displaced human residents, but to the possibilities for human thinking itself. Yet there can be no doubt that our decision to sit quietly meditating on our breath or poetry involves many difficult practical choices. To sit still in the midst of the restlessness of the technological world is as much—indeed, is more—a deliberate action than rushing out the door brandishing a placard. Simply sitting and reading Heidegger implies a host of practical judgments. To put aside books on integrated business management and be bothered with Heidegger's ontological questions at all runs counter to the self-assuredness and instrumentalism of the latemodern world. And remaining open to these questions, if we choose to be so bothered, is difficult amid the hurly burly of technological life.[87]

Contrary to Dreyfus, I consider that "ecological destruction, nuclear danger, consumerism, and so on" is distressing in a way that is simultaneously and inextricably ontological and corporeal.[88] The literature of radical ecophilosophy attests to this being so. My concern about the accumulation of carcinogenic pesticides and heavy metals in the tissues of my children is at once a concern with the technological diminishment of human possibilities and a concern with the practical task of living in more sane, more care-full ways. Certainly my preoccupation with the well-being of my children could be narrowly construed as a mere instrumentalizing concern with the survival of my genes. Similarly ambivalent are alternatives to harmful, unsustainable practices offered via the ecomodernist drive for ecoefficiency. If I can afford them, I can choose from alternatives such as genetically engineered pest resistance, the substitution of timber in my house, and of lead in petrol and paints, by more sophisticated synthetic products of industrial laboratories. However, history has shown the propensity of such solutions to create new sets of problems, for which new sets of technological solutions are soon required. This is, after all, the dynamic of technological profligacy that defines modernity. There is thus much weight to Dreyfus's argument that to attempt to solve our problems in this way is to move another step further down the path to fully technologized forms of life that obliterate the possibility of our encountering our relational selfhood. But where does this leave us as we negotiate the ambiguities of daily life? If I choose to reduce toxicity in my family's diet by the collection of rain water, by turning my backyard and local public land over to organic forms of food production, by adopting simple passive design methods to reduce the risk of termite damage, by cycling to avoid the combustion of fuel, or by bartering for the vegetable-based paints made by a neighbor, am I necessarily falling prey to a death-defying desire for control? Conversely, are philosophers who spend long hours meditating on Hölderlin or the Japanese tea ceremony, and who unquestioningly accept the (time-saving) presence of latemodern technosystems in their everyday practices, thereby released from the oppressive ontological grasp of technology? I think not.

Genuine Practice and our Inner and Real Core

The problematic distinction that Dreyfus makes between our primary ontological distress and its secondary material symptoms derives, in my view, from the distinction that Heidegger drew, after the war, between our essential nature, our "inner and real core," and our everyday latemodern lives:

> We can use technical devices, and yet with proper use keep ourselves free of them, so that we can let go of them any time. We can use technical devices as they ought to be used, and also let them alone as something which does not affect our inner and real core. We can affirm the unavoidable use of technical devices, and also deny them the right to dominate us, and so to warp, confuse, and lay waste our nature.[89]

This remarkable passage from his 1955 *Memorial Address* effectively draws Heidegger's explanation of technology full circle. Beginning with his critique of instrumentalism, through his description of technological ontology and the destining of being, this passage returns us to the instrumentalist promise of technology in the form of the comportment of releasement and openness. Of course, Heidegger would present this movement as a spiralling upwards towards the heights of essential questioning. And I do not seek to deny that there is considerable merit in seeing reflection as a spiral movement that returns us to places we have not been before. Nonetheless, his argument is that by adopting our place as artful and meditative dwellers, technologies become instruments once again: we can set them aside at any point. This is nothing less than a restatement of the instrumentalist assertion that although technologies define the material form of our lives, our minds are free to define the moral and ontological form of our lives. In asserting that, provided we preserve our core, the receptiveness of our thinking, we can live in the world of technology yet stay always beyond it, Heidegger elides the simple fact that modern technosystems are designed precisely so that we cannot put them aside at any point. Just how do we "let go at any time" of the technosystems of money? Just how do academics refuse to let computers dominate and lay waste the practices of education now that students born in the computer age are unable to conceive sentences without keyboards and university bureaucrats have restructured campus life along digital lines in an effort to maximize the production of competitive educational product?

I see in Heidegger's meditative instrumentalism vestiges of Cartesian dualism of theory and practice, despite the fact that he claimed that "the distinction, stemming from metaphysics, made between theory and praxis . . . obstructs the path toward insight into what I understand to be thinking."[90] Asking, "does thinking remain only a theoretical representation of Being and of man; or can we obtain from such knowledge directives that can be readily applied to our active lives?" Heidegger responded that the "answer is that such [genuine] thinking is neither theoretical nor practical. It comes to pass before this distinction. Such thinking is, insofar as it is, recollection of Being and nothing else."[91] Further, genuine "thinking is a deed. But a deed that also surpasses all *praxis*."[92] So how do I think Heidegger perpetuates the dualism of theory and practice? To answer this question, I want to retrace some of the territory we have covered in this chapter, exploring more closely the hyperseparation Heidegger imposes between (1) the essence of technology and technological artefacts; (2) *Techne* and modern technology; and (3) meditative thinking and latemodern practices. I shall consider each of these in turn.

(1) The essence of our technological world is, for Heidegger, technological ontology. He thus presents the essence of technology as arising in the philosophical canon more than two thousand years before it became evident in technological forms of life. Because of Heidegger's obscure location of this essence within the history of being, we can easily lose sight of the basic struc-

ture of his claim, which is to explain technology as the application of a prior form of theorizing. In a fashion strikingly similar to the instrumentalist assumption that technology is applied science, Heidegger argues that technology is applied metaphysics. Not surprisingly, he considers that our only genuine response to modern forms of life is therefore an alternative ontology, what he calls genuine thinking. Despite all of his disparaging remarks about modern technologies, Heidegger retreats behind the claim that his concern is not and has never been technological things, creating for himself, as one who thinks in an unthinking unworld, an "intellectually hermetic environment."[93] Heidegger gives us no cause to think that our everyday practices will be technically much different in the epoch beyond technology.

(2) It is only in old forms of life that Heidegger considers that building and thinking belong together in dwelling. Meditative thinking (*Logos*) and the practices of building (*Techne*), such as those centered on a bridge or farmhouse, listen each to the other and both to being through dwelling in a world.[94] But in the technological world, the realm of everyday action is entirely, unambiguously truthless, says Heidegger. Those who emphasise how Heidegger exposes the saving power of insignificant things in everyday practices typically point only to old things.[95] Dreyfus tells us that "we must foster human receptivity and preserve the endangered species of pretechnological practices that remain in our culture, in the hope that one day they will be pulled together in a new paradigm."[96] Leaving aside the not unimportant issue of what could qualify as a pretechnological practice—and one need only step inside the latemodern studio of a master cello maker to be confounded by the thorough imbrication of old and new—we are left to conclude that modern practices are unambiguously and unredeemably claimed by technological ontology. In *Being and Time*, we learned that the workshop of the old craftsman expressed ambiguously both poetic skill and instrumentalist calculus. But we are asked to accept that technological practices themselves can never be poetic in and of themselves. According to Heidegger, the only essential building, the only *Techne* possible in the technological world is that which takes place in genuine language, *Logos*: "For thinking is genuine activity, genuine taking a hand, if to take a hand means to lend a hand to the essence, the coming to presence, of Being. This means: to prepare (build) for the coming to presence of Being that abode in the midst of whatever is into which Being brings itself and its essence into utterance in language."[97]

(3) We thus confront Heidegger's paradoxical assertion that the turning away from *praxis* is the only genuine action open to us in the modern world. "Why? Because [for Heidegger] what human activity, *praxis*, has become in the modern technological epoch is itself a challenging-forth," asserts Richard Bernstein.[98] Heidegger considers our essential core as thinkers to be preserved not through *praxis* but through the "intention to pit meditative thinking decisively against

merely calculative thinking."[99] This claim is notable both for its wilful, combative overtones and for the dualistic polarity Heidegger creates between technological (inferior) and nontechnological (superior) thinking.[100] The heights of thinking to which Heidegger seeks to retreat leaves his argument prey to the same modern philosophical disregard of everyday life encouraged by the rampant instrumentalism of our age. While instrumentalists assume technological life's essential neutrality, Heidegger assumes its essential emptiness. In his 1991 essay *Heidegger's Silence?* (1991), Bernstein establishes, to my satisfaction at least, that Heidegger's postwar thought is founded on the attempt to "collapse *praxis* into *Gestell* [technological truthlessness], or to displace *praxis* by thinking."[101]

Heidegger's Displacement of *Praxis*

Heidegger's notoriously failed attempt to involve himself in the pragmatic affairs of his time in a bid to wrest his homeland from the dislocation being wrought by industrialization, is integral to his later philosophical displacement of *praxis*.[102] Ironically, the saving power that Heidegger saw in the upsurgance of German folk traditions between the world wars turned out, in the rise of Nazism, to have been an intensification of the technologization of the German people. Badly scarred by his experience, Heidegger retreated in the mid- to late 1930s back to what Hannah Arendt charitably described as his residence in true thinking.[103] Thenceforth, he maintained a silence on the events of the war and, stimulated by intense study of Nietzsche, made plain his view that the entirety of political and everyday life is enframed by the unthinking procedures of instrumental reason.[104]

Jacques Derrida is one of many to label Heidegger's postwar silence about the Nazi atrocities as "horrible, perhaps inexcusable."[105] Yet Heidegger broke this silence in the unpublished 1949 lecture from which *The Question Concerning Technology* derives, and his commentary is arguably more horrible than his silence. In this lecture Heidegger claimed that industrial agriculture is "in essence the same as the manufacturing of corpses in gas chambers and extermination camps."[106] Fifty years on, this comparison is chilling. Its impact in Bremen four years after the end of the war can only be guessed at. Perhaps equally shocking is Heidegger's 1945 defense of his infamous Rectorial Address of 1933/34 in which he claimed that everything is now enframed by the "universal rule of the will to power. . . . [W]hether it is called communism, or fascism, or world democracy."[107] In accepting the National Socialists' offer of the Freiburg Rectorate—which he portrays as the noble act of leaving his abode in thinking to help his people embrace genuine thinking—Heidegger would have us accept that he "risked the attempt to save and to purify and to strengthen what was positive" in National Socialism.[108] He takes his failure in this attempt not as a reflection of his own shortcomings, and certainly not as a refutation of his efforts in thinking, but as a reflection that genuine thinking is now all but impossible in "the midst of the consummation of nihilism."[109]

The apolitical comportment that Heidegger seeks to defend, a comportment that is "blind and impervious to mundane suffering and misery, victimization and mass extermination," is disturbing in the extreme.[110] I agree with Bernstein that Heidegger's condemnation of metaphysical humanism and his analysis of the obliteration of the conditions of *praxis* can be read as intended "to 'justify' his silence about the Holocaust."[111] Such a subversive reading of these texts needs to be tightly focused, however. The ontological critique of modern technology pioneered by Heidegger seems to me vitally important in advancing our understanding of the everyday character of our latemodern lives. What is specifically called into question by his postwar silence is his otherworldly representation of the "saving power." Heidegger offers us a false choice by juxtaposing an unauthentic everyday world with poetic meditation upon the possibility of a cosmic, epochal *Gestalt*.[112]

LIVING QUESTIONS

Heidegger conflates Socrates' question of "How one should live?" with that of "How must we think?"[113] He exaults in the poetic ambiguity of the essence of technology but denies the ambiguity of our latemodern experience in the technological world. In contrast, I consider that there are continuous and countless possibilities for setting ourselves upon a cultural journey toward forms of life in which we can encounter our ecological and social reality as other than that which must be ordered. To set out on this journey to and through the margins of the technological world, we have need of not only the saving power latent within genuinely poetic thinking and acting, we have need of the saving power of practical moral judgment in our everyday forms of life. Before moving on to consider the possibilities for moral wisdom within our technological world, however, we also have need of Zimmerman's wise counsel:

> If one approaches Heidegger's writing as a guide for action, one will find such a guide, but it will probably be the reader's own aims re-expressed in Heideggerean terminology. Most of us are caught up in a web of egoistic self-understanding that seeks to manipulate reality in ways that promote our ideals and goals. Given this state of mind, we ought not to be surprised if, after reading Heidegger, we come away full of ideas about how to replace the 'bad' old scheme with a 'good' new one. Our understanding of and impulse to action, however, need critical examination.[114]

Zimmerman focuses our attention squarely on the question of what constitutes genuine activity in our technological world. I agree that we need to be careful in our questioning of Heidegger's displacement of *praxis* to recognize that latemodern life is increasingly enframed within a frenetic productivism. Our distress at the debasement of our ecological world is dismembered from its ontological sources and is reconfigured as the urgent need for ever more eco-efficient action. Our understanding of sustainability is thoroughly shaped by the urgent desire to produce solutions amidst crisis. We who have known nothing other than the technological challenge of an inhospitable world, we who have been babysat by television sets and spoken to by telephones from birth, cannot be too careful in scrutinizing our desire for certain, unambiguous answers to necessarily ambiguous problems.

But we are right, in light of Heidegger's conflation of fascism and late-modern democracy and his subsequent adoption of an expedient meditative instrumentalism, to be very cautious of his description of the genuine activity of thinking. We can take up Zimmerman's call to expose our impulse to productive action to critical examination, but also add that Heidegger's impulse to genuine thinking equally needs critical examination. Thus I shall inquire in chapter 6 into the relation between productive activity (*poiesis*) and moral life (*praxis*) in the technological world. I shall argue that a truly practical understanding of our selves and of our world undermines the dualistic dislocation of our lives into productive and relational modes of agency.

CHAPTER SIX

DISORIENTING MORAL LIFE

INTRODUCTION

MY INQUIRY INTO the contested moral meanings of sustainability has as its principal subject matter the lived reality of the technological world. I am convinced that any philosophically rich understanding of our ecological crisis must address itself to the deformation of everyday forms of life as much as to the deformation of theory. But this does not mean that I seek to dismiss the importance of metaphysical questions. On the contrary, my concern is to bring to light the relationship between our dominant practices and our dominant ontological narratives. Thus, in the preceding chapter, I took up Martin Heidegger's refrain that humans are "thrown" into existence as worldly beings. That is, to have a worldview is first to occupy an embodied position within a world from which viewing is possible. It is through our technological and dialogical engagement with a world of others and things that our social and ecological reality can be grasped as meaningful. This ontological grip, this world-revealing, is necessarily partial and particular. Partiality does not, however, imply a limitation upon or a loss in human wisdom; and particularity is an inescapable feature of moral experience.

Worldhood is inescapably part of personhood. The moral upshot of the modernist denial of this fact—a denial encoded by technological ontology into the very structure of our technological world—is the subject of the present chapter. We shall see that the stripping of both ontological narratives and technological experience from our understanding of the subject matter of moral life has left us profoundly disoriented in our search for the good life. To blindly build a world deformed by the dualism of means and ends is at once to build a world in which moral language can only be disordered and moral life disoriented. It is to degrade *praxis* to production.

We live, in our technological world, in the bright light of apparently unprecedented freedoms to express our individual preferences. Latemodern practices present a glamorous foreground of vibrant options against a drab, uninteresting background of neutral machinery. Yet, "notwithstanding the flexibility of everyday technological practices and the enhancement of moral agency that they make possible," Ian Barns justly points out that at a "deeper level the continuing trajectory of technological change is towards a continuing commodification and instrumentalisation of human life."[1] Resting on the shared, public, and taken-for-granted background of devices, latemoderns are generally at liberty to choose from a diversity of cultural, religious, sexual, political, educational, therapeutic, and other lifestyle options, provided, that is, that we do not overstep the legal lines drawn by liberal democracy to protect the similar private freedoms accorded to others.

To appreciate this juxtaposition of background and foreground, we need only consider the ways in which industrial agriculture is bringing about a standardization and homogenization of agricultural practices. Modern technological advance in agriculture has drastically reduced the diversity of productive species, landscapes, and rural communities, and it continues to do so. Industrial agriculture has established ecologically moribund, machine-dominated monocultures around the globe. Yet we consume the resultant agricultural produce as a stupendous array and a dazzling celebration of traditional world cuisines and their uncountably many hybrids.[2]

Consider, also, that the households of a single latemodern suburban street are far more likely to share the presence of microwaves, for warming our cuisine of the moment, than they are to share lifestyle preferences. While neighbors may appear to disagree on fundamental moral and spiritual matters, this disagreement generally doesn't penetrate their cooperation in the proliferation of credit cards, reticulated sewerage, and transgenic vaccinations. It is perhaps characteristic of this world that the decision my partner and I took some years ago to live without a television set has prompted far more comment, puzzlement, defensive rebuttal, practical obstacles, and mutual recognition of marginalized others than has our decision to be unmarried parents.

This is not to say that ethical debate has fallen silent in the technological society. Quite the opposite; ethical debates prosper. Concerns about the conduct of journalists, accountability of corporations, treatment of domesticated animals, use of the Internet by pedophiles, and high rates of suicide amongst homosexual teenagers rightly make for passionate discussions about ethics. Yet it seems that the more we seem compelled to take a stand in ethical debates, discovering and giving expression to our values, the less we seem able to take account of our peculiar, deformed practices and the public world they produce. Increasingly we find expert ethicists, ethics committees, and ethical codes of practice in most areas of social debate. But, "[f]or all the proliferation of medical ethics, business ethics, and the ethics of investment," asserts Joanna Hodge, "what marks contemporary life is an absence in place of a sense of what human living is about."[3] The more discussion about ethics thrives, the more it seems detached from any meaningful social and practical negotiation about how our forms of life might be changed. "As it ponders social choices that involve the application of new technology," observes Langdon Winner, "contemporary moral philosophy works within a vacuum. The vacuum is created, in large part, by the absence of widely shared understandings, reasons and perspectives that might guide societies as they confront the powers offered by new machines, techniques, and large-scale technological systems."[4]

The levitation of moral discourse against the gravity of everyday experience is nowhere clearer than in well-intended but unconvincing books like John Hart's *Ethics and Technology* (1997). Hart tries to bridge the apparent gulf between humanity and technology—an appearance maintained by the dualistic structure of the technological world—by appealing to us to add values to technology so that it can be directed to ethical ends:

technology + tradition = transformation

> The union of ethics and technology can transform the world. The appropriation of social ethics by innovators, entrepreneurs, citizens, and communities can lead to the development of appropriate technologies and to the creation of good jobs that help to provide for people and preserve the planet.[5]

The fallacy of externalization, displayed so clearly here, is nothing less than the externalizing of the conditions of moral experience from moral discourse, and it has ensured that there has been little discussion amongst moral philosophers and lay public alike about what terms new technological developments demand of our moral lives. The instrumentalist hope that technology can be harnessed to some set of values that exist independent of it has ensured that technological culture has been unable to make sense of Hans Jonas's observation that technological "powers to act are pushing us beyond the terms of all former ethics."[6]

Heidegger saw that the technological world, in its unprecedented ability to conceal the partiality of human knowing, is rendering our experience progressively less coherent. Our world has a powerful capacity to obscure its origins in the locality and particularity of human practices. We have come to revere above all else our technological proficiency, but in the process we have come to understand ourselves as instruments for sustaining productivity, rather than as bearers of moral lives. The more our universe is becoming scientifically intelligible, the less our embodied relationships with the world around us seem intelligible and welcoming. The more our instrumentalist epistemology aspires to universality, and the more our vision is crammed with devices, the less we understand our social and ecological correlationality. The less we understand the reciprocity of self and world.

As I see it, *moral life* names the fact of our practical concern to live good lives, while moral discourse arises from our need to understand how this good is nourished and threatened by the world in which we encounter it. Moral experience is necessarily embodied and ontological, and it is necessarily concerned with technology as world-building. The substance of moral life, and the truly practical reasoning that illuminates it, is the concern of chapter 7. However, there is still ground to cover before we can locate the practical moral possibilities that remain open to us in the technological world. In section I of this chapter, I describe the truncated nature of modernity's understanding of moral life that establishes technology as the *de facto* ultimate good in our world. In section II, I offer a critique of the resultant conflation of all forms of practical reason with the impulse for technical control.

I MORAL SCIENCE

James Gouinlock prefaces his excellent book *Rediscovering the Moral Life* (1993) with the following observation: "Moral philosophy needs reorientation. As professional philosophers engage topics in the field with ever more sophistication, their analyses recede ever further from vital subject matter, and their conclusions wither on the vine of inconsequence."[7] Although Gouinlock does not directly engage with issues of technology, I believe that his argument about philosophy's neglect of the material conditions of moral experience clearly establishes the problem that technological ontology poses for moral thought. If our eyes are closed to the conditions and character of our forms of life, how are we to understand and talk about our moral experience? As remarkable as it seems, modern moral philosophy's response to this question has been to ignore it altogether by excising from its deliberations the reality of our moral lives. It has denied the reality "that moral *life* brings forth and incorporates moral philosophy: it is not derivative of it," claims Gouinlock. "Perhaps it is a truism to say that moral philosophy is a function of the moral life, but one would get little hint of it by reviewing the efforts of philosophers over the last fifty years. *A crippling lack in recent theory is its drastically truncated subject matter.*"[8]

Although Gouinlock's criticism is broadly true, we must also acknowledge the increasing number of latemodern moral theorists who seek to establish situated practice, rather than universal reason, as the central subject of moral inquiry. In chapter 7, I shall rely heavily on the positive contribution that latemodern discussions about moral *praxis* can make to our understanding of truly world-disclosing, practical epistemologies.

"GOOD" LOGIC?

The Enlightenment turn in philosophy was more than anything else a turn toward the view that atomistic human consciousness is the foundational moral subject. This was a turn that inevitably, although only gradually in practice, saw theology and moral philosophy part company. From Immanuel Kant, Jeremy Bentham, and David Hume to Henry Sidgwick, George Moore, and Alfred Ayer to Richard Hare, Kurt Baier, and John Rawls and on into the present, modern moral philosophy has been preoccupied with founding moral discourse on a detached, epistemological footing.[9] Despite the diversity of moral theories encompassed within this lineage, it represents a coherent tradition, what I call "moral science," in its shared fidelity to technological ontology and instrumentalist epistemology. Moral scientists have sought to recover ethics from the partiality of tradition, myth, and religion so as to see morality in its ubiquitous and verifiable essence. Moral scientists conclude that the best way to conceive of morality is to aspire to the categorical, universal certainty of the natural sciences in making detached, principled observations about moral judgment and logical descriptions of moral language. The influential positivist Jean Piaget gave this aspiration pithy formulation in 1932: "Logic is the morality of thought just as morality is the logic of action. Nearly all contemporary theories agree in recognizing the existence of this parallelism—from the *a priori* view which regards pure reason as the arbiter both of theoretical reflection and daily practice, to the sociological theories of knowledge and ethical values."[10] Framed in this way, competing formulations of fact/value, is/ought, cognitive/noncognitive dualisms have long held center stage in academic ethics.[11]

Moral science truncates our understanding of moral life through its dualistic construction of fact as absolutely separate from, and primary to, value in human reasoning. This construction has two profound effects on moral thinking. First, it encourages an Archimedean, detached stance with regard to the everyday contexts of moral life. Moral scientists are exhorted to follow Sidgwick's lead in giving voice to the "point of view . . . of the Universe."[12] Second, moral science seeks to develop theories that are universally conclusive.[13] In this endeavor it has produced two apparently hostile species of theory that we can broadly call absolutist and relativist.[14] The former theories aim to prove the existence of absolute, foundational rational moral principles. The latter seek to prove that morality springs from nonrational sources (chiefly sentiment, intuition, and religion) that lie beyond, or more precisely, beneath, objective description and rational deliberation. Despite the apparently antithetical nature of absolutist and relativist claims, they are, Gouinlock asserts, the Janus faces of the same assumption:

> Partisans of both camps, as well as neutral observers, suppose that the purpose of moral theory is to *try* to arrive at universally valid moral principles. . . . The absolutists purport to establish moral principles that are both precise and certain: precise in what they require in the way of conduct, and certain in that their validity is thought to be established in a manner that withstands rational controversy. . . . Relativists are those who are convinced that there are no intersubjectively binding constraints on the determination of moral norms. They give up normative moral theory as a lost cause.[15]

And while this debate continues to rage in the corridors of philosophy departments, relativism—and the moral subjectivism to which it gives rise—holds sway in the world at large.

Moral Subjectivism

After Virtue (1981), Alasdair MacIntyre's watershed critique of modern moral philosophy, documents how the seeds of subjectivist disorder sown by Cartesian dualism, with its underlying technological ontology, grew and produced their bitter fruit in the twentieth century in the "intuitionism" of Moore and subsequently in the "emotivism" of Ayer and Charles Stevenson.[16] Despite logical positivism's postwar decline in favor, and despite the resurgence of absolutist stances such as Rawls's neo-Kantian constructivism, subjectivism has become the dominant feature of public moral discourse.[17] MacIntyre argues that moral language and judgment are now considered little more than "expressions of preference, expressions of attitude or feeling. . . . [M]oral judgments . . . are neither true nor false; and agreement in moral judgment is not to be secured by any rational method, for there are none."[18] We need not look far, he suggests, to see how the superficiality of the language of values and preferences is leading public debate toward perspectives that see moral difference as "a confrontation between incompatible and incommensurable moral premises."[19] As a result, moral discourse is increasingly displaying the disturbing characteristics of scepticism and nihilism. Confronted by intractable moral debates in which each disparate moral stance offers competing yet equally valid frameworks that claim to be true in their own terms, it is easy to be unconvinced that any of these frameworks get at the truth or, and more devastatingly, to deny that there is, after all, anything resembling moral truth at all.

In truncating moral commitment to the realm of personal conviction, an individual's moral sensibility is seen to be relative to their own subjective desires, disposition, and experience. Expressions of moral commitment can point to little beyond the preferences of the individual, and we are thus often sceptical about the moral utterances of others. The declarations of others as to wherein the good lies typically fail to build a bridge to our own realm of preference; there seems little reason why we should give such declarations credence. Why should we trade our subjective view for that of others if we share with them no inherently morally laden spaces? Adrift in a sea of individual moral realms, we are prey to the nihilist temptation of considering that the "goods" held dear by individuals are nothing more than make-believe, the projections of insecure egos unable or unwilling to comprehend that there is no right and wrong, no good and bad, no action or form of life that is in any way better or worse than any other. Anne Manne's question is to the point: "So here we are, floating without a map on a raft of different choices, rudderless and alone in a sea of freedom. . . . [W]hat islands of obligation, moral constraint, and restraint still exist?"[20]

According to MacIntyre, the subjectivist disintegration of modern moral discourse came about through the gradual, historical expunging of the epistemological force of the premodern, theistic moral *telos* that provided an ontology of the human through the eschatological metaphysics of eternal salvation and damnation.[21] In MacIntyre's account, this ontology of the moral unfolding, the ultimate moral point, of human life lay within a lineage of Occidental moral traditions in which the essence of moral personhood was understood as the cultivation of virtuous character. MacIntyre traces this lineage back to its pre-Christian roots, explaining that, for Homer, "a virtue is a quality which enables an individual to discharge his or her social role," while for Aristotle, the New Testament, and Aquinas, "a virtue is a quality which enables an individual to move towards the achievement of the specifically human *telos*, whether natural or supernatural."[22] This specifically human *telos* conferred upon moral discourse the ability to expound social rules for right conduct in terms of a larger, transcendent order, thereby furnishing everyday situations with a unifying sense of wherein the ultimate good of a whole human life lies.

The Disappearance of "The Good"

Modern moral philosophy is built upon the attempt to do away with shared ontological narratives in moral discourse, and for this reason moral teleology is now most commonly discussed in the dry terms of utilitarian calculus rather than in terms of what a human life is about. Michael Sandel has well explained the plight of politics under the reign of moral subjectivism. As he sees it, the core thesis of latemodern liberalism is this:

> a just society seeks not to promote any particular ends, but enables its citizens to pursue their own ends, consistent with a similar liberty for all; it therefore must govern by principles that do not presuppose any particular conception of the good. What justifies these regulative principles above all is not that they maximize the general welfare, or cultivate virtue, or otherwise promote the good, but rather that they conform to the concept of *right*, a moral category given prior to the good and independent of it.
>
> This liberalism says, in other words, that what makes the just society just is not the telos or purpose or end at which it aims, but precisely its refusal to choose in advance among competing purposes and ends.[23]

Communal visions that articulate what is good in life, what we are acting for or toward, have atrophied. Any landmark or signpost in the latemodern moral landscape seems as good as any other. And so, as Nietzsche foresaw, while nineteenth-century modernity clung to the comforting illusion that Christian morality remained intact despite the demise of its metaphysics, latemoderns face the daunting prospect that soon no socially commanding moral features will be in view at all. For MacIntyre this prospect leaves us with a stark choice: "*either* one must follow through the aspirations and the collapse of the different versions of the Enlightenment project until there remains only the Nietzschean diagnosis and the Nietzschean problematic, *or* one must hold that the Enlightenment project was not only mistaken, but should never have been commenced in the first place. There is no third alternative."[24] On the basis of this alarming conclusion, MacIntyre pronounces that there are "no remedies" for "the condition of liberal modernity." He insists that if we wish to preserve the human moral condition, we must seek wherever possible to sustain those instances of "life of the common good" that have hitherto resisted the "disintegrating forces" of modernity.[25] According to MacIntyre, there is nothing good at all in or about modernity.

Like MacIntyre, Sandel, Charles Taylor, and other moral theorists of communitarian bent, I believe that the elevation of an instrumentalist epistemology of right conduct over a shared teleological ontology of the good life provides a fundamentally distorted picture of moral life. To take a striking image from Taylor, modern moral discourse lies mutilated and cramped on the Procrustean bed of concern "with what it is right to do, rather than with what it is good to be."[26] But, nonetheless, I find MacIntyre's quasi-theological and harsh diagnosis of latemodernity oppressive, and I find the return to tradition that he advocates reactionary, totalizing, and misdirected. While modernity is deeply destructive of ecological and social relationships, it is also a deeply ambiguous phenomenon and cannot be simply dismissed as an unequivocal failure. More than anything, perhaps, modernity is deeply destructive of its own ennobling impulses. Thus, unlike MacIntyre (and, as I showed in chapter 5, unlike Heidegger) I believe that the healing of the technological world's deformation must begin from within, just as is the case in the healing of our own bodies. While our freedoms to shape our world are undeniably bounded, there is much we can do, as is the case with our bodies, to build up the health of our world.

A valuable countervailing aspiration to the nostalgic longing for premodern traditions that afflicts some communitarian thought is emerging from the situated ethics of feminist theory. Like communitarianism, much feminism provides a trenchant critique of moral science. Yet, emphasizing the way in which thicker, small-scale communities of the past, commonly founded on patriarchal family structures, kin-based hierarchies, and the stifling mores of face-to-face politics, often suppressed difference, moral theorists like Marilyn Friedman call attention to the liberatory potential of our chosen social relationships.[27] And I agree: chosen communities are vitally important in latemodernity. We must be careful not to denigrate their orienting power, for they are without question a sustaining feature of contemporary life. Most of us build islands of intentional community, expressing our preference to find commonality with the subjectivist goods of others whom we like and respect rather than drown as a result of nihilist apathy in the loneliness of absolute freedom. Any adequate response to the moral subjectivism of our times, then, must explicate the irreducibly relational character of moral experience in a way that remains open to the possibilities for moral life to be oriented around both traditional (or at least pregiven) and chosen relationships, relationships to other people, to other worldly things, and even to otherworldly things. For our essential relationality in a world is a correlationality within a coevolving social and ecological mosaic. Moral life is always oriented to a reality that includes but transcends human-to-human relationships. We live, thank goodness, in a more-than-human world. Any meaningful counter to the claims of moral subjectivism needs to establish that moral discourse, including moral science, is necessarily ontological discourse: it orients us to the wider reality within which our practices embed us.

Moral Ontology

In *Sources of the Self* (1989), Taylor contends that "[o]ntological accounts have the status of articulations of our moral instincts. They articulate the claims implicit in our reactions."[28] Such accounts articulate our reactions to the commanding reality around us as we encounter it in everyday life. Further, our encounter with the world is at once an encounter with ourselves. We learn what is good about and for us as we learn what is good about and for the world around us. We necessarily, inescapably make what Taylor calls strong evaluations about the things around us from the moment we seek to claim our mother's nipples as our own. These evaluations have the status of pre-understandings. As will become clearer in the next chapter, these evaluations are drawn out, ordered and reconciled into frameworks for cognitive deliberation through narratives that call forth the defining features of human existence. It is through narratology, rather than through epistemology, that our largely inarticulate sense of the good is aligned with wider cultural visions about the good life.[29] Narratives of this sort encode what Taylor calls moral ontology.

Moral ontology articulates those supreme, final goods, our love of which "empowers us to do and be good."[30] Our love of good things gathers together our moral instincts into a coherent and practical if largely inarticulate vision of how we should aspire to live in our world. The deliberative, reasoned concern with what is right rests inescapably on the narrative background by which we are oriented in what Taylor calls moral space toward the good life that we desire to live.[31] A shared view of the world is always contingent upon a shared evaluation of the lived reality of our world of practice. A worldview is inescapably contingent on a world evaluation that is itself contingent on direct human experience of a world worth caring for.

"From the publication, say, of Ayer's *Language, Truth, and Logic* in 1937 until quite recently, our moral philosophers have had nothing more edifying to do than to scrutinize the English language to see whether its moral words are indeed irreducible," contends Gouinlock. "Gone are the absorbing questions of the moral implications of the nature of man and the world; gone is any distinctly moral motive for investigating those great subjects."[32] The absorbing questions about our self and our world to which Gouinlock refers have disappeared to the extent that moral ontology has been suppressed in ethics. And there is, Taylor reminds us, "a great deal of motivated suppression of moral ontology among our contemporaries."[33] This suppression is on proud display in Hare's claim that there are no "ontological questions which are neither about the logical properties of words nor about matters of empirical fact. . . . [I]f moral philosophy can be done without ontology, as I think it can, that makes it a lot easier."[34] The divorce of ontology from moral philosophy may indeed make moral philosophy easier, but it also thereby makes it largely irrelevant to our moral experience and to the worlds we build. Their suppression of moral ontology leaves moral scientists "constitutionally incapable of coming clean about the deeper sources of their own thinking." In fact, suggests Taylor, "the very goods which move them push them to deny or denature all such goods."[35]

Moral Ontology and Technology

In a most peculiar way then, the ontological narratives of technological progress deny their own inherently evaluative character, asserting that our world of practice makes no strong claims upon us because it is simply an instrumental resource—a standing reserve. The theistic *telos* that gave shape to medieval moral discourses has been steadily usurped and displaced by the technological *telos* that now dominates our practices. Technological machinery has become the *de facto* horizon, what Taylor calls a "background of intelligibility," by which our lives are oriented to questions of the good, despite the fact that these questions have all but vanished from moral discourse itself.[36] In the Enlightenment promise of technology, we see a shift of the locus of moral *telos* from ontological narratives to technological structures. The domination of modern technosystems in our forms of life is increasingly realizing in practice, and thereby making redundant in theory, the epistemology of moral science. The technological background of means is becoming increasingly universalized, predictable, and detached from our moral instincts. Resting on this unquestioned background, this lived *telos*, and provided we acknowledge the prescriptions of what we ought to do established by moral science, latemoderns are free to devise, modify, and express their own nonrational version of a good way of living. Yet, the more the question of "How should I live?" is stripped of its worldliness, the more it is rendered superfluous by seemingly unstoppable technological trajectories, and the smaller seems the legitimate domain of moral questions. Ethical debate has retreated almost entirely into the foreground of technological life, there staking out its claim to provide objective rules for right action as well as prescribing the social conditions that best accommodate subjective values.

What is remarkable about the latemodern horizon of the good is that the more technology has defined our forms of life and dominated our reasoning, the more this horizon has vanished from view. Inevitably, the more our vision of the good recedes from view, the more frantically we build a world in which means and ends are dislocated—a circular dynamic that only accelerates the processes of technological proliferation. The more technology proliferates, and the more our world is alien to us and opaque in our reflections, the more it becomes a formless background upon which we can only hope to display our own individual, unique goodness. While we can and do establish important and partially orienting moral relationships in the technological foreground of ends, the fact remains that the background of means relates us to our ecological and social worlds in ways that are almost completely unquestioned. Barns is right, then: the recovery of moral agency entails nothing less than "a recovery of moral ontology, not just at the level of abstract philosophy but at the level of technological practice."[37] That is,

> to maintain the conditions of human dignity, justice, freedom, democracy, and the like, requires not just simply the re-assertion of popular control over technology. More deeply, it requires a deeper reflexive recognition of the significance of technologies in articulating and maintaining the discursive ordering of social life. It also entails a posthumanist quest for the ontological conditions of selfhood, and in particular the renewal of practical reason as the primary mode of human agency, including the renewal of those moral traditions which articulate the deeper sources of our humanness.[38]

But before we can speak of the possibilities for the "renewal of practical reason as the primary mode of human agency"—and thus before we can speak of the possibilities that practical rationality may be able to disclose within everyday practices for the building of more sustaining worlds—we need to understand more completely how our capacity for practical reasoning is presently suppressed by the *telos* of technical control. We need to consider the epistemological implications of our supression of moral ontology.

II FROM PRACTICAL TO TECHNICAL REASON

It is interesting, if not ironic, that two of Heidegger's most influential students, Hans-Georg Gadamer and Hannah Arendt, have forged philosophical paths that enable us to see the technological deformation of *praxis,* and of wisdom that belongs to it, as well as seeing the unhelpful displacement of *praxis* in Heidegger's own critique.[39] Gadamer, in his hermeneutic treatment of the phenomenological method, has made clear that the core deformity of practices in the technological world is that they subordinate doing to producing and concomitantly degrade practical thinking to technical calculus.[40] For her part, Arendt, in an attempt to develop a neo-Aristotelian politics of *praxis*, emphasised that our technologically produced homelessness, our world-alienation, arises from the inability of modern productive activity to illuminate our world as a "web of human relationships."[41]

What is taken for practical reasoning in modernity is, in fact, ever more sophisticated forms of technical reasoning. This technical, productivist rationality, however, is sophisticated only in the sense of being closely aligned with principled, theoretical reason. Understood as the technical application of objective rationality, the Enlightenment notion of practical reason meshes together what Aristotle called *episteme* and *techne* (theoretical and technical reason), yielding what I have called instrumentalist epistemology.[42] Production has progressively moved out of the sphere of skilled, embodied knowledge passed from practitioner to novice through long experience and lived tradition. It has instead become aligned with the dualistic metaphysics of scientific explanation. Gadamer shows us that "the more strongly the sphere of application [of knowledge] becomes rationalized, the more does proper exercise of judgment along with practical experience in the proper sense of the term fail to take place.[43]

The rationalization of practice, as Bernard Williams explains in *Ethics and the Limits of Philosophy* (1985), is nothing less than the rationalization of reasoning itself: "The drive toward a *rationalistic conception of rationality comes* . . . from social features of the modern world, which impose on personal deliberation and on the idea of practical reason itself a model drawn from a particular understanding of public rationality. This understanding requires in principle every decision to be based on grounds that can be discursively explained."[44] As Williams points out, this requirement cannot, in fact, be met. For, as I shall argue more fully in chapter 7, the deliberations of moral life do not embody first principles so much as they respond to the claims made upon us by a world of good things—claims that can only be articulated in the context of our experience and, even then, that remain always at least partially inarticulate and beyond cognition. The instrumentalist model of practical reason is founded on the impracticable attempt to separate, absolutely, theory and practice, and, Williams observes, it thus presents a "false image of how reflection is related to practice, an image of theories in terms of which they uselessly elaborate their differences from one another."[45] Put this way, it seems remarkable that so many philosophers have for so long persisted with such a useless exercise in pedantry. Yet this false image of practice has been adopted by modernist philosophers because, buttressed by their quasi-scientific pretensions, they have considered their theories to be quintessentially practical. And modern thinking has indeed become preeminently practical if the measure of practicality is taken to be technological control.

Moral science has sought to remove from practice the unruly complexities of situation-by-situation judgments, thought to be based on an erratic, idiosyncratic, and nonrational store of experience and habit, by imposing an all-encompassing form of reason. The aspiration for practical moral principles purified of the contingencies of lived experience, and founded instead on universally applicable rules, is nowhere clearer than in Kant's *Critique of Practical Reason*. Philosophy, concludes Kant, is the gatekeeper of the theoretical wisdom that *directs* our actions toward moral ends: "In a word, science (critically sought and methodologically directed) is the narrow gate that leads to the doctrine of wisdom, when by this is understood not merely what one ought to do but what should serve as a guide to teachers in laying out plainly and well the path to wisdom which everyone should follow, and in keeping others from going astray."[46] Lewis White Beck, in his translator's introduction to *Critique of Practical Reason,* notes that Kant was not, as many have claimed, establishing a dichotomy between pure and practical reason. Rather, he was "trying to show that pure reason can be practical and must be practical if morality is not an illusion."[47] For Kant, reasoning is practical only to the extent that it can be disengaged from the bias of our necessarily local embodiment. Kantian reason, in the realm of morals, produces practical imperatives that would be binding if we were fully rational beings and not always at risk of being led astray by irrational, that is, impractical, desires.[48] Kant provided what he considered to be conclusive proof that "in ethics what is right in theory must work in practice."[49] This holds good in political as in personal life, Kant tells us, such that the persistence of mutual distrust between nation states, for example, does not make the theory of universal justice wrong; it makes the actions of states thoroughly impractical.[50]

The Procedures of Instrumental Reason

Following Taylor, we can categorize modernist notions of practical reason, of which Kant's is a preeminent exemplar, as naturalistic or procedural.[51] Procedural explanations share with Kant the view that "we do not give the name *practice* to every activity, only to that accomplishment of an end which is thought to follow certain generally conceived principles of procedure." And they thus share his conclusion that "the value of practice depends entirely on its appropriateness to the underlying theory."[52] This conclusion Taylor characterizes as the primacy of the epistemological, that is, the "tendency to think out the question of what something *is* in terms of the question of how it is *known*."[53] It is not simply that our actions must be held to objective account, however. The procedural claim is far more demanding, as Frederick Will exposes, for practical reason, and even the essence of moral experience itself, is conceived of as existing independently of practice:

> Adhering to this conviction, a long line of modern philosophers through the years persevered in the search for resources in the governance of practice that are identifiable as rational and independent of all established institutions and accepted practice. The plausible general loci of such resources were clearly *experience* and *reason*, employing the latter term in its more narrow common usage to signify rational *thought*. But these familiar resources, as they are normally employed, are thoroughly permeated by social practices. It was therefore proposed that by careful intellectual analysis we might eliminate from them their conceived unessential and invalidating social elements, arriving at, as a purified form of experience, pure sensation, and a purified form of thought

> engaged solely with the relations of what were conceived to be very intimate, personal intellectual resources, namely, our own 'ideas'.[54]

As agents of ideas, we bring to bear on our practical engagements the force of Cartesian scepticism. This scepticism, founded on the ontological cleaving of mind and body, holds within it a deep anxiety, a "dread of madness and chaos where nothing is fixed."[55] Cartesian anxiety about the chaotic indeterminateness of everyday life leads naturally to the longing for a form of disengaged agency in which modern agents can set about the task of reordering everyday practice according to the first principles of pure reason.[56] We thus have the strange result that the Cartesian assertion of mind's radical separation from materiality leads not towards the mountaintop contemplation of the ascetic, but to the engineer's preoccupation with pragmatic affairs. The paradox of pure reason's productivism leads to the disorienting stance of detached engagement.[57] Paradoxically, the modern becomes preoccupied with the pursuit of ever greater physical security and comfort, while insisting that the body is merely the mechanical vehicle of an essentially noncorporeal humanness. We may legitimately ask: why has pure mind, detached as it is supposed to be from its worldly reality, become increasingly occupied with technological matters? Wouldn't rampant technological proliferation be more characteristic of a mind engaged in, rather than detached from, the material reality of worldly things?

Insightful responses to questions of this kind, such as that of Jürgen Habermas, recognize that our epistemological scepticism is founded on the claim that we "only know an object to the extent that we ourselves can produce it."[58] Attending to Arendt's essay *The Concept of History* (1954), we can add that this scepticism necessitates the subordination of history to engineering. Arendt reminds us of Vico's conclusion that "[m]athematical matters we can prove because we ourselves make them; to prove the physical, we would have to make it." Vico, one of the "fathers" of modern historiography, searched for truth by turning away from the foreignness of physical phenomena and towards history. He did so in the belief that, as they are the makers of history, "historical truth can be known by men." But, Arendt points out, Vico "would hardly have turned to history under modern conditions. He would have turned to technology, for technology now does what Vico thought divine action did in the realm of nature and human action in the realm of history."[59] The actualization of technological ontology is thus also the transcendence of history. The technological world is fundamentally ahistorical.

The "cult of the future . . . has turned us all into prophets." And, adds Wendell Berry, "if we are living for the future, then history is on our side—or so we are at liberty to think, for the needed proofs are never at hand." In the technological world, then, sustainability inevitably becomes the technocratic task of projecting our mastery onto the future, rather than the task of sustaining a present of meaningful, historical things. Sustainability inevitably becomes subsumed within what Berry calls "the dream of the future-as-Paradise."[60] Technology, as the rationalization of practice, commands our allegiance more than any other thing in our world. Technology detaches us from the past that is encoded in our present and draws us ever more into a preoccupation with what lies just beyond our world, in the future.

The transcendence of history is the inevitable result of the construction of politics on lawful principles independent of any human experience, relationship, or tradition. Habermas argues that the elevation of procedures over relations in epistemology has given rise to an impoverished understanding of politics as a deductive exercise. This procedural politics replaces "instruction in leading a good and just life with making possible a life of well-being within a correctly instituted order."[61] Epistemological proceduralism divorces politics from ethics, where politics is understood as the application of procedures for maximizing social order, and where ethics is understood as the obligation of private individuals to respect this order, and their concomitant right to the free expression of their evaluative judgments within the confines of this order. Habermas describes three unwelcome consequences of the procedural deformation of political activity: first, there is an attempt to establish the "conditions for the correct order of the state and society as such;" second, the practical application of knowledge becomes simply the technical problem of aligning behavior with this correct order; and third, our self-understanding is dominated by the claims of "human sciences."[62] Habermas rightly declares that in the political order of sustained production, "we are no longer able to distinguish between practical and technical power."[63]

Technology and Freedom

The Enlightenment promise of technology constructs judgments about the good as private and subjective, as standing largely outside of the relationships of everyday life and outside of the procedures of politics. But the modern fetishism of technology cannot be understood if it is construed as a blind, vicious urge for control. The worship of technology was oriented, at least at first, Taylor tells us, by the final good of individual freedom:

> So instrumental reason comes to us with its own rich moral background. It has by no means simply been powered by an overdeveloped *libido dominandi*. And yet it all too often seems to serve the ends of greater control, of technological mastery. Retrieval of the richer moral background can show that it doesn't need to do this, and indeed that in many cases it is betraying this moral background in doing so.[64]

Taylor argues that the rich moral background of freedom has been steadily debased as the narratives of moral ontology have been suppressed by "a deeply wrong model of practical reasoning, one based on an illegitimate extrapolation from reasoning in natural science."[65] As we saw in chapter 5, this flawed epistemology arises from a form of ontology, a technological ontology, that denies altogether that value is to be found in a purposeless reality. To the extent that technological ontology is a moral ontology, it locates the supreme good of human life in the task of securing an indifferent nature to our purposes, taking control of the conditions of our freedom and the prospects for our survival.[66]

In *The Ethics of Authenticity* (1991), Taylor argues that the ultimate good of individual autonomy oriented the modern challenge to religious and political tradition through the epistemological claim that every mature human agent, regardless of social position and religious conviction, is endowed with the virtue of universal, objective reason. This emphasis on an ubiquitous, evenly distributed, and thus inherently democratic reason lent itself naturally to the view that moral judgment was most likely to be authentic when purified of social mores, traditional beliefs, and inherited practices. The hold of church and aristocracy on European life was clearly weakened by the new freedoms of rational, detached agency. Industrial practices proliferated and rapidly replaced longstanding ecological and social relationships.[67] The mechanization of nature and of social life brought forth a powerful unease that manifested during the Romantic turn in aesthetic expression.[68] This turn did not, however, displace the Enlightenment push for the freedoms of universal reason so much as it located the ultimate good of freedom inward. The foundling industrial citizens of Europe began to search for human richness within themselves, rather than within a shared world. The world became the realm of fact, the self became the home of truth. The truth of human experience was firmly joined to the idea that "each of us has an original way of being human."[69] Being authentic to one's self thus became of foundational existential importance. If I fail to be true to myself, concludes a latemodern, "I miss the point of my life; I miss what being human is for *me*."[70]

The Enlightenment freedom to think rationally became, by degrees, the freedom to explore one's own essential inner originality. Such freedom was seen to be threatened by "the pressures towards outward conformity" imposed particularly by the church. But it was also seen to be threatened by the pressure of "taking an instrumental stance toward myself" being imposed by the new detached objectivity of the Enlightenment.[71] "Authenticity involves originality," describes Taylor, "it demands a revolt against convention."[72] This revolt from convention, however, took the form of a retreat away from the public order and into the private realm of subjective preference. Thus, although the ideal of authenticity holds within it a critique of instrumentalism, the retreat from a shared world that it encourages tacitly reinforces the political procedures that give the public order over to the *telos* of technological efficiency. Technological progress and the culture of moral subjectivism dance together: the advance of technology takes place with the retreat of moral discourse away from our correlational world and into the self.

Self-Production

Moral inquiry is increasingly internalized into the task of self-understanding and self-development, rather than that of world-understanding and world-building. The building of the technological world is the lived actualization of the ontological disengagement of human consciousness and material things postulated by Descartes. Once the building of this world began in earnest, the task of thinking became that of grappling with the design specifications necessary to produce a rational world in the image of pure, transcendent mind.[73] In this new technological world, natural objects appear meaningful only to the extent that they embody scientific principles and thus stand as precursors for the genesis of rational organization. Even other humans begin to appear meaningful only to the extent that they are also producers of ideas and organizers of nature.

In the latest high-tech walking treadmill, Internet software, and frozen dinner, we engage in walking, communicating, and eating in a way that appears to free our mind from the dreary mundanity of everyday life. We need not walk through our neighborhood, with its taint of heavy metals and muggers, for we can exercise in our bedrooms, our ears plugged with the latest Walkman. We need not communicate with our local grocer, papergirl, or even our own family, for at a "terminal" our mind can engage, in a space free of bodily inhibition, with whom we choose, wherever they happen to be; to a remarkable extent we can even choose who we will be in this virtual engagement. Nor need we cradle seed in our palm or watch anxiously for rain or be rich in culinary arts to be ensured that our frozen dinner will be waiting for us at the end of the day.

The latemodern ideal of authentic selfhood was constituted by the love of the freedom of individual self-expression. However, with the degradation of our social and ecological relationships, this freedom has become principally a freedom to produce and to consume objects, information, and "values." Politics aspires to become the procedural order within which each individual is given the freedom to explore their essential originality. But in this order, technology proliferates under the guise of neutrality, and the questions of the good recede from public view and from the range of language. The only authentic goals in this public order are technical ones. The only authentic goal of private life is to direct our moral awareness inward, cultivating fidelity towards ourselves before all others. Latemoderns consequently participate in a culture of self-*poiesis*.[74] Self-expression has become self-production in a world meaningful only to the extent that human production creates and sustains it. In this world, "practical wisdom can no longer be promoted by personal contact and the mutual exchange of views among citizens." It is not just that our primary exchange is with devices rather than with people, continues Gadamer, it is that "many forms of our everyday life are technologically organized so that they no longer require personal decision."[75] Our everyday practices no longer demand practical decisions. They demand material decisions whose defining focus is acquisition, rather than relationship.[76] The procedures of instrumental reason sever our experience of worldhood from our knowledge of personhood and invent, thereby, a vacuum that only productivist fervor can fill.

This is a vacuum in which the possibilities of our moral *praxis* become subordinated to the reality of our technological production. If we are to recover in our social theory the truly practical possibilities that remain open in the technological world, we must recover skills of practical thinking that can break the hermetic seal imposed on our theories so that our experience of our worldly relationships can flow back into our moral thinking. We require skills of judgment that can lead us from the foreground of self-production towards the background of world-producing means that relate us to our wider social and ecological reality. In chapter 7, I shall present the craft of moral life in this technological world as beginning with the embodied activity of developing powers of skilful judgment that can draw means and ends back towards and into each other in our experience and in our discourse.

CHAPTER SEVEN

RECOVERING PRACTICAL POSSIBILITIES

INTRODUCTION

I MAINTAIN THAT within all of our technological forms of life there exist possibilities for change in the way that we see the things that make up our world, and thus for change in the ways in which we build our world. There exists within all of our practices a tension between what the technological world enframes and what it does not, and, therefore, it is possible to probe towards the boundaries of our world from any point, from any practice, within this world. And we can take heart from the fact that the more the moral *telos* of sustained productivism takes hold in the technological world, the more evidence of the noninstrumental in human nature abounds. Our television news bulletins are turgid with dry empirical representations of the market. But the advertisements that punctuate such news about the latest consumer confidence index or the falling gold price tell quite another story. From teenagers leaping out of aeroplanes on surfboards in a sugarlust for soft drink to the latest car promising escape from the gridlocked city into wilderness to the deodorant that ensures us an enduring community of friendship, we can remember the ephemeral ground on which the concreteness of our technological activity is built. By engaging more fully with the reality of our lives, by inhabiting more completely our practices, by exploring more completely our volition to act at all, lived remembrance of what lies beyond the desire to produce security is always possible. The latemodern possibilities for genuine activity are to be sought within the technological world, not beyond it.

What I wish to elaborate in what remains of this book, then, is my perception of the never-ending possibilities for gradual yet profound cultural transformation that exist within our latemodern practices. It is within these possibilities that I place my modest hope that we may begin building worlds more capable and more worthy of being sustained than the technological one in which we find ourselves. I want to describe at least some of the epistemological conditions within which this hope may take root and grow strong in public discourse about sustainability. In so doing I part company from received moral theory and the project of ecomodernism that it informs in my conviction that these possibilities are disclosed in forms of practical discourse that cannot be circumscribed by instrumentalist forms of reasoning. I neither reject our technological world as unauthentic nor embrace it as the only rational way to live. Rather, I consider that healing of our world's deformation requires that we resist the suppression of our practical moral reasoning by instrumentalist epistemology, recovering sight of ourselves as relational beings.

In striking out away from the one-dimensional arguments established between the totalizing critics and the enraptured advocates of our technological world, I follow the invaluable contemporary project of rehabilitating practical moral reason as a substantive philosophical concern. In addition to the efforts of Charles Taylor and Albert Borgmann, efforts to which I pay close attention in this chapter, this is a project pioneered by scholars as diverse as Richard Bernstein, Lorraine Code, Hubert Dreyfus, Carol Gilligan, Jürgen Habermas, Alasdair MacIntyre, Nel Noddings, and Bernard Williams.[1] Much of this rehabilitation draws upon the traditions of practical philosophy that spring from Aristotle, although feminist contributions to this effort, in particular, offer a crucially important critique—and, in places, a damning one—of the substantial political limitations of many of these traditions.[2] Unlike some in this list, however, my primary goal is not to explore the role of practical reason in the everyday moral dilemmas so beloved of moral philosophers, important though this concern is.[3] I am interested in the world of practice in which our everyday moral choices take place, rather than in those choices *per se*. In my view, these world-structures frame the moral domain. They give it its shape, in terms both ontological and practical. My interest in practice is thus not with virtuous behavior, however that may be defined, but with our self-understanding of ourselves as bearers of moral lives searching for the good in the technological world.

In the first section of this chapter, I offer a substantive, in contrast to a procedural, explanation of the nature of our practical reasoning. I present practical reasoning, the rationality of relationality, as that disciplined thinking capable of guiding our search for the good in a particular world of practice. In section II, I follow Borgmann's lead in arguing that substantive explanations of practical reasoning must be extended to challenge the longstanding philosophical lore that our doing and our making, our relational and productive activity, our moral experience and technology exist in isolation from one another. The search for moral orientation in the technological world is nothing less than a practical craft of skilled judgment about, and experimentation with, the ways technology constitutes our self-defining and world-defining relationships.

I THE RATIONALITY OF RELATIONALITY

Under the polarizing forces of dualism, the term *practice* is now so radically separated from and subordinated to *theory* that it is effectively defined through negative contrast with theory. In the reasoning of moral science, theory is pure cognitive activity and practice is pure bodily action. Practice, separated from thinking in this way, seems either devoid of theory (habitual, intuitive, instinctive behavior) or seems contingent upon the application of theory (rational, scientific, productive behavior). If Immanuel Kant stands at the head of the Enlightenment attempt to develop epistemological procedures for practical thinking, Aristotle stands at the head of attempts in the philosophical canon to explicate practical reason in its lived substance. Bernstein contrasts the low, dualistic sense of practice with the high sense of practice described by Aristotle.[4] In *Nicomachean Ethics,* Aristotle describes how relational action (*praxis*) calls forth forms of knowing that are neither technical and pragmatic (*techne*) nor abstract and universal (*episteme*), but that are preeminently practical and moral (*phronesis*).[5] The *telos* of *praxis* is "not primarily the production of an artefact, but rather performing the particular activity in a certain way, i.e., performing the activity well: *'eupraxia'*."[6] Performing social activity well is not dependent on the guidance of abstract theory but involves the cultivation in lived experience of perceptive and prudent exercise of practical judgement. In this high sense, *praxis* belongs always to ontological narratives that tell of the well-being of the world in which it takes place. *Praxis*, our relational action, and the practical moral wisdom which belongs to it, brings us into self-defining relation with our world.

The high sense of *praxis* has largely slipped from our modern grasp.[7] Bernstein reports that this loss has "been the source of innumerable confusions, even among philosophers."[8] The history of modern philosophy is burdened with the attempt to deny or transcend the local and dense entanglements, the worldhood, of human life. The lived duality of theory and practice that the technological world makes paradigmatic of our experience has meant that, even where modern philosophers have attempted to develop a nondualistic conception of practical thinking, they have largely been misunderstood. "[A] man has not to do Justice and Love and Truth," John Dewey reminded philosophers over a hundred years ago, "he has to do justly and truly and lovingly. And this means that he has to respond to the actual relations in which he finds himself."[9] In the building of the technological world, vigorous common sense of this sort has been well and truly buried under the weight of epistemological procedures. The recent excavations of feminist theory show how deeply these procedures have also buried the wisdom of the mother and other agents whose practices have been centered in care rather than in technological control.[10] Bernard Williams is right: it is peculiar that modern moral philosophers have presented the acquisition of moral maturity as the procedure of shrugging off the guiding hands of caring others in an attempt to become self-sufficient, Archimedean observers of a distant social and ecological reality.[11]

The Substance of Practical Reason

Understood substantively, practical reason is that discipline of thinking that can only develop, be expressed, and be passed on through embodied relationships.

Practical reasoning embodies the truth of Hans-Georg Gadamer's claim that "moral knowledge must always be a kind of experience."[12] Practical reasoning aspires not to universal descriptiveness, nor to the rule-bound governance of practice, but to an ever more rich awareness of our moral lives, a richness that arises as moral understanding and moral experience draw together. Practical reasoning arises out of the reality of human love in a world of good things. While "[p]ersons, rights, obligation . . . are concepts at the center of one way of thinking about morality," points out Raimond Gaita, "[h]uman being, human fellowship, love, and its requirements are concepts at the center of another."[13]

Taylor describes practical reason as "a reasoning in transitions." As such, it does not aim for absolute conclusiveness, it is inherently comparative. This is not to say, however, that it is inherently relativistic; practical reason takes its bearings always from the broader character of a discrete, identifiable world of practice. In the context of a shared materiality, practical reasoning aims at making the comparative judgments by which we live a "transition which we understand as error-reducing and hence as epistemic gain."[14] Such comparative judgments become part of biographical and communal narratives that encode our trust in ideas and beliefs that have proven in our experience to foster better, less error-ridden judgments, that are better in the sense that they increase our ability to sustain the good in our lives and in our world. Thus, explains Taylor, practical reason was understood by the ancients as being inherently normative: "To be rational was to have the correct vision, or in the case of Aristotle's *phronēsis* [practical wisdom], an accurate power of moral discrimination. But once we sideline a sense or vision of the good and consider it irrelevant to moral thinking, then our notion of practical reason has to be procedural."[15]

Theoretical and technical knowledges are of course necessary for the flourishing of human life, but they are not sufficient for it.[16] They lack the capacity to produce wise action. They lack what Taylor calls the power of correct vision, or what Borgmann has called "world-articulating and world-explaining significance."[17] By reasoning practically, we are enabled to make sense of theoretical and technical knowledge as it is applied in our forms of life. Our practical reasoning takes serious account of the claims of principled thought on our actions, but not in a way that makes the embodiment of these principles the point of action itself. As Hubert Dreyfus and Stuart Dreyfus emphasize, "ethical practices can function perfectly well without abstract universal principles of rightness being invoked, while principles of rightness are totally dependent upon the everyday practices for their application."[18]

Practical reason is perspectival, and the inquiry it fosters is inherently experimental, although not in the sense of seeking factual confirmation of a predetermined hypothesis, as in deductive, apodeictic reasoning.[19] The experimental character of practical reason arises from at least five attributes that are inherent to it. First, practical reason displays a profound openness to the inexhaustible novelty of our practical entanglements. Second, practical reason has regard for the modest limits of human comprehension amidst the vastness of our experience. Third, practical reason accounts for forms of human knowing and human wondering that do not find rational expression. Fourth, practical reason acknowledges the irrepressibility of both human benevolence and human malevolence. Fifth, practical reason affirms that our powers of practical moral judgment are learned, developed, and passed on through our participation in a shared world of practice.

It is not surprising, then, that dualistic philosophical traditions thoroughly misunderstand the nature of practical reasoning, for it is essentially nondualistic. As a consequence of this misunderstanding, modernist philosophers "have written as if practical judgments were a degenerate form of theoretical judgements."[20] They have been unable to see that practical inquiry is of a character at once pragmatic, ontological, and moral. As Jim Cheney and Anthony Weston have argued so clearly in their advocacy for a shift from epistemology-based ethics to world-affirming ethic-based epistemologies, modern philosophy lost sight of the ground of human valuing. "Our task is not to "observe" at all—that again is a legacy of the vision of ethics as belief-centered—but rather to participate," they remind us.[21] That is, *"ethical action is first and foremost an attempt to open up possibilities, to enrich the world."*[22] Like Cheney and Weston, I consider that practical reason discloses for us the ways in which practical possibilities for enrichment "*surround us at all times*."[23] However, the moral scientists that have had the ear of modernity for so long have denied the ways in which worldly wisdom and moral practice belong together and they have, in the process, largely lost sight of our moral experience. This denial has been actualized in the building of a technological world that constructs practical means and moral ends, world and self, as a technological duality, and that consequently disorients moral experience.

Unlike the instrumental reason that designs the means required to produce a given end (of, say, the strongest bridge), practical reason cannot know the *telos,* the end, of our moral activity prior to or outside of our participation in the activity itself. There is no absolute blueprint for practical reason, although we do without doubt build up and draw upon a tacit store of exemplars from past experience.[24] The point is, however, that our moral ends cannot be understood from a detached, Archimedean perspective.[25] Our moral ends are, as Dewey formulated it, only ever ends-in-view.[26] They can only ever reflect the line of sight afforded by our embodied and correlational location. Practical reasoning arises out of the ways in which the ends, the goods, that we seek are not external to our material context but are constituted through our participation in this materiality. Ultimate moral goods are not realized as the completion of a practical task, but are realized as the ongoing achievement of cultivating and sustaining the good in our lives. It is through the ongoing participation in practices that the idea of the good life gains clarity, eloquence, and expression. As MacIntyre makes clear, there is an internal, mutually constitutive relationship between narratives of what it is good to be and the everyday practices through which we live. Which is to say that "the end cannot be adequately characterised independently of a characterization of the means."[27] MacIntyre sets out the meaning of this high sense of *praxis*:

> By a practice I . . . mean any coherent and complex form of socially established cooperative human activity through which the goods internal to that form of activity are realized in the course of trying to achieve those standards of excellence which are appropriate to, and partially definitive of, that form of activity, with the result that human powers to achieve excellence, and human conceptions of the ends and goods involved are systematically extended.[28]

The goods we seek, including those final constitutive goods that Taylor contends empower us to "do and be good," belong to our practices.[29] The ultimate, final good that Aristotle proposed as the end of all moral *praxis*—the state he called *eudaimonia*, or well-being—has been wrongly understood in much modern philosophy as the individual's quest for happiness.[30] The internal relationship of means to end that both Aristotle and MacIntyre describe, however, is not a relationship internal to an individual's life experience, it is internal to a world of practice. Well-being correctly refers to a state that is sustained by participation in social practices and that issues forth from virtuous behavior to infuse public life.[31] In the "ancient and medieval worlds, the creation and sustaining of human communities—of households, cities, nations—is generally taken to be a practice."[32] Virtue in human character, whatever a particular historical community takes this to be, emerges not as the acquisition of individuals, but as the attribute of practices that reveal a world in which virtue is possible. The ultimate end to which practical judgment aims is that of sustaining the well-being of the communal life of the *polis*. It is characteristic of internal goods, MacIntyre believes, "that their achievement is a good for the whole community who participate in the practice."[33] Moral ends are not, Patricia Benner emphasizes in her philosophical study of the public practices of nursing, "limited to a 'private feeling' or psychological state. They are grounded in the shared public, historical world of being related to one another in social groupings, common meanings, mores, practices and skills."[34]

The ultimate good of virtuous life is encoded within a world of practice. Moral *praxis* is that part of our agency that is an end in itself, for it is action that sustains the well-being of our world.[35] Among the goods that are internal to practice are the goods of prudent judgment and practical wisdom themselves.[36] The reciprocity—the communion, even—between the practical means we use and the power of practical reason to illuminate the ends we seek cannot be broken without both the ends we seek and the moral nature of the means available to us becoming unintelligible.

The Riches of Relationship

Practical reason does the work of aligning our everyday actions with our intuitions and narratives about the good life. The substance of practical rationality is nothing less than a vision of the good that is worked out, endlessly and inexhaustibly, within the partiality and ambiguities of embodied practice. Unlike scientific or technical reasoning, practical reason orients our practical decisions towards the horizons of moral ontology by which we understand what is good in and about a world that is more-than-self and more-than-human. According to Taylor, to "make practical reason substantive implies that practical wisdom is a matter of seeing an order which in some sense is in nature. This order determines what ought to be done."[37] Practical wisdom takes its bearing from the moral order that is encoded in our practical relationships with significant things and others. Practical wisdom provides insight into the ways in which these practical relationships are deeply implicated in our understanding of who we are. The reward of practical reasoning is neither the systematization of theory nor greater technological efficiency. Its reward is the nurturing of wisdom about what is truly valuable in our social world by encoding it in the narratives and practices of everyday life. This wisdom animates our stories far more powerfully than any extravaganza from Hollywood.

In its essence, practical reason responds to the question of what it is, above all else, that our actions should seek to sustain. Practical reason provides small, transient spaces of clarity amidst the irremovable ambiguity of human experience. Rather than attempting to "gain authority over ambiguity by getting hold of its controlling conditions," as does instrumentalist reason, Borgmann affirms that practical reason responds to the claims made upon us in particular situations, by particular people and particular things.[38] That is, the "experience of a thing is always and also a bodily and social engagement with the thing's world"[39]

Our practical understanding of what is most precious in our world provides the framework within which we orient our moral judgements. Taylor reminds us that the "goods which command our awe must also function in some sense as standards for us."[40] On the basis of our love for what is finally and decisively good, we make strong evaluations in our practical judgments. These are evaluations that "stand independent of our own desires, inclinations, or choices, . . . [for] they represent standards by which these desires and choices are judged."[41] A practical situation never arises before us as a random conglomeration of objects from which we can detach ourselves. Our cognitive awareness of this situation emerges through the filter of narrative and experience by which we make strong evaluations about wherein the highest good lies and about how it is to be sustained through our actions. Our practical thinking is claimed precognitively, or, rather, infracognitively, by our love of, and thus our desire to sustain, particular things in and about our world.[42] In the situations I confront in everyday life, for example, I make strong evaluations about the magnificence of a vastly aged forest or the deep satisfaction of growing food for those I love or the joy of listening attentively to my children or the incomparable relief of having friends physically close by when trauma strikes. Strong evaluations of this sort subsequently orient and guide my practical decisions about what I ought to do.

The Rationality of Relationality

In its general character, practical reasoning is the rationality of relationship. It brings to light the moral character of our essential relationality as members of human and biotic communities. It brings to light our essence as children, siblings, parents, lovers, neighbors, friends, combatants, leaders, followers, carers, sufferers, teachers, and students. It brings to light our essence as sensuous, omnivorous, mortal, and reproductive inhabitants of the living earth. To reason practically is to live in ways that celebrate our worldhood as the "implicate order," the undivided flowing movement, within which our selfhood is enfolded.[43] It is the lived celebration of the intimate holographic embrace of world and self. This perennial wisdom is sustained for us by the prose of Robert Browning:

> This world's no blot for us,
> No blank. It means intensely and means good.[44]

And this is, of course, merely to state what was such common sense to the original cultures of lands like Australia and America that it needed no discussion, no debate: it is a wisdom that gained expression with their every practice. There is much to think about in David Abram's observation that "Kant's writings could not be translated into Navajo or Pintupi."[45]

Human practice is the drawing toward and into our selves of worldly things: things living and nonliving, artefactual and ecological, human and nonhuman, earthly and heavenly. In the beat of sunlight on our skin, or in the remarkable timelessness of the water in our glass, or in the ingenious elegance of the bicycle sprocket harnessing the power of our legs, we are engaged in a shared, eloquent reality. However, even to affirm relationships in this generic way is to retain the vestiges of abstract reason. It is not that I reason practically from the general categories 'parent', 'son', and 'friend' toward my experience as the father of Angus and Mara, the son of Bevis and Mary, the friend of Ian and Patsy. Practical reasoning works always from the specific, the embodied, the experiential, toward the generic.[46] Practical reasoning is a discipline of care-full attention to the demands placed upon our self-understanding by our lived embrace of specific people and specific things in specific places. Moral wisdom does not lie in discerning what is absolutely right to do, but in practically nurturing and sustaining the relationships we have with particular people and places which bind us to a wider shared world rich in good things. So, whereas for instrumental reason the written text is the paradigmatic exemplar of learning, for the practical reason it is the intimate experience of others and things. It is, as the wonderful poet and essayist Mary Oliver has put it, "the intimate, never the general, that is teacherly. . . . Time must grow thick and merry with incident, before thought can begin."[47] I thus think Abram right. The public environmental ethic for which so many of us yearn in the technological world, one that cherishes our more-than-human world, can only come through "renewed attentiveness to this perceptual dimension that underlies all our logics, through a rejuvenation of our carnal, sensorial empathy with the living land that sustains us."[48] I would clarify this claim, however, with the affirmation that human attentiveness to the perceptual dimension of sensorial experience encodes its own logic; it develops in us, over time, the capacity of a practically reasoned, a practically wise, understanding of our relationality.

Technology and Relationality

In affirming that moral wisdom has its roots in and is thus nourished by its world, we must note that, while it is instructive, Aristotle's high sense of *praxis*—as political and moral activity that is neither primarily theoretical (*theoria*) nor productive (*poiesis*)—cannot be appropriated in our latemodern world in any direct, literal way.[49] Although Aristotle appreciated that *praxis* belongs intimately to the *polis*, he discounted the ways in which technological activity is integral to the moral character of the social world. Reflecting the social stratification of his classical world, Aristotle saw the statesman as the paradigmatic actor of *praxis,* the philosopher as the paradigmatic actor of *theoria* and the craftsman as the paradigmatic actor of *poiesis*. For Aristotle, these (patriarchal) social roles and the agency they embody, "are generically different."[50]

Whatever the merits or otherwise of the Athenian *polis*, this distinction is entirely misleading in our technological world. We need explanations of our inherently relational activity that can encompass the role of technology in producing, that is, in building and revealing, moral worlds. We require an understanding of everyday practice that illuminates the ways in which relational and productive activity interpenetrate and inform each other. Doing and making belong together in human agency. Neither can be reduced to the other, nor can either be understood in isolation from the other. While modern devices often displace our experience of relationality, they have not and cannot obliterate the possibility that we will be able to recover sight once again of the interconnections between our practical judgments and the well-being of our world. Thus while I accept, in general terms at least, the Aristotelian account of practical reason, I agree with Borgmann that we must treat with great scepticism the philosophical lore that doing and making, *praxis* and *poiesis*, represent two discrete classes of human agency:

> The segregation of doing from making, of morality from production, goes back to Aristotle and has been carried forward into the modern era by Immanuel Kant. It has been a comforting tradition for philosophers and the public alike, for it suggests that the modern developments that entirely upset and destroyed traditional production left morality intact, if not in its particulars, then, at any rate, in its general cast as the theory and practice of human conduct.[51]

The dualism that first cleaves doing and making and then subordinates doing to making is central to the emergence of technological ontology in Occidental philosophy and in wider material culture. This dualism hyperseparates morality and production, rupturing our awareness of technology as world-building. And, of course, the actualization of technological ontology that begins with Descartes and Bacon is nothing less than actual subordination of *praxis* to *poiesis* in our everyday lives.

Practical explanations of our world need to move from the specific to the generic in a way that keeps visible the character of our technological world. They should be able to shed light on the moral, ontological and epistemological significance of our technological structures, thereby making more intelligible what is at stake in the everyday practical choices about technology that remain open to us. Disclosing what is at stake for our moral lives in these judgments is an embodied, rational craft that lies at the heart of any possibilities for building worlds that are more sustaining of our ecological and social reality.

II THE CRAFT OF MORAL LIFE

What is required, if we are to counter the productivist fervor of latemodernity, is a recovery of the rationality of relationality to the center of our everyday practices. We are in need of a deep political awareness that it is not only rational to value material security, it is also rational to want to belong in a welcoming, sustaining world. We must recover those skills of ontological thinking about our practical judgments that can illuminate and guide us toward preserving what is good in and about our world. "A morally trenchant conversation about the good life requires, then," Borgmann articulates with great clarity, "that the pattern of our actions which is disguised and diffracted in the prevailing moral discourse is itself made the moral issue."[52] There is nothing more rational, yet more neglected in latemodernity, than inquiring into the symmetries that exist between the character of our lives and their material setting. This practical inquiry will confound conventional theoretical boundaries, because, as Borgmann makes clear, its central task is to see our lives in their constitutive essence:

> To bring out the contrast between these symmetries is a task that is at once ontological, moral, aesthetic, theological, and political. It is ontological in raising the problem of what is real. It is moral in directing us to the very substance of human conduct. It is aesthetic as it involves us in the question of what human works are centrally enchanting and illuminating. It is theological because it leads us to the issues of grace and divinity. And it must become political and make us consider our responsibility for the common order. Either we see this task in all of its dimensions or we will miss it altogether.[53]

Seeing this task in all of its dimensions is the particular capacity of the rationality of relationality, for it is inherently paradigmatic: it explores the connections that hold us in a particular social and ecological space. It responds to the specific substance of our practical engagements in such a way so as to illuminate the systemic character of the world in which these engagements take place.

However, even in the work of those, such as Taylor, MacIntyre, and Williams, who advance so greatly our understanding of practical epistemologies, we still find unworldly, almost protean, descriptions of practices that ignore their technological constitution. The crucial next step all too often neglected by philosophers is to relate the procedural suppression of practical thinking to the technological deformation of action by the material structures of latemodernity. In this neglect, points our Borgmann, what remains "unexamined all the while is the power of products, of the material results of production, to shape our conduct profoundly."[54]

ACTION, STRUCTURE AND AGENCY

As we saw in MacIntyre's definition of practice as "any coherent and complex form of socially established cooperative human activity," moral theorists have generally emphasized practice as activity, rather than as enduring structure. MacIntyre's account is an extrapolation from Aristotle's insight that virtues develop only through the achievement of goods that are internal to practices and which are, thereby, internal to relationships between practitioners. But what is not sufficiently emphasized here is that virtue is also internal to human relationships with things and, further, that relationships between practitioners develop within a matrix of things. To alter this matrix, even in small detail, is also to alter the nature of the spaces within which social relationships can take place, not to mention our relationships with other inhabitants of the living earth.

In his essay *Social Theory as Practice* (1985), Taylor provides a convincing description of the mutually constitutive relationship that exists between ideas and practices in practical reasoning. Taylor argues that truly practical social theory is necessarily founded upon a reflexive self-awareness that theorizing is itself a type of practice. Self-reflexive theorising makes it apparent that "part of what is involved in having a better theory is being able to more effectively cope with the world."[55] Conversely, better practices may result through the explication of the implicit self-descriptions that are encoded in practices and which participants absorb as part of the lived reality of their world. This task of explication is, as Taylor sees it, the task of practical reason. Such practical theories "do not just make our constitutive self-understandings explicit, but extend, or criticize, or even challenge them. It is in this sense that theory makes a claim to tell us what is really going on, to show us the real, hitherto unidentified course of events."[56] Practical theorizing, then, works back from the understandings encoded in the world of practice in which it takes place, making these understandings at least partly visible and partly open to question once again. Visible and questionable, these understandings become the subjects for new theories that shed fresh light on our practical world of problems and solutions. Thus: "What makes a theory right is that it brings practice out into the clear; that its adoption makes possible what is in some sense a more effective practice."[57] At root, practical social theory explicates, questions, and transforms the self-definitions encoded in our world. Further, suggests Taylor, "because theories which are about practices are self-definitions, and hence alter the practices, the proof of the validity of a theory can come in the changed quality of the practice it enables." This change in quality is not a change toward greater scientific veracity or a change in technological efficiency—or, at least, it need not necessarily require these kinds of change—it is change that "enables practice to become less stumbling and more clairvoyant."[58] Good social theory improves our vision of life's moral horizons.

Even Taylor's impressive description of the features of thick social theory, however, retains a somewhat abstract, or at least a rarefied, notion of practice. Take, for instance, his digression on diachronic causation in *Sources of the Self* (1989):

> The kind of ideas I'm interested in here—moral ideals, understandings of the human predicament, concepts of the self—for the most part exist in our lives through being embedded in practices. By "practice" I mean something extremely vague and general: more or less any stable configuration of shared activity, whose shape is defined by a certain pattern of dos and don'ts, can be a practice for my purpose.[59]

A "stable configuration of any shared activity" will indeed serve us well as a general, if bland definition of practice. The emphasis on shared activity produces examples of practice such as "the way we discipline our children, greet each other in the street, determine group decisions through voting in elections, and exchange things through markets."[60] More specifically, shared activities that Taylor considers to be causal to the emergence of the Enlightenment ideal of disengaged agency include:

> [R]eligious prayer and ritual, of spiritual discipline as a member of a Christian congregation, of self-scrutiny as of the regenerate, of the politics of consent, of the family life of the companionate marriage, of the new child-rearing

> which develops from the eighteenth century, of artistic creation under the demands of originality, of the demarcation and defense of privacy, of markets and contracts, of voluntary associations, of the cultivation and display of sentiment, of the pursuit of scientific knowledge.[61]

This is an interesting list, but, as a result of the way it is framed, it also obscures the moral significance of technological artefacts. Each one of these examples is intelligible only to the extent that a reader is able to call forth the material context in which these activities take place. "A Christian congregation" may evoke images of the world-disclosive Cathedrals of Europe, products of massive social organization, huge sums of money and cutting-edge medieval technology. It may evoke the printing presses that came somewhat later and showered the word of God upon people at large. Or, perhaps, it calls forth the dowdy, almost secular, suburban brick chapels of latemodernity, centres of worship provided more amply with car parking than with ornate vestments. Similarly, the "new child-rearing" cannot be conceived without reference to artefactual contexts. In the latemodern world, we need to think substantively about the suburban environments that fragment family life into its now predominantly nuclear form and that support the historically recent phenomenon of secular schooling and the even more recent industries of child-care. Nor can we understand how our children are reared without being attentive to the presence of multitudinous factories producing even greater multitudes of inconspicuous artefacts, such as milk formula, plastic nipples, disposable nappies, cots, prams, stuffed animals (curiously enough, the majority of whose living counterparts were slaughtered long ago), television programs, videos and books, that display how the technological society cares for, shapes, and orders its young.

I agree absolutely with Taylor—invaluable diagnostician of the frail condition of contemporary moral discourse that he is—that "one of the laughable, if lamentable, consequences of modern moral philosophy is that you have to fight uphill to rediscover the obvious, to counteract the layers of suppression of modern ethical consciousness."[62] I would suggest, however, that in addition to fighting uphill to rediscover the good, we also have to fight uphill to recover the obvious fact that without the introduction of technological structures into moral theory, the subject matter of practical reason is only partially understood.

Taylor's discussion of the central political practices of modern liberal democracies is a case in point. As part of this class of practices Taylor includes activities such as "elections, decisions by majority vote, adversary negotiations, the claiming and according of rights, and the like."[63] The act of casting a political vote would appear to be paradigmatic of *praxis.* It is, in theory, pure democratic, relational action. But in our technological world, this action is structured within ossified histories of action that have revered above all else the capacity of instrumental reason to reorganize practices according to the dictates of optimally efficient production. The technological background of means frames the substantive political character of the democratic *praxis* of voting. To exercise a democratic vote means something very different in a national election of millions than it does in an extended kin group founded on intimate face-to-face relations. The structure of our practices makes possible a *polis* based on vast and often random aggregations of people, and it simultaneously makes marginal any form of political community founded on face-to-face relations. It is theoretically significant therefore, that latemodern liberal-democratic political practices are embodied in large-scale technosystems of information production that make possible such things as electoral boundaries, voter lists, and the like, and of information dissemination that present candidates to us and dissect the results of the popular voice. As my pen moves over the page of candidates, it is true that I am exerting my free rational judgement, but I am doing so in a highly delimited and unique world which enframes the nature of this judgment. Even simply marking crosses on paper with pencil invokes the political reality of a state-of-the-art pulp mill situated in an international industrial ecology park and the drone-like activity of workers in a large Southern pencil factory. My action that seeks to sustain wise governance in my community is patterned by forms of life that greatly constrain the forms governance can take.

Relationality and Productivity

Much talk of practice belies the fact that our experience of practice resides ambiguously between production and product, action and materiality, method and subject. A practice is a drawing together and a transposition of human, natural, and artefactual things into enduring structures. A practice is not simply a cultural site for action, it is a material ossification of histories of action that constrain, and yet invest meaning in, present forms of action. Our activity, our *praxis* and *poiesis*, build the structures in which future activity must take place. This future activity will in turn transform these structures, building, eventually, new worlds. There is thus a need in our account of the rationality of relationality to affirm the coevolution of relationality and productivity. We should also be aware, however, that there is a fine line between affirming this coevolution and following attempts such as Heidegger's to conflate the two in his account of disclosive production. And this line is truly a watershed. With our focus now firmly upon the substance of practical reason, it is worth looking further at what I called, in chapter 5, Heidegger's displacement of *praxis*.

It is one of the great ironies of Heidegger's account of technology that he mirrors, and thereby only reinforces the apparent inevitability of, the latemodern subordination of *praxis* to production. And why does he do this? Because, as Dana Villa shows us, he understood *praxis* to be, in its essence, "a form of radical *poiesis*;" that is, a form of disclosive artful skill.[64] Heidegger argued that it is artful technology in its disclosive essence, rather than deliberately social action, that builds the ethical bond that secures humans, that allows humans to dwell, in their world.[65] Practical moral reasoning, as the wisdom that inheres in *praxis*, is thereby subsumed by Heidegger in his explanation of the knowledge that inheres in our technological engagement with things; our technological bringing-forth of a world.[66] The degeneration of *techne* into technology is for him thereby the degeneration of *praxis* into production. His complete lack of interest in ethics arises out of his view that the technological world is inherently amoral because its practices are utterly arelational.

I see a contemporary variant of this conflation of *praxis* and *poiesis* in Hubert Dreyfus's neo-Heideggerian account of *phronesis*. Admittedly, unlike Heidegger, Dreyfus does not dismiss discussion of ethics. Indeed Dreyfus is an influential advocate for a situational ethics that acknowledges that moral judgments are based in experience, experimentation, and tacit exemplars rather than in abstract principles.[67] Yet, although I have in this chapter endorsed just this sort of account of practical reason, I have serious concerns with Dreyfus's phenomenological treatment of practical skill, a treatment that draws heavily from early Heidegger.

Dreyfus emphasizes that moral expertise is acquired in ways not dissimilar to expertise in playing chess or driving a car. In his view, practical wisdom is "a case of skill applied in everyday life." In the 1991 interview from which this remark comes, Dreyfus goes on to draw Heidegger's mantle around him in describing his idea of skill:

> I don't know whether it is Aristotle or just me, Stuart [Dreyfus, brother and co-author] and Heidegger. We need a notion of skill which cuts right across these two [*techne* and *phronesis*]. Skill is what you need in order to do your art and craft, getting things done with things. And skill is what you need for dealing with other people and running a government. I don't know what the word for skill would be in Greek, because there doesn't seem to be enough words to go around. But it looks like, in fact, *phronesis* and *techne* are two subcategories of skills.[68]

Dreyfus develops this idea further in his 1997 paper with Charles Spinosa, *Highway Bridges and Feasts*, and the earlier *Disclosing New Worlds* (1995) with Spinosa and Fernando Flores.[69] These authors argue that we are able to respond to the solicitation of the latest in highway "spaghetti junctions" by using a car cellular phone to reschedule a meeting while in transit. We are told that, using the technological equipment available to us, we are able to respond just as skilfully, just as poetically, to the gathering of the highway bridge as premoderns were able to respond to the solicitations of any of the old bridges. According to Dreyfus and Spinosa, the flexible way in which we are enabled to respond to the call of a balmy, inviting afternoon by shifting our meeting from town to lake is an example of "how Heidegger thinks we can respond to technological things without becoming a collection of disaggregated skills."[70] We are offered another example in a similar vein in *Disclosing New Worlds*:

> Frequently, as soon as one member of a family starts using a cellular phone for business purposes, it expands into the family. A working wife with a cellular phone will find that she may be able to schedule a last-minute lunch with her husband. When the couple are going out to dinner or a movie, they can call the sitter from the car. Dead time can be used not only for calling customers but for calling children and spouses. Pretty soon the practice of using the cellular phone is appropriated by the family. But the whole style of the office does not come with it, rather what primarily enhanced efficiency in the office furthers togetherness and warmth in the family.[71]

I do not wish to deny that artefacts such as mobile phones may enhance human relationships. Nor do I see any point in objecting to the obvious fact that a meeting by a lake, attended in the comforting knowledge that the carer minding our children can contact us at any time, is likely to be far more pleasant than if it had been held in a dull seminar room. But the attempt by Dreyfus and his colleagues to reduce the ontological possibilities of technology to purely local phenomena leaves their account of moral expertise prey, in my view, to the very metaphysical subjectivism against which postwar Heidegger railed. The local worlds that Dreyfus and his colleagues point to are, to my mind, not worlds at all in any meaningful, ontological sense. The disclosure they describe does not shed light on the patterning of language and technology that shapes our worldview. These local disclosures illuminate only the foreground of commodious freedoms that devices make possible. While we may indeed be able to exercise choice expertly in this foreground, and while this expertise does indeed have some moral significance, the triviality of the examples provided by Dreyfus and his colleagues makes clear that the technological substructure of our world remains hidden in this account of skill. These examples trivialize the systemic, widely patterned implications of massive technosystems such as car-based transport and telecommunications.

The proposition that a carload of Northern entrepreneurs, chatting on their cellular phones in air-conditioned comfort, have subverted the dominance of a productivist understanding of reality because they now race down a highway (through agricultural land soaked in acid-rain?) toward a lakeside meeting (about the prospects for marketing laboratory-grown produce?) is simply untenable. In the end result, this kind of affirmation of the skill of *phronesis* is effectively indistinguishable from the affirmation of instrumentalist freedoms offered by arch-Prometheans. Dreyfus and his colleagues provide a welcome attempt to understand "human beings as neither primarily thinking or desiring but as skilful beings."[72] But, as Borgmann counters, "[c]onsumption rules production, and desire rules consumption. 'Disclosing New Worlds' is right in disputing the claim that humans are primarily thinking beings. But neither are they, in the prevailing order of cultural justification, primarily or finally skilful beings."[73]

Toward a Thick Theory of Practice

A substantive theory of the craft of moral life needs considerable political depth if it is to avoid rarefied or trivial explanations of the inextinguishable possibilities of world-revealing *praxis*. What such theory needs to do, I suggest, is to address the peculiar bifurcation of our latemodern world of practice into a foreground of apparently autonomous human agency and a background of apparently autonomous machinery. The technological world makes real the segregation of doing and making only postulated by philosophers. Certainly, in the foreground of ends, we consciously make and debate choices about our social relationships, and in this foreground we are indeed engaged in reflexive, skilful, phronetic *praxis*. But in the background of means our relationality is subordinated to our productivity. Joanna Hodge reminds us that in the technological world "human beings are placed in relations above and beyond our control."[74] Therefore, "viewed as a theoretical structure technical relations may be the actualizing of metaphysics, but as a lived experience they are a reification of ethics."[75] The background of machinery on which our lives rest is nothing less than the reification of an instrumentalist understanding of moral life.

In the foreground of ends, ethics thrives as we question the merits of euthanasia, the implications of divorce for the mental health of children, and the moral grounds on which we may defy the law in coming to the aid of a friend. These are undoubtedly crucial questions. And, to be sure, there is nothing to stop us debating these questions on a cellular phone, by satellite conferencing or in a space shuttle. But we ask and answer these questions largely oblivious to the world of means to which they belong. We debate the merits of abortion over a plate of fried eggs from factory chickens, perhaps in a lakeside café, being careful not to stain clothes that in all probability were produced through the ravages of industrial cotton farming and Southern factory labor. Ethical deliberation, judgment and debate have not vanished, but are increasingly constricted to the foreground of liberal freedom. They are, further, increasingly unable to keep visible the dualistic, anthropocentric, subjectivist, and productivist qualities of so many of our relationships that render our world ultimately unsustaining.

Public and academic discourse about *praxis* is routinely restricted to the technological foreground of ends, while the background of world-building means is virtually given over to the imperatives of maximum sustained production. The more technological ontology is actualized in our practices, the more the nature of everyday practices as a mosaic of worldly relationships becomes opaque in our experience and in our discourse. Our world is built in a way that makes the conditions of our moral agency invisible. We are limitlessly innovative in increasing our power as producers and consumers. We display a remarkable solidarity in our shared fidelity to artefacts like highways and cellular phones, but we seem largely unable to influence the trite, superficial, and often exploitative nature of our relationships with the things and others around us.

I do not accept the argument that skilled appropriation of commodities can rehabilitate the Enlightenment promise of technology, although certainly such skill is important. But neither do I accept the argument that our capacity for ontologically disclosive *praxis* has been obliterated by technology, although it has certainly become weak and diffuse. I agree with David Strong that we live "neither in an iron cage nor in an arena of unconstrained choice; we inhabit a possibility space where some moral choices are being made."[76] If we are to reason practically about the spaces of moral possibility that we inhabit, we must first ground our thinking about practices in their actual, material context. "Talk of practices remains inconclusive," observes Borgmann with characteristic insight, "until it hits the catalytic layer of tangible reality. . . . Social theory, catalyzed by material reality, gains access to the reciprocal relation between material reality and social practice."[77] I thus hold that the production of worlds (*poiesis*) and the revealing of worlds (*praxis*) are irreducible dimensions of human agency. If we are to understand the nature of our moral lives, building and revealing should not be conflated, polarized or made mutually exclusive.

Reclaiming the Question of Sustainability

I began part II of this book with the observation that technology is the focal fact of our latemodern lives. This fact is twofold, encompassing the phenomenon of world-building and the experience of correlational practice. My aim in this chapter has been to argue for forms of disciplined, rational inquiry into technology that move from our experience of correlational practice, the experience of discovering our selves through our practical engagements with others and things, toward the cultural phenomenon of world-building. Stephen Ross, a philosopher who has taken up the idea of substantive practical judgment with originality and clarity, gives voice to what our experience confirms unstintingly:

> Technology is neither uniquely threatening and dehumanizing nor uniquely beneficial and empowering, but is practice itself in its large-scale contemporary and political forms. It contains the panoply of practical entanglements, the dense and specific materialities, that characterize practical judgment, especially in those far-reaching political forms that are most autonomous, with lives of their own. The issue of technology, then, is the generic issue of practical query—not one problem, but inexhaustibly many, the question of how we are to be able to work toward the future, given that our forms of practice determine and transform that future, enrich and despoil it. It is a question of whether practice is possible where control is impossible.[78]

Like Ross, I have argued that technology is practice itself. I have argued that technology is best understood constitutively as the correlational engagement with reality whereby humanity is oriented or disoriented toward what is ultimately, finally good in their world.

The prospects for building more sustaining worlds rest, ultimately, with a recovery of our practical experience. Sustainability is nothing less, in late-modernity, than the craft of moral life. As I see it, this recovery requires three steps. First is the step of recovering in everyday practice an awareness of the ways that our practices are deformed by the structural bifurcation of means and ends. Second is the step of recovering in moral discourse the world-disclosive force of practical epistemologies capable of shaping this awareness into public declarations of the sources of our moral orientation and disorientation, the sources of our moral enervation and sustenance. Third is the step of recovering the possibilities that exist in everyday practices for sustaining what it is that seems most commandingly good in and about our shared world. The first two steps have been the principal concern of my discussion in part II. I shall now take up the challenge of describing at least some of what the last step demands of us in part III.

This recovery of moral possibilities is simultaneously a recovery of the ideal of sustainability from instrumentalist discourses and technocratic agendas. In the course of this text, I have travelled far from the prosaic policy discourse that I took as my point of departure in chapter 1. I hope, nonetheless, that the question of sustainability has remained visible, even while I ventured into the esoteric terrain of Heidegger's ontological analysis and the convolutions of moral philosophy and epistemology. Part III returns us to where I began, contesting sustainability. But I return having reclaimed the question of sustainability, thereby reconfiguring it, as the question of what possibilities exist for practical inquiry into the moral character of technology. The question, that is, of what it means to sustain technology.

I have in this chapter tried to sketch the epistemological conditions of this recovery. I have tried to clarify the difference between technical and practical power, and I have sought to explicate the need to liberate our self-awareness of our relationality from its subordination to the narratives of self-production. But, in the end, such theoretical clarification only makes the path of recovery ahead more visible. It is unable to take even a first step forward by itself. In chapter 8, then, I return to the question of sustainability through a narrative account of my own personal search for moral sustenance in the deformed world of technology. In chapter 9, I shall more directly contest the ideal of sustainability, and I wish to do so in a way that bears relevance to policy, but I will do so in a way that acknowledges that my concern about sustainability begins and ends with the practical experimentation with technology that gives expression to my yearning, and the yearning of many in my world, to live in more sustaining ways. I give expression to my conviction that there is room to experiment within our existing technological practices. For I am convinced that there is room amidst the on-rush of technology for us to move to the edge of the technological world and to thereby catch sight of the possibilities in our moral lives for building more sustaining worlds.

III

The Search for Sustenance

Now I re-examine philosophies and religions,

They may prove well in lecture-rooms, yet not prove at all under the spacious clouds and
along the landscape and flowing currents.

Here is realization,

Here is a man tallied—he realizes here what he has in him,

The past, the future, majesty, love—if they are vacant of you, you are vacant of them.

—Walt Whitman

CHAPTER EIGHT

A WORLD WORTH CARING FOR

INTRODUCTION

WHAT TRULY SUSTAINS US? Why? And how do we know?
What are we to sustain above all else? Why? And how may we do so?

These were the questions I introduced in chapter 3 as basic to any contestation about the cultural meanings of sustainability. Our ability to comprehend and respond to these questions arises directly out of our richly ambiguous experience of the things and others that inhabit the world that sustains us. Questions of this sort exist antecedently to cultural contestation about sustainability. They hold open the discursive spaces in which these contests can take place. Although they are rarely asked in the stripped-down form I have framed them in here, these questions probe the anxieties about and the visions of the future that arise with the latemodern evidence of ecological and economic unsustainability.

The spaces held open by the questions of sustainability accommodate both pragmatic objectives and cultural critiques. They support the need for both incremental and systemic reform of the technological world without collapsing one mode of reform into the other. These spaces accommodate our urgent practical need to live in more ecologically sane and economically just ways with our desire to deepen our cultural understanding of how we came to live in such unsustaining ways in the first place. It is my hope that within these spaces we will be able to probe, discursively and experimentally, through the weak points of internal contradiction that are to be found throughout our technological world and out onto paths of world-building that lead toward more sustaining worlds.

In this chapter, I draw questions of sustainability closer to direct, pragmatic concerns. I want to offer a personal, practical response to these questions. In so doing I shall try to give substance to Albert Borgmann's claim that the commodious freedoms afforded by latemodern technology "are alluring, but they are not sustaining."[1] Yet I shall not attempt to provide a comprehensive, global discourse within which to articulate "the" morality of sustenance. I don't believe such a discourse can guide us in experimenting with the possibilities that exist in our everyday practices. What I shall do is give expression to my experience of what I find ultimately sustaining. This form of expression is necessarily partial and particular. It necessarily aspires to being emblematic rather than comprehensive. The experience of which it tells is shot through, no doubt, with my own imperfections, and it is colored more by what it lacks that what it encompasses. Yet, in its humble boundedness, this experience is undeniably real, and it tells of a world shared by others and worth caring for.

Encouraged by Borgmann's declaration, "[l]et everyone speak in the first person, singular and plural," the discourse I enter into is one that bears witness to embodied agency.[2] While this approach is unlikely to make any impact on the floor of the United Nations General Assembly, it is possible from within such a discourse to point up systemic, paradigmatic features of our technological world. It is possible to resist the institutionalized schizophrenia that would sever our worldly relations into technical and practically unreasonable categories of 'public' and 'private'. It is possible to draw our experience into wider cultural visions and thereby to be informed and sustained by the particularity of our relationships in the political task of joining with others in caring for a shared world. I shall attempt this act of cultural composition in the next and final chapter, seeking to close the circle on an inquiry that began with the pragmatic, worldly concerns of policy.

"When it comes to bringing home to us the full appeal, the real force of correlational coexistent things," says David Strong, "philosophy and theory need to give way to narrative and poetry."[3] I agree, and in this chapter I relinquish the burdens of argument in search of the disclosive force of narrative. Borgmann reminds us that the "only reality author and reader can be sure of are traces of ink on the page. These marks, no matter how real, would be forever silent were they not embedded in a communal context wherein they invite and instruct the reader to recall or call forth a certain reality."[4] It is this invitation to call forth your view of a shared world that I extend here through a reflection on my embodied location in the technological world and through a celebration of those correlationships in whose embrace I have been sustained through this study.

I STORIES

> "We need some knowledge of the rocks beneath . . ."
> Oh yes indeed, oh yes, indeed we do.
> The furnace of an old volcanic breath
> survives and culminates in me and you.
> . . . .

"During the Cainozoic lava-flows
these ranges were built up." They wear away.
We perch upon them now in half a doze
sitting with gently folded hands today;
containing all prehistory in our bones
and all geology behind the brain
which in the Modern age could melt these stones
so fiercely, time might never start again.

Judith Wright—*Geology Lecture*[5]

I live in a plain, single-story, short-legged wooden house.[6]

I live in the embrace of the *Jarrah* tree.

Jarrah tongue meets *Jarrah* groove to form a floor, to panel hall and lounge. *Jarrah* beams support a roof of corrugated iron. *Jarrah* 4-x-2s hold lath and plaster walls within, *Jarrah* weatherboards without. The whole rests on *Jarrah* stumps sunk into the sandy ground. This house, four years younger than my father, has withstood decades of summer heat, expanding and contracting; creaking and protesting; sighing and singing. In swirling winter storms the whole appears to shake, its center warmed by *Jarrah* disintegrating in the fireplace.

I grow food, looking to the land, to my blooming and buzzing friends, to teach me the way in the ancient sandy soil of our small backyard. More sand than soil, more lime than anything else. A dry ocean floor. I grow food, drawing contentment as well as nutrients from the soil, inside a valley carved into the soft limestone by a river long-gone. A soil devoid of the rich deposits of volcano or glacier.[7] A soil without rain from November to March, when the hot sand burns a white foot. A soil out of which huge gum trees—*Tuart* mostly, but also *Marri* and *Jarrah*, as they have been called for thousands of years by the local peoples—once towered, shading the ground.

In this soil Captain Fremantle founded the attempt to produce an Anglo-Eden, although a land more unlike the British Isles it is hard to imagine. And now, generations later, I live the life of technological urbanity in the town of Fremantle, participating in the technological world that is consuming the earth, and consuming the past and future of this place. Yet I live in this place atop and within limestone, embraced by *Jarrah*. This place is now my home. It is where I am at home, where land and people sustain me. It is a home full of wonder that I want to help sustain.

At home, amidst technology, it is still possible to draw food from the tired soil tenderly, cooperatively. At home, amidst technology, it is still possible to hear the stories *Jarrah* tells, in houses and in forests. At home, amidst technology, it is still possible to hold our children close, letting joyful havoc undermine a little of the regimentation of our world.

It is still possible to sustain our home-places, but it is a struggle, fraught and confusing. In the technological world we stand with one foot on the throat of the past, the other on the throat of the future. It takes conscious effort and wise deliberation to shift our stance, to let our past and our future once again breath vigor and depth into our present. It takes time, enthusiasm, focus, mistakes, teachers, mystery, and joyous communion with others to develop in the craft of giving and receiving sustenance in the technological world.

Jarrah and limestone, more than any other blessings from the land, have given form to the Port of Fremantle, at the mouth of the Swan River. They are the raw material out of which a European settlement was hewn from a nonplace, a nullity. A significant nullity, however. The point at which a nameless river poured out of unknown, wild antipodean land into a vast ocean. A nonplace, latitude 35°6'20" South, longitude 118°1' East of Greenwich.[8] A gateway, more than a place, to timber, to agricultural land, to gold, to the sea lanes of global trade. The gateway of the colony of Western Australia.[9]

Limestone and *Jarrah* have sustained the port town of Fremantle for one hundred and seventy years. And yet, of course, nature's blessings were never raw. The rawness was in the eyes of settlers—Bacon's children—who yearned for a promised land. It was never a nonplace. It was not nameless. The *Nyungar* peoples knew it for uncountable generations in the meeting of *wadarm*, *kwangkan,* and *bilu*; of sea, sandplain, and river.[10] The pelican knew it in the taste of the mulloway. The waves massaging these beaches, the river's endless journey past its limestone banks, *Jarrah's* red strength: all aspects of this place tell stories of many worlds. Worlds of seashell scuttle, death, and compaction. Of Gondwana wandering across the face of the earth. Of icy inactivity. Of now-extinct marsupial rhino. Of the first human arrivals in this land, turned black by the sun some sixty thousand years ago. Of the formation of Swan River Limestone, only ten thousand years ago. Of *Nyungar* babies, music, headaches, and feuds.

This is a place where vast crowds of the white-tailed black *Ngulya* squabbled amongst the massive branches of the *Tuart* and amongst the grey-green needles of the tall, thin *Condil*. The little *Woorening*, its wing-tips touched with gold, lapped the juice from the overflowing orange flowers of *Magnite*. The squat broad-trunked *Wonnil* added a waft of peppermint, the weeping branches of *Kabbur* their fire-liberated stink of urine. For uncountable generations, human hands plucked the pendulous fruit of *Dumbari* and roasted the tubers of *Tjunguri* in this place. The leaves of *Balga* roofed huts and its resin attached shaft to spearhead. Human wounds were wiped clean with the woolly fibre at the base of *Jeeriji* fronds. Canine *Yakany* hunted ringtailed *Ngurra,* and all fled before the great crashing sigh of blazing *Tuarts*. This is a place where the larrikin laugh of the *Kulbardie* has been sustained from then to now.

Then: plant so tightly embraced weary soil that much of this flora ventured no more than a few kilometers from the coast. The plants of my home place belonged so deeply to their depleted soil that they thrived better with scarcity than with plenty. A couple of valleys behind ours, *Tuart* and *Marri* began to give way to the claim of *Jarrah*. In the granite escarpment that marks the end of the coastal plain, twenty kilometres to the east, the plant community had an entirely different politics and many different members.

Now: my house rests at one of the lowest points of the limestone valley that runs parallel to the coast on the southern side of the river mouth. Its front door faces west, ready to receive the breeze that rushes, mid-afternoon, off the Indian Ocean. This breeze, The Doctor, is the healing joy of our town on summer afternoons. It cools the limestone walls of the grand buildings, the cramped terrace cottages, the warehouses, and the prison. It soothes those sitting in the sidewalk cafés and ushers fishing boats back home. It moves through the spirits that dwell in the land, spirits that call still to the *Nyungar* to return home.[11]

II JUXTAPOSITION

Could we just once encompass the stars in a circle of love,
Just once induce our heart to emit the pulse
That would dissolve its casing and travel to the rim
Of space, then could we stake out our claim
To hope. The vectors of faith point not only forward,
Toward the future, but inward,
Where those whom we have driven to extinction
Still rise with dripping faces from the dark,
To forgive us.

Freya Mathews—*Mysticeti Testament*[12]

Fremantle. A town of limestone and Jarrah, of salt-laden Doctor and river beauty. Yes. A town, also, of cable and conduit. Of cars, mobile telephone towers, brass sculpture, operating theaters, railway lines, neon signs, fashion labels, aeroplane shadows. The home of plastic in its inexhaustible polyformity, and of synthetic wastes in their persistent corrosiveness. The home of air-conditioning plants, video stores, and the pill. A port for Internet users as much as for visiting warships, racing yachts, and the grotesque prison ships bound for Saudia Arabia, bearing sheep much as convicts were once borne in the opposite direction. A town whose square is graced by the Church and Oak of England rather than by the provincial *Tuart*, itself once a cathedral to an uncountable insect congregation. A town now thoroughly absorbed into the metropolis of Perth.

The long-local peoples and plants have been torn from this place and the scars covered with buildings, bitumen, and lawn. This place is the home now of the industrious European honeybee, the laconic local bee having long given up the struggle. The home now of European rat and mouse, dove and fruit fly, grass and mollusc.[13]

Fremantle is the home now of libraries. Of books that store wisdom and platitude, tales and technical specifications. Books that hold onto the stories of

Homer. Books that despoil the much more ancient poetry of names like *Condil* and *Kabbur* on the Procrustean bed of scientific discovery, recording the tortured presence of *Allocasuarina fraseriana* and *Jacksonia sternbergiana* (a tribute to the culture of egoism more than to botanists Fraser and von Sternberg).

A town of global citizens, more than of inhabitants of the Swan Coastal Plain. A town where the *Nyungars* that remain must do so under the suffocating force of the whitefella's, the *Wadjalas'*, law—a force so appallingly strong that some incarcerated *Nyungar's* prefer suicide. A port town where now few wharfies live, their drinking holes shouldered aside by wine bars and bistros. A port into which the Southern and Eastern European immigrants of the 1950s and 1960s breathed cultural life, so that now the Holy Mother blesses the fishing fleet, and olive trees thrive all around. Poor people escaped from the postwar chaos of their homeland to Fremantle, worked hard, and many made good. And, in making good, the "Italian has been able to educate the Australian," puncturing some of the Anglo-Saxon's reserve, as Nunzio Gumina, owner of Fremantle's iconic café, *Old Papa's*, knows better than most.[14]

Fremantle is now a bohemian town of yoga studios, street acrobats, markets, extraordinary musicians, joyous festivals, and political rallies. Restaurants offering Sicilian gusto and Milanese style must now compete with Chinese pragmatism, Japanese delicacy, Thai fragrance, vegetarian good-vibe, and McDonalds. They vie for the favors of the global palate.

The environmental movement is strong here, too. Many of us in Fremantle choose to buy organic produce and second-hand clothes, vote green, clean our houses with vinegar, find our Christmas presents in the Wilderness Society's shop, compost, recycle, and bicycle. Here you can also find, if you know where to look, housing cooperatives, permaculture community gardens, wind turbines, solar passive architecture, integrated pest management, and lifestyles of relative simplicity.

And this is also a town recently taken to real estate speculation and the lucrative industries of rustification. Now most of the limestone warehouses store not wool and wheat but the wrought iron and glass fantasies—apartments, no two the same, no space too functional—of the global professional seeking a unique life-style option with more earthiness, less factory produce, more vivacity, less suburban burden to it.

A local manifestation of the global economy instantly recognizable in its technological affluence to the coursing tourists who graze contentedly upon its surface of difference. Tourists who take a tour of the city by horse-drawn buggy between visits to aboriginal art shops and the prison, whose limestone walls loom above the oval. A prison where inmates still placed their bucket of excrement on the landing for morning collection when I arrived in Fremantle in 1990. Now a prison of memories where concert goers and tourists mingle with the still remaining hangman's noose.

A sand plain on which many recent blows-ins, like me, have been gratefully snagged. A river mouth where Captain Fremantle was blown in six generations ago to found a city, Perth, whose population has grown to over thirteen hundred thousand fossil-powered, television-watching, lawn-watering global citizens: a great mass of people drawn from all over the earth to live the Great Australian Dream of suburban freedom around the Swan River.

Fremantle is a vivacious place, justly envied by the growing number of people in Perth who can no longer afford to live near its newly gentrified center and who are pushed by the invisible hand of the real estate market away from coast and river into the suburban hinterlands. Fremantle is, for most, an easy, comfortable place to live. It supports an apparently random and fluid conglomeration of lesbian lawyers, born-again Buddhist carpenters, pagan stockbrokers, Italian grandmothers, Timorese children. A conglomeration of lives where the global intimacy of electronic bulletin-boards command as much loyalty as those on the walls of the Town Hall.

Fremantle is an urbane place that combines latemodern glamour with rustic charm. But it isn't, local writer Ken Kelso affirms, "merely nostalgia embodied in architecture of a life of cafes in which to sun yourself or blather on. It's a living town. That means there are contradictions, some fierce contradictions."[15]

The Doctor blows now as before, joyfully puffing away the day's heat while bringing with it the smog that accumulates during the morning in a dense mass behind Rottnest Island (ensuring that sales of medication to the swelling ranks of Fremantle's asthmatics continue to be profitable).

The surface of the river glitters increasingly with pleasure craft (and the shores crawl with mansions) while its contaminated depths are increasingly lifeless.

The street-life of the cappuccino strip is ever more convivial as more discover its delights (much to the displeasure of the locals), while the poverty of the Southern peasants who grow the world's coffee beans is as crushing as ever.

The local cats and dogs live the good life, lavished with love, boutique canned cuisine, and hospitals, while the once-local *Chuddich* and *Wyolie* inhabit only photograph albums.

The vegetables and meats in the supermarkets look wonderful, unblemished and waxily bright, while they taste of less and less (and more and more like each other) and come laced with pesticides, hormones, and antibiotics.

The private University, recently established amongst the warehouses of the West End, testifies to that burgeoning export, educational product, while the high rates of teenage drug abuse and suicide testifies to the aimless disorientation of many young people.

The ambience of Fremantle's old limestone and *Jarrah* buildings is rapidly being restored (indeed, is being made more original than it ever was) while the little ancient *Jarrah* forest that remains is tumbled down (or is redefined as a lure for tourist dollars), and the demand for stone continues to scar the earth.

The shop windows celebrate the return of beautiful handicraft and antiques (that admittedly return only for a premium), while many of Fremantle's old people cower, neglected and justifiably frightened, behind security mesh.

A juxtaposition of alluring surfaces and troubling depths. Surfaces glamorous, malleable, and diverse, but often thin and substanceless. A juxtaposition of fear for the warming earth and coal heat trapped in the overhead wires. An abundant place where many of us yearn to belong respectfully—with each other, the land, the *Nyungar's* wisdom, the world's poor and oppressed—yet where we belong always with one eye on real estate prices, the other making sure that our doors and windows are secure. Fremantle is nonetheless, as Kelso says better than I can, a place that "seems to meet a great need in some of us. There's been just as much self interest and greed, I guess, in Fremantle as anywhere, but the accident of time and circumstance has left the town with reminders of a more communal way of thinking."[16] It is a place where a great many of us continue the search for the good life of good work.

III GOOD WORK

Good work finds the way between pride and despair.
It graces with Health. It heals with grace.
It preserves the given so that it remains a gift.
By it, we lose loneliness:
we clasp the hands of those who go before us, and the hands of those who come after us;
we enter the little circle of each other's arms,
and the larger circle of lovers whose hands are joined in a dance,
and the larger circle of all creatures, passing in and out of life, who move also in a dance, to a music so subtle and vast than no ear hears it except in fragments.

Wendell Berry—*Healing, IV*[17]

Fremantle is the product of unflagging endeavor, the product of generations of hard, diligent work. In his 1839 manual for emigrants to Western Australia, a mere decade after settlement, Nathaniel Ogle called on prospective settlers to apply themselves vigorously to the task of possessing this "smiling" land:

> Taking into account the climate, extent, and position, it may be looked on as being among the finest portions of the habitable world, now given by Providence, a free gift, to those who find the old world too difficult an arena in which to encounter the vicissitudes of life; or to those who, actuated by a high and noble impulse, avail themselves of the offer of their Creator to go forth and possess themselves of this smiling land, and there to increase and multiply, and enjoy the fruits of their industry.[18]

Freed from the constraints of a homeland caught between tradition and machine, the new colonizers "prepared a path in the wilderness, which those who now follow will find leading to a promised land,—not a land of idleness, but of uniform labor."[19] In this promised land, "the anxious father of a family exchanges the corrosion of doubt and anxiety for the future provision of a quiet and secure feeling, that God has placed before him in a beautiful region, so vast that thousands of years will elapse before it feels the same pressure of numbers and want of occupation as the old and noble country."[20]

Fremantle was indeed not a place for the idle. It has a long and proud history of backbreaking work. A history of production so enormous that the pressure of numbers and want of occupation have already arrived, and have settled down to stay. This is a history most clearly evident in the plight of the aptly named "lumpers." Later dubbed "wharfies," the lumpers loaded bags of wheat and unloaded barrels of rum on Fremantle's docks on the strength of their backs. No less did the bodies of the women of the waterfront ache as they strove to feed their families in the harshness of Fremantle's sun, the blandness of the soil, and the unforgivingness of its early economy.

The labors of the port of Fremantle have been ceaseless. In 1835, only six years after Captain Fremantle's arrival, twenty-four ships docked at Fremantle's rickety jetty with over four thousand tons of cargo. Just two years later, Fremantle sent seventy-one tons of whale oil back to England.[21] Building of the famous Long Jetty began in 1873, and by 1883 this Jetty, made of Swan River Mahogany, as *Jarrah* was then known to the settlers, extended over three thousand feet into the Indian Ocean.[22] In 1889, the year Heidegger and Wittgenstein were born on the other side of the earth, destined to begin the recovery of philosophy from Bacon's dream, the Fremantle Lumpers' Union was created out of the need to resist the already legendary power of the port's elite mercantile families, Fremantle's own Merchant Princes.[23]

In 1897, when the present day harbor was opened, the largely redundant Long Jetty became the scene of promenading and fishing. But within two years the gold rush had slowed, temporarily, immigration declined, and Fremantle was awash with unemployed and starving workers. The shipping companies imposed harsher working conditions on an already disgruntled workforce. The Fremantle Lumpers' Union embarked on the colony's first, and still one of its most momentous, strikes.[24] The nineteenth century closed on Fremantle promising an uncertain, crisis-ridden future.

Much remains the same on Fremantle waterfront. In early 1998 the wharf was once again paralyzed by a bitter strike, with workers and community members lined up against security guards and their dogs.[25] The original cargo sheds remain, although they now house tourist attractions, markets, and art exhibitions. Ships still unload goods coming from, and load goods going to, all parts of the globe. But now it is the dip and sway of gantry and crane that does the work of lumping. Fewer and fewer wharfies move ever greater quantities of goods, their managers diligently comparing Fremantle's hourly container rates with Hong Kong and Rotterdam. The story of the tough, fearless, fiercely loyal brotherhood of the lumpers lives on, but mostly through Fremantle's Australian Football League team, *The Dockers*.

Compared with the lives of our founding fathers and mothers, work in Fremantle is now much less gruelling, mundane, and debilitating. The present technologies of Fremantle afford less hardship and more choice than even a generation ago. Sweat still pours, but it does so mostly on soccer fields and in aerobics classes. Injuries still happen, but a bevy of lawyers awaits an unscrupulous boss. The commodious surfaces of Fremantle ensure that most do not go hungry or without shelter, although some surely do, for Fremantle's new-found affluence is the product of competition and private acquisition, not sharing, while the social marginalization of *Nyungars* and other long-oppressed groups remains strong.

The new pressures imposed by the central obligation of contemporary life, the obligation of choice, were anticipated by the American Youth Commission in 1942:

> The new, thin, dangerously unsustaining element before us, in which we will be called upon in peacetime to live, is free choice, in many more lives than ever before, as to what we shall do in the several hours of each day when we are under no compulsion from urgent material necessity.
>
> . . . The choice set before more ordinary men and women than ever before is between the flabby, passive, easy, effortless use of time which makes of human life a tale told by an idiot, signifying nothing, and the opposite use of time in coherent, persistent effort to handle creatively the raw materials available to us in human existence.[26]

In the last sixty years, Fremantle has made great progress toward the goal of free choice. Much of the harshness of life in Fremantle has gone. Life is softer. Smoother. Gentler. Easier and more pleasurable. And, let us be clear, life is fairer. In the last sixty years, much (yet not all) of the institutionalized sexism, racism, and general prejudice against those deemed different to middle-class norms of respectability has gone from places like Fremantle.

But gone too is much of the shared focus, resourcefulness, and local character of places like Fremantle. Much of the oppressiveness of our society remains, but it has been steadily submerged out of our experience through being encoded in global technosystems that continue to treat the earth and the majority of human cultures as so much raw material. Gone is much of the point of having choices, for the point of choice is inexorably becoming choice itself; a lesson advertisers have come to learn too well. In Fremantle, as in the rest of the technological world, we rightly celebrate the overt demise of harshness and of prejudice, but we also face the bewilderment of pointlessness. Depression, agitation, and anxiety flavor the dreams of many.

The disburdening facility of our technologies makes the flabby use of time ever more likely, often separating us from those forms of skilled proficiency that bring satisfaction and resourcefulness. We are ever more distanced from those forms of cyclic, everyday work that provide a space within which we can locate ourselves within the cycle of birth to death to birth. Good work is, its exemplary champion Wendell Berry explains, the "name of our proper connection to the earth . . . for good work involves much giving of honor."[27] It is simple but not easy work. It is the work of choosing to cultivate fidelity to people and place. It is the work of sustaining a public intimacy.

The simple but skilled work of growing spinach and oranges, for instance. Of learning to read the moods and habits of the long-local beings. Of repairing our houses, in the process sharing tools and life's traumas with neighbors. Of spending time with our young, allowing the eddies of their primordial energies to move us with their mystery. Good work is endangered in the technological habitat, attracting little prestige, political affirmation or economic reward. The endangered practices of good work are becoming ever more skeletal and isolated from each other. The extinction of good work progresses steadily under the guise of liberation.

In Fremantle, as in many other places in the technological world, the extinction of good work is not passing unheeded. Many swim against the current of technological progress to recover good work to safer ground in their lives. Just where the technical need for good work has been overcome by commodious amenity, many display a practical need to participate in good work. There are a great many practical experiments into what contemporary forms the giving of honor can take.

But the existence of good work is fragile. Evidence of good work is inscribed around the edges of Fremantle's technological core. Good work no longer defines the character of our relationship with the land or with each other. The imperatives of production rule with the impunity of a feudal lord, urging us onward to control rather than to share our home place. Despite the affluence of our town, despite its commodious surfaces, the future seems as perilously crisis-ridden as ever. Uncertainty moves freely through the walls of the citadel of "progress."

IV SPACIOUS EMBRACES

> Wood is hospitable,
> even eatable,
> subject to destruction;
> the only stuff to live in
> for destructible people.
> Who'd live in steel and plastic,

corseting their lives
with things not decently mortal
We're perishable, house,
but nourishing.
Judith Wright—*Habitat, III*[28]

The crisis posed by our planet's struggle in the jaws of technology has come to our attention as a global problem, a problem with the globe. It has come to our attention largely through mass media and largely as a fading of the glamorous surfaces on which our lives rest. It is a crisis which appears as some external calamity bearing down on our lives, as a brightly colored sail is faded by a harsh sun. But the truth of this crisis lies inward, down in the hidden, contaminated depths of our lives. The desecration of the future bubbles to the surface inconspicuously in every act we make in places like Fremantle.

Homelife is always crisis bound. Homelife is founded on the daily, never-ending struggle to feed and shelter ourselves, our kin, and our friends; to foster our and their health and vitality; to fathom what it is to be a friend; to strengthen each other in the virtues of tolerance, compassion, and love; to be resourceful in the face of the transience of material achievement, preparing one another for our coming deaths.

So there is crisis at home. All sorts of everyday, normal crises. The crisis of falling under the spell of another virus, our body craving respite from a seemingly endless assault. The crisis of Megs, a resident chook, taking on the posture of a duck, rear end bearing down under the pressure of some unknown pain. The crisis of a wallet stolen out of the lounge room, of a bully at school, an exorbitant bill, a thoughtless neighbor.

Yet now the ordinary crises of homelife are bound up within the crisis posed by the technological consumption of the planet. This is a crisis recorded in the leaded paint flaking off the *Jarrah* weatherboards of my home. A deadly snow which flows into the soil, into the plants and groundwater, lodging in the nervous systems of Fremantle's children. This house's *Jarrah* stumps record this crisis also in their organochlorine fortifications. These deadly chemicals not only make the ground under our house forbidden to childhood escapades for decades to come, they were sprayed indiscriminately throughout our valley forty years ago, in an (unsuccessful) effort to halt the march of an ant, so that DDT now inscribes its unwanted signature in the eggs of chooks and in those of the native birds. The walls of the shed tell of the crisis of extracting asbestos from the earth. Asbestos mined in the deadly boom-bust town of Wittenoom, now a ghostly backdrop for lawyers serving compensation claim and corporate rebuttal back and forth through the courts. Asbestos fibers which The Doctor doubtless peels off a neighbor's roof and supplies, free of charge, through our windows. Fibers coming to a fatal rest in my lungs, perhaps. Who can say?

The present struggle between planet and technology is contained within the very fact of a *Jarrah* house. My house was built out of the flesh of trees that grew strong before Europeans disembarked at the Swan River. It was built out of the lifeforce of seed that fell to ground before Descartes' umbilical cord was cut. Living in the embrace of *Jarrah*, the need to sustain the *Jarrah* forests is sharp and strong. Sharp and strong is the need to sustain these ancient places so that we can walk with our children within them. So that birds and animals can nest in their hollows, soil can be held firmly to the earth and rain can be induced to fall on the surrounding farmland. So that we can remove old trees to make exquisite chairs to send, with our blessings, into the future. So that we can continue to make houses within whose embrace we can be at home in this place.

Yet there is so little original forest left anywhere in this land that it is now too expensive for most of us to build a house out of *Jarrah*. Even if we can afford it, this timber is too young to emulate the rich glow of the settler's furniture. For generations *Jarrah* was felled and burnt to make way for agriculture. For generations *Jarrah* has been logged and wasted as if there would be no end to its bounty. There was no need for this bounty to end, had it been sustained by loving action, although it seems that it now surely will.

But, in any case, so the logic of our economy would have it, the depletion of *Jarrah* is no loss, it is an opportunity for new forms of industry, for the fresh breath of innovation. After all, *Jarrah* houses lack the flexibility of brick and metal in keeping up with the race of fashion. *Jarrah* houses are draughty. They let the air in, defeating the attempt to condition it. They shift position with the soil, creating gentle slopes here and there in the floor. They let sound in and out, defying attempts to shut the world out of homelife.

A *Jarrah* house creaks. Burns. Changes shape with the weather. Its window frames call for the effort of applying beeswax. Its weatherboard needs continual loving attention. It is, like us, edible. And fallible. (The thump, thump, thump coming from this house in the wild May storm of '94 was found to be the rhythmic rise and fall of roof on front veranda posts, the roofing nails having been pulled free from *Jarrah's* exhaustible grip.)

Despite the fact that the market would wrench us free from *Jarrah's* embrace, encouraging us to move forward, to go places, to progress, those of us who love the good use of *Jarrah* know that within the limits it imposes, just like those imposed by our body and by our planet, a *Jarrah* house makes available an abundance of space in which our lives can be spread out. The limits of *Jarrah's* embrace give definition to our sense of what it is to be at home. Without the restrictions imposed by the embrace of things, life lacks the depth of intimacy.

This house was built in 1934 as the family home of Alice and Jack: she a local farm girl; he a Danish sailor who jumped ship in Fremantle. They raised four children in this place. They raised children who have now raised children becoming parents.

This house was well built. It was intended to demonstrate the skills of a local craftsman. This house is, like Jack is reported to have been, without ornamentation and with some cussed ways. Unlike Jack and Alice, this house persists and Dale, my partner, and I bought it after their deaths. One among many impersonal real estate transactions. And yet now we are becoming part of a history of fidelity to and in this place. When I sit in this house to write about sustainability, I am conscious of its span of the past. Nearly sixty years Alice and Jack lived in this place with each other—a history of stability that lives on still in some of our elderly neighbors, although, in our time here, most of the old people have gone.

Our street's postwar stability, with its taint of life's incarceration by rigid mores, has given way to a latemodern generation rejoicing in their newfound liberty. But this is also a generation deeply worried about the unsustainability of the present. Worried about its consumption of the future. A generation wanting sustainability—willing, even, to pen voluminous pages on the subject—but wary of practicing actual fidelity to people and places.

When we first arrived, I began removing Alice's knotted old rose-bushes, casting aside fanciful exotica in my commitment to native and "productive" plants. But the longer I live here, the more I cherish those roses that remain. It is to these plants that Alice's children and neighborhood friends gravitate when they visit. Smelling one of the blooms, they recall how the short old lady was lost from sight amongst these bushes. It is these roses we now offer to our neighbors to celebrate a birth, to brighten a hospital room, to soften a funeral.

Now Alice's roses are embedded within the story told by our garden. A story written in the limey soil, atop an ancient limestone ocean floor. The lawn is gone, replaced by a busy jumble of herbs, vegetables, fruit trees, and native plants. A jumble that has inexorably spread, spilling over onto the verge, crossing the street, heading north and south along the other side, becoming the street's garden where neighbors mulch the plants side-by-side. Our garden has cultivated in me a public possessiveness for my limestone valley so that now I work with others to heal the public land. Linking arms together, many of us now encourage the long-local beings back into their homeplace and hold open spaces in which the *Nyungar* dreaming that gave birth to this place can be reinscribed on the land. Together we are teaching each other how we might grow food ecologically in this place, and we experiment with ways to decouple livelihood and profit.

My days invariably start in the back garden. I usually wake early and wander cup in hand along the sawdust paths between the limestone rocks I have collected from around the valley. Crunching snails, checking the pumpkins, greeting the chooks, chatting to the fruit-trees, looking for tadpoles in the pond, resetting the fruit-fly traps. I have walked hundreds of miles on these short paths, covering the same ground again and again. Grooving myself into this place. And often I just sit. Looking over and over again. Intently. Sitting in the hammock. Sitting at the table under the almond tree. Sitting, watching as my son chases butterflies, hunts for lizards, scoffs mulberries, and picks the salad for his lunchbox. Watching as my daughter totters toward me, somehow moving past and not into the outstretched thorns of Alice's Roses.

V PARENTHESIS

> For parents, the only way
> is hard. We who give life
> give pain. There is no help.
> Yet we who give pain
> give love; by pain we learn
> the extremity of love.
>
> Wendell Berry—*The Way of Pain I*[29]

Our houses and our gardens encode histories, they tell stories, which are in need of being sustained. So too do our children. Children can bring us into intimate contact with the future that projects out of the stories encoded in our present. To be bodily engaged in caring for children is to bring the abstract talk of future generations alive, to have it animate our everyday practices with a yearning to build and sustain a good world.

In our short stay in this house Dale and I have become parents. Like the roses, our children have received and given care in this place: Angus, born in hospital during our first year here; Mara, born in the lounge room in our fifth. Just as my questioning about what I most wish to sustain in and about my world has realized the present inquiry, so too it has realized my commitment to stay at home with my children. The practice of fatherhood forms a lived narrative to which this book is largely *parent*hetical. The pain and love bound up in being related to my technological world as a parent—as a father to Angus and Mara, as a parent with Dale, as a fathering son to my and her parents, as a parent among other parents, and as a parent on a planet struggling in the jaws of technology—is the underbelly of vulnerability that nourishes the protective spine of scholarship. It is a narrative easily split by the knife of analytic precision, but in its fragility it is undeniably real.

Reaching (and breaching?) the extremities of love has been an everyday occurrence in this family. Many afternoons, I have been seated in the study trying to write on the global development crisis, to hear Mara and Angus giggling or screaming, sharing a fierce battle of wills just as easily as they share a storybook. Many afternoons I have found myself out of my chair and headed to join the fun or the fight before my studious self has comprehended that it has been seduced by the flowing allure of life yet again. As much as possible I have sought to understand writing this text as one among many of my everyday, ongoing domestic tasks. In all of these practices, I seek to be so full of care for the things bequeathed to my present by the past that the future will be able to take care of itself.

From shortly after his birth, Dale and I carried Angus in a cloth sling looped over a shoulder, held fast by two steel rings, for long periods each day. I carried him in this way for almost three years, and then for another year or more in a pack on my back. As a newborn, Angus used to sleep slung sideways across my chest, his ear pressed to my heartbeat, the cloth pulled up around him to shield him from the world. As an inquisitive toddler, Angus used to sit on my hip to greet the world. Joined in this way we would cook and clean together (our four hands performing a crazy duet), tend our garden together (my hands planting, his uprooting), travel on buses together (me looking out the window, he enchanting those sitting behind). In the embrace of the sling, we would converse and play with others—Angus at first shy, burying his head in my armpit, and then, eventually, with confidence, his arms waving in front of my face. And he would sleep, hour after hour, tied to my chest as I walked about the shops or as I sat at home reading.

This child-lumping was hard work, as good work often is. And, as good work often does, its self-fulfilling character prepared both of us for the time when Angus would begin to carry himself and his many other burdens. This sort of good work started all over again with our daughter, Mara, who as a newborn liked to sit in the sling facing forward, staring out at things with her legs drawn up in front of her, and will soon have reached its end.

Feeding from their mother and being carried in the sling were Mara and Angus's respite from the vast novelty of life. The chests of both of their parents gave, for a time, strength to the nets they spun to haul the world to themselves. For a time, Dale and I were at the center of the net by which they drew sustenance from the world. Later, when language and friendship became their self-defining sources of strength, breastfeeding and being held in the sling remained as a source of solace when the world loomed menacingly before tired, sick, hurt, or confused eyes.

There are, to be sure, as many paths to understanding mothering and fathering as there are possible practices through which these identities—these narratives about practical tasks—can be constituted. Child-lumping, in its myriad historical forms, has given rise to technologies as primal as the axe (not that male historians have given much thought to technologies of this kind, the only ancient slings commonly recorded in history books being the mini-missile launchers of the Egyptians). But the technologies and techniques of child-lumping are nonetheless central to most human forms of life.

The ornate and beautiful cradleboards of the Chippewa people of Canada's Great Basin were called "our third mother" by the infants who rode in them on the back of their "second mothers," the earth itself being their "first mother."[30] Laced firmly into a furred cocoon, these infants stared out at the snowy world about them, drawing sustenance from the ecological, biological, and technological dimensions of their mothering world. Whilst in the deserts of Australia, where baby-carrying technologies of any sort were mostly and understandably avoided, infants of only a few months balanced miraculously, without being held, on the shoulders of their kin.

The concrete ground of the technological world is ideally suited to the wheels of a pram. This world beckons for us to give up the work of lumping altogether. It is easier and more efficient to remove our young from our embrace and to place them in the care of a pram. While few children grace their prams with the title of mother, many sleep peacefully in their care. And then there are a great many techniques and technologies for helping "tame" our children, to domesticate them into the rhythms of technological life. So, now, many of our babies travel about town in their prams and sit at home in front of their televisual storytellers, their mouths full of plastic nipple. (A form of life only lightly ruffled by the astonished headlines of last year which told of research linking the excessive use of dummies, or pacifiers, to lower IQs, a story ideal for the editorial wit, "dummies make for dummies." Where would we be without research which can show us that a mouth full of plastic may not be as stimulating to the development of minds as caring attention, vocal and bodily converse, and visceral comfort?)

In this world, the daily practice constituted by a simple strip of cloth has sustained profound engagement between this father and his children. It has enabled me to more easily substitute for the presence of their mother, so that she and I could share the burdens of paid work while our children were still "on the breast." It is a practice around which Dale and I have been able to reduce our need for a great many of the technologies and techniques offered by the technological world to parents. Holding our babies close, it has just seemed natural to go largely without cots and prams and milk formula and playpens and television and the rest. Holding them close, it simply made sense to give birth at home. Holding our babies close we have longed to be around them, and so we have learned to live on less money, in the process stumbling onto the pleasures of resourcefulness and prudent, selective frugality. Holding our babies close, we have involved our children as partners in all parts of our lives. Holding our babies close, we have wandered the neighborhood meeting other babies, forming groups simply for the purpose of play. Holding our babies close, Dale and I have been joined with each other and with others in ways that baffle and tire yet simultaneously reward and sustain us anew.

Our slings have been made locally by a Fremantle mother, yet, certainly, they are the product of deeply unsustaining latemodern industries of cotton, dye, and steel production. And, certainly, their use brings with it no guarantee that the lot of parent or child will be improved. Nonetheless, related to my world through a strip of cloth, I feel like I have been sucked deep into this world, beneath its surfaces, through others, through things, through past, through future, and into myself.

Embraced by a *Jarrah* house and embracing my children. Inside a limestone valley and telling my story in the limey soil. It is within this commerce of sustenance, rather than within the commerce of consumption, that I have begun to live a little differently in the technological world. It is through this small difference that I have drawn nearer to the edge of the practices of consumption, for I have seen a little of what consumption cannot offer.[31] I have seen something of the ways in which intimacy and fidelity have a worldly character: a public as well as a private character. Indeed, I have seen that within the commerce of sustenance the boundaries between public and private are constructed in ways that are partial, multiple, and overlapping. Within this commerce I have seen how technology can be a craft of reproduction rather than a technique of production. I have seen only a little of all this, but enough to want to keep looking.

CHAPTER NINE

SUSTAINING TECHNOLOGY

INTRODUCTION

IN THE PREVIOUS CHAPTER I sought to disclose something of my experience of the commerce of sustenance. I did so in the belief that the broad qualities of this experience are shared widely. Certainly, the specificity of my experience is shared only by those in my local community and even then is shared only by a few in this community and in its full detail by no one else. My experience reveals something of the appalling profligacy of my world. Surely, only a world of such reckless extravagance could produce books like this one: a surfeit of food, assured medical help, an army of machines, a steady supply of excellent red wine; this text is underwritten by the immunity of abundance. It is underwritten undeniably by the historical and continuing disenfranchisement and destruction, the death, of the earth's astonishing array of cultures and ecologies.

Yet at the same time this text tells of a yearning to rewrite the unjust terms of my affluence. It shows that abundance is not the same as freedom; consumption is not the currency of happiness. It speaks not to a universal audience—is there any such thing?—but directly to those who share my unsustaining world. It is my hope that the essential qualities of our yearning for and experience of embodied sustenance call forth certain systemic, paradigmatic features of our technological world. I believe that through communicating something of this experience with each other, we will be able to put into words more of our ambivalence about the constriction of our practices and the concomitant constriction of our thoughts by the logic of efficient production. Through public affirmations of our sources of sustenance, I believe that we will also be able to put more of this ambivalence into our social forms of practical judgment, that is, into our policy making. In what follows, then, I make the move in practical reason from disclosive reflection toward cultural contestation over the ideal of sustainability.

The commerce of sustenance exists at the margins of latemodern life. It is pushed from the center of our daily experience by the commerce of consumption. Our experience of what is truly sustaining is compressed within those small spaces of private choice in a way that has little impact on the overall character, the core policies, of the technological society. The commerce of sustenance is thus fragmented and dispersed. The conclusion of the argument I have been developing here is that the moral challenge posed by the ideal of sustainability, its undoubted capacity to call us forth to respond to the deformation of our world, is firstly that of bringing this commerce out into the arenas of public discussion. It is then the challenge of seeing how isolated examples of the practice of sustenance can be crafted together into the tentative beginnings of new cultural worlds.

This conclusion has been the guiding *telos* of this book. It is the culmination, not the product, of my inquiry. I am not, therefore, suggesting that my argument is conclusive. I have not sought the power of proof. The power of its conclusion is of a quite different order. Its power is practical rather than deductive, and it is particular rather than universal. The conclusion that the moral life-blood of sustainability is our yearning for sustenance has the power to make my own practice a little "less stumbling and more clairvoyant," to recall Charles Taylor's description of practical social theory.[1] My hope is that it may also be powerful for many others.

The qualities of the commerce of sustenance are many and various; my own experience brings to light only a few, and even then, of course, only from one angle. Further, these qualities stand juxtaposed to the abundant commerce of consumption in an infinite variety of ways. Seeking the resonant possibilities of emblematic rather than comprehensive discourse, I wish to explore the qualities of five sites in which sustenance and consumption stand together in my own experience. I shall explore the juxtaposition of future and past, public and private, production and reproduction, technique and craft, theory and practice. These are sites in which I have begun to build a little differently, in the process moving a little further toward the edge of our technological world. Like David Abram, I think that "perhaps we may make our stand along the edge" of our civilization, where we can be open "to the shifting voices and flapping forms that crawl and hover beyond the mirrored walls of the city."[2]

It is my hope that the public recognition of these and other resonant sites in our latemodern experience may enlarge the spaces within which the ideal of sustainability can be contested in practice. It is my hope that this public recognition will spill over into a sustaining *ethos* of mutual encouragement in the task of experimenting with technology at the edges of our latemodern practices where sustenance and consumption meet. This is an *ethos*, an ethic of sustainability if you will, in which we recognize and celebrate each other's efforts, no matter how limited their extent and no matter how foreign their form, to engage in the giving and receiving of sustenance. This is an *ethos* in which our intimate experience of being embraced by and of embracing particular fragments of our shared world can form the basis of clasping hands with others in the good work of building more sustaining worlds. This is the good work, the joyfully prudent work, I call sustaining technology.

The idea of sustaining technology is ambiguous in as much as it moves our thoughts fluidly back and forth between the need to identify technologies that sustain us and the need to identify what it is that we wish to sustain through technological skill. It is open-ended, for it defies resolution by holding open a space of important questions within which our thinking about technology can move beyond the pervasive explanations of neutrality or autonomy. The idea of sustaining technology recalls Ivan Illich's description of convivial tools as those that encourage us in developing a graceful playfulness within—with and in—our embodied relationships in a world worth caring for.[3] It is vital to understand, however, that the power of conviviality is not conferred upon us by external objects. The good work of sustaining technology is the work of encouraging ourselves and each other through the reality of our technological being to nourish the human capacity for grace, play, and care.

The five sites of sustaining technology I describe here do not exist discretely from one another: in the perplexing multiplicities of practice, each site is bounded by the others. Further, these sites cannot be understood dualistically. In them, sustenance and consumption stand together, not one over and against the other, a juxtaposition that does not become visible unless technology is acknowledged to be constitutive of our humanity and of our world. To see technology as an external tool at the service of our humanity is to be blind to the nature of sustenance altogether. In this final chapter, I explore these sites looking for ways in which the questions of cultural sustainability may begin to lead us beyond the technological world. I shall point out, in the process, how the answers of technocratic sustainability are disorienting our path-building toward more sustaining worlds.

I FUTURE AND PAST

> I imagine this past and this future as two vast balloons of time, separated from each other like the bulbs of an hourglass, yet linked together at the single moment where I stand pondering them. And then, very slowly, I allow both of these immense bulbs of time to begin leaking their substance into this minute moment between them, into the present. Slowly, imperceptibly at first, the present moment begins to grow. Nourished by the leakage from the past and the future, the present moment swells in proportion as those other dimensions shrink.
>
> David Abram[4]

The questions of cultural sustainability, such as those with which I opened chapter 8, are inherently temporal. They call on us to reflect on our experience as beings who live in what Hannah Arendt so perceptively called the gap between past and future.[5] They call on us to reflect on what it means to live in a temporal opening in which the forces of the past push us into the future, and in which the forces of the future push us into the past. A space in which, as Abram so beautifully puts it, past and future nourish us on all sides.

The questions of sustainability call on us to investigate the ways in which the technological world seeks to lift us out of this gap, out of the place of our temporality, locating us ahead of ourselves, as it were, in the future. I see great wisdom in Wendell Berry's contention that the technological world is founded on an almost occult yearning to possess the future in the attempt to make our present more hospitable: "What has drawn the Modern World into being is a strange, almost occult yearning for the future. The modern mind longs for the future as the medieval mind longed for Heaven. The great aim of modern life has been to improve the future—just to *reach* the future, assuming that the future will inevitably be 'better.'"[6] But now a great many of us are being drawn by the ideal of sustainability to question what it means to long for the future. What does it mean to have the present at the flick of a switch? What does it mean to preserve our present, recording it for some time in the future? What does it mean not to be in the present? Where, in fact, are we? Are we "in" time? Do we still "have" time?

In the world built through the commerce of consumption, we have become disoriented. We trample much of what is good in our past and in our future because technological efficiency is thought, wrongly, to provide us with the sustenance for which we yearn. As Martin Heidegger saw, the practice of skilled craft certainly does bring us into closer, more disclosive contact with those things that we wish to sustain in our world, but this has little to do with efficiency. The cultural questions of sustainability spring from a totally different source from that which produces the answers of technocratic sustainability. The former founds our temporal experience on the search for things in the world that nourish in us a love of life. The latter founds our temporal experience on a frantic race toward security and away from death.

The commerce of consumption founds our experience on what Albert Borgmann calls the "hyperreal glamour" of commodities that have no past and that tell no stories: "To be disposable, hyperreality must be experientially discontinuous from its context. If it were deeply rooted in its setting, it would take a laborious and protracted effort to deracinate and replace it. Reality encumbers and confines. Disposability and discontinuity are marks of hyperreal glamour, and glamour, in turn, is the sign of perfect commodity."[7]

Commodities do not engage us with the stories of the past, for the past is inherently worldly, and it is the shared world of our shared mortality from which technology seeks to liberate us. Commodities, regardless of their ecoefficiency, distract and disorient us from the world around us. More sustainable commodities may produce better statistics about environmental impact, but they nonetheless create a world of glamorous experience in which human life is inclined to oscillate between the pointlessness of sullenness and the erratic waywardness of hyperactivity. Evidence of global ecological crisis has thus increasingly provoked nihilistic indifference and technocratic optimism rather than practical steps toward healing our world and our selves.[8]

The hyperactive technobabble of ecomodernists like Gregg Easterbrook is disturbing but instructive. As I indicated in chapter 1, Easterbrook advocates that we make use of gene technology to create a New Nature through which humanity can transcend oblivion:

> At present human beings do not know whether upon death consciousness is lost or preserved in some higher form. For the sake of our own souls and those that have come before, we should hope an afterlife exists. But if it does not, then people should make one.
>
> . . . .Today men and women look back on their forebears of distant centuries and view them with sadness as benighted creatures that lived out crude lives with constant material suffering and in ignorance of the most basic facts of the world around them. Someday our descendants may look back with greater sadness at us . . . seeing us as the last benighted human generations that on physical death went to oblivion, rather than having their consciousness continue living, as for millennia people have supposed would be the fitting progression after temporal life.[9]

Ecomodernists, like Easterbrook, see the future of technology as exciting, as limitlessly creative. They emphasize the supple resourcefulness of latemodern technology in responding to the problems of modernity. Ecomodernism reconfigures ecological crisis not as a challenge to the commerce of consumption but as a challenge to the hyperingenuity of engineers and economists. Yet, ecomodernists do not limit their optimism to merely technical matters. They display a peculiar form of techno-ethical optimism. Ecomodernisation is founded on the assumption that the commerce of consumption provides moral and political as well as technical solutions. The techniques of Sustainable Technology are represented as freeing human life from the tyranny of ecological and economic limits. The development of greater ecoefficiencies is presented as nothing less than an ethical strategy that will produce at least three new global ethics. These are, first, an ethic of care and a policy of caution directed toward future human generations; second, an egalitarian ethic and a policy of equity directed toward the global community; and third, a stewardship ethic and a policy of global management directed toward planetary stability. But despite these noble objectives, ecomodernists are unable to be reflexive about the essentially contestable moral commitments that animate their pursuit of efficiency. Their discourses are founded on the epistemological assumption of their essential cultural and moral neutrality. The project of ecomodernization unwittingly acts out a set of moral answers to the basic questions of sustainability that runs counter to, and that thus defies, its ennobling impulses.

I have no wish to deny that the hyperactivity of ecomodernism can achieve much, and indeed must achieve much, in curbing the worst excesses and inefficiencies of global technosystems. The project of ecomodernization keeps alive the hope, although it provides no certainty, that coming regimes of global management may be more equitable and more ecologically astute than in the past. But if these discourses are permitted to displace our experience of the temporality of sustenance, then the *telos* of human and nonhuman life will only become ever more bound up in the narratives and practices of efficient management. This *telos* is inherently ahistorical: it is projective rather than reflective, and it is abstract rather than embodied. It is not that ecomodernism denies human care for its world in its drive to overcome the oblivion of death. It is, ironically, that it seeks to express this care within what Abram so rightly calls the oblivion of linear time:

> Any approach to current problems that aims us toward a mentally envisioned future implicitly holds us within the oblivion of linear time. It holds us, that is, within the same illusory dimension that enabled us to neglect and finally to forget the land around us. By projecting the solution somewhere outside of the perceivable present, it invites our attention away from the sensuous surroundings, induces us to dull our senses, yet again, on behalf of a mental ideal.[10]

Through the discourses of ecomodernization, the ideal of sustainability is rapidly and sadly becoming just such a dulling force.

If the *telos* of ecomodernization were to be fully realized, the past would cease to live in the present. Through it our present is being defined by the technological attempt to design a known future. This prospect, no matter how distant it may be, is already working itself out in latemodernity. The earth's present-history as a vast brewery of ecological and cultural diversity is truly becoming a past as we build a present-future of ecological and cultural homogeneity. The goal of ecomodernism is first to colonize and then to engineer the future. Borgmann reminds us that the "distinctive discourse of modernity is one of prediction and control. In the teeth of severe cultural and moral crises, we continue to use it as if it were the sole alternative to sullen silence." As a consequence, we "vacate our first-person place and presence in the world just when we mean to take responsibility for its destiny. Surely there is deprivation and helplessness in this."[11]

The commerce of sustenance stands in bold contrast to the enervating narratives of mastery through consumption, of salvation through productivity. My experience of sustenance, my sense of what it is that sustains goodness in my life and in my world, encourages me in the view that moral care for the future can only emerge out of our desire to take actual, practical care of what seems most valuable in our present. It seems that by holding my children close or by entering personally into the cycle of seed and vegetable with fellow cultivators or by prudently nurturing the strength of my *Jarrah* house, good things will flow forward in time and keep flowing well beyond my own time.

It seems to me that what seems worthy of being cared for in the present, what seems worthy of being sustained, only does so through being embedded in historical webs of narratives and practices that give expression to this worth.[12] It is in this sense that I consider moral life to be a practical craft, rather than a science of technical calculation. Practical moral reason "originates not with the question, What should I do? but with the question: What is going on?" David Toole's point here is thus that moral life belongs to an epistemology that is temporal not through being descriptive or predictive, but through being worldly: "Questions about how one ought to act, if they are to be questions of substance, must be accompanied by an interpretive activity that requires one to ask epistemological questions: What sort of person am I? And what sort of world do I find myself in?"[13]

To be a parent or a farmer or a builder is to have the possibility of taking cues from those who first cared for us, those who grew our first meal, those who built the room in which we were born. The healing of sustenance comes from the fact that, in its commerce, "we clasp the hands of those who go before us, and the hands of those who come after us." The "little circle of each others arms"—arms of every living sort—is the arena of moral life.[14]

In my experience, the ideal of sustainability is nothing less than an expression of the primal desire to belong within, that is, to be nourished between, a past and a future that are meaningful and good. This is nourishment present in the corn kernels sown in early spring for our summer meals to come, this year and on into the future. This is nourishment present also in the songs which bind the movement of life, a ceaseless reproductive movement from seed to vivification to cessation to seed again, into social stories about the human mortal journey from seed to seed. "My great corn plants," the Navajo chanted, "Among them I walk/I speak to them/They hold out their hands to me," and, looking at the small patch of corn in my Fremantle backyard, I feel I know exactly what they meant.[15]

II PUBLIC AND PRIVATE

> If any one phenomenon distinguishes the start of a new era and a post-condition, it is this: humanity's newly achieved ability to effectively destroy its own sustenance. The world has become post-immortal; not in the sense that life on the planet is necessarily mortal but rather that there is no longer any assurance of its immortality. We live in a world-at-risk, where life has become contingent upon our own actions.
>
> Bent Flyvbjerg[16]

Rather than the ability to nourish itself and those within it, the technological world displays the distressing ability to destroy its own sustenance. While the techniques of ecoefficiency may be able to make this world more sustainable of a technocratic social order, they are unable to make it sustaining of our moral experience. The enervation of moral life is the real meaning of unsustainability. Instead of being nourished by leakage from past and future, we live in a leaky present: a present insignificant in the face of the past achievements of technological modernity and in the face of the growing fears for its future.[17] We live, also, in a social world that denies our inherent sociability. Unsustainability and unsociability cultivate each other. In part, our world destroys its sustainability because it denies the publicness of our selfhood. It has turned the sustaining public practices of food production, healing, and learning into the industries of agriculture, medicine, and education. It has traded the worldliness of sustenance for the privacy of sustained acquisition.

Moral life now takes place within a shared, public world-at-risk. This is not simply a factual observation. It is a normative observation that bears on the character of contemporary moral life. Sustainability, the ability of giving and receiving sustenance—what I have been calling the commerce of sustenance—has inevitably become a dominant preoccupation in this world-at-risk. The discourses of sustainability have resonant power and broad appeal in a world consuming the future after which it longs. These discourses give vent to our distress at the despoliation of human and biotic communities. These discourses are heavy with fear of personal and planetary injury and death. In the North, these discourses give vent to the bewilderment created by the dictates of pointless consumption. In the South, they give vent to the pain and anger created by institutionalized poverty and cultural disintegration.

The discourses of sustainability are rich with ambiguity. They juxtapose and hold in tension our desire to take control of a dangerous situation, a dangerous future no less, and our desire to step outside of the on-rush of technology, moving gratefully back into the gap between a meaningful past and a meaningful future. This rich ambiguity has been steadily leached, however, by approaches to sustaining development that seek to recover the ability of sustenance through the perfection of production. Since the early 1980s, ecological crisis has been increasingly interpreted as a threat to human survival that can only be countered by redesigning nature to make its central *telos* the ability to secure the future of the technological society. Ecomodernism aims at nothing less than an "extension of the technological paradigm to the global scale where the earth itself is seen and treated as a device." And if "that technological totalitarianism comes to pass," Borgmann assures us that "life will take on an essentially secure, trite, and predictable cast."[18]

The commerce of consumption demands that the earth itself become the ultimate commodity, the ultimate product. The devices and techniques of productivity thrust forward, away from the past. The dynamic of production is linear, heading always toward a better, more efficient product. We produce commodities, forcibly taking from the raw material around us and giving back mostly waste. We produce commodities not through cooperation with each other or with our world, but through competition. We compete, observes Berry, in the mistaken belief that it is only through the motivations of competition that we can be drawn out of the reach of the ever-present dangers of idleness and mediocrity: "We have been wrong to believe that competition invariably results in the triumph of the best. Divided, body and soul, man and woman, producer and consumer, nature and technology, city and country are thrown into competition with one another. And none of these competitions is ever resolved in the triumph of one competitor, but only in the exhaustion of both."[19] In this exhaustion we sap the conditions of sociability, neighborliness, and public fellowship. Trust has inevitably become a problem for modernity, notes Adam Seligman:

> With the realm of value relegated to the sphere of the private, it becomes increasingly difficult to represent the collective whole, the realm of the public (especially when we consider the moral or value-laden aspect of every representation). In this sense and very tersely, we may note that the loss of honor as a category of public value represents the triumph of the private. Contrariwise, such a radically constituted private cannot support itself, cannot, in fact, represent itself: this is the situation at present.[20]

It is uncomfortable to find our world untrustworthy. It is, in fact, insufferable, and we respond by ordering our private worlds through bonds of community based on preference. This helps, of course, but the practices of everyday life remain suspended in the bitter medium of distrust.

The commerce of sustenance holds out the hope that we may once again begin to build the conditions of trust in our shared world. In the commerce of sustenance, it is mutuality, not competition, that defines our experience. Sustainability requires that the giving and the receiving of sustenance be bound inextricably together. One encourages the other. They belong together as partners in a dance, Berry's verse reminds us, "to a music so subtle and vast that no ear hears except in fragments.[21] To give sustenance is to be sustained. It is to receive. To be graced with sustenance is to develop the capacity of giving gracefully. The production of sustenance is thus always a reproduction. Things of sustenance hold within them the seed of their beginning and the seeds of new beginnings.

III PRODUCTION AND REPRODUCTION

> Times beyond remembering, I have seen such moments: summer falling to fall, to be followed by what will follow: winter again: count on it.
>
> Mary Oliver[22]

Human experience is inherently temporal, inherently social, and inherently reproductive. Each of us carries the seed of the future within us. Each of us is the product of the union of seed in the past. We are truly the product of union, not of competition. We are truly reproductive beings before we are productive beings.

Good work is, in its essence, reproductive. As a father holding his babies close in our technological hyperreality, I have become more receptive to the wisdom accumulated in the caring practices sustained in our patriarchal culture by the reproductive work of women—practices long devalued in Occidental history and steadily marginalized in the modern era by the drive for certainty in a perilous reality. Wrapped in my children, enclosed in *Jarrah*, held in place by limestone, I have caught sight of the latemodern possibilities that exist for practicing a more reproductive masculinity. I have glimpsed the hallmark of reproduction in spaces I previously thought to be the rightful domain of production. The practices of farming, for instance, must be reproductive, if they are to display sustainability. So, too, forestry and fishing are in essence reproductive work. So, too, is the transport of walking or the skill of building or the alchemy of the kitchen or the lingering transience of music. So, too, is the art of friendship and fidelity. So, too, is the work of fathering.

The good work of reproduction may give rise to a product. But often this work offers no product by which its value can be judged. Often it produces no outcome, for this good work is its own outcome. Take my unfolding decision, that is, one that crept up on my reasoning as a response to my children's embodied emergence into my life, to take up for several years the hard work of carrying my children close to my body in a cloth sling. This hard work produces no identifiable product. According to the commerce of consumption, the end of this work is transport, an end that can be obtained with much less effort by using a device such as a pram. Yet I have found that the good work of child-lumping has engaged me with my babies in a multiplicity of ways that the idea of transport cannot encompass. My children have been sustained by the muscles of my back and legs, and soothed by the languages of texture, contour, warmth, gait, mannerism, vibration, heartbeat, muscular tension, and smell. I have been sustained by their wholesome trust. In receiving their trust, I have felt worthy of their trust. In receiving their trust, I have valued the trust of others more highly.

The work of child-lumping is tiring, even potentially harmful, and it is obviously not for every fathering body. No more is it for every child. But in being tired by doing good work, I have found that strength flows back all the more freely. I have found in the reproductivity of child-lumping something worth working for, over and over again, endlessly. Something that can never be secured and possessed, but something that only exists through the never-ending activity of renewal. In the reproductivity of child-lumping, I have come to experience more powerfully and more directly the richness of trust. And now, with every extra kilogram my daughter gains, I pay more attention to what other forms the practice of trust could take in my life.

Carrying my children I have daily taken long walks around my neighbourhood. In a world of cars and planes, walking is, of course, an inefficient form of transport. It is slow. As we walk, time rushes by. Walking is, perhaps, an efficient form of exercise, but then it is more efficient on a treadmill than on a street. But, unlike the transportation produced by car or plane, walking is a form of reproductivity. Only by long walks do our legs develop the capacity of long walks. To remain healthy we must walk again and again and again.

Walking about in the weather, I have come to understand more of the reproductivity of the seasons, a reproductivity you can count on. That you can trust. Moving at the pace of conversation, the walking pace of comprehension, I have begun to listen more completely to those eloquent forms around me. Arriving exhausted at a mountain hut, breathing deep the joyous release as my backpack slips from sore shoulders, freeing my puffy feet from their boots, I have been sustained by being drained. Rinsed clean with sweat, emptied by walking all day, and yet filled anew beside a campfire, that most wonderful technology.

IV TECHNIQUE AND CRAFT

> A curious alliance: the cold impersonality of technology with the flames of ecstasy.
>
> Milan Kundera[23]

The idea of development exercises a strong grip on the latemodern mind. Our public institutions strain at the bit of the past in their headlong rush into the future, pursuing technological and economic development. The technological society rushes as fast as it can after development, hoping in the process to lift off the ground into a space where there is only ecstasy and no pain, only birth and no death, only growth and no decay.

Our private lives are also increasingly dominated with the concerns of personal development. To develop in this sense is to produce. It is thus also to consume. We produce the future and we produce our selves, and we consume the resultant products. We consume our future and we consume our selves with the productivity of consumption. Development, in this sense, is to take confusion and turn it into ambition and to take insecurity and turn it into self-assertiveness.

Within the private and public forms of the commerce of consumption, development is thought of as a technique. It is thought to be the product of strategies that are assured of achieving pregiven, quantitative parameters of improvement.

Development, in this sense, is inconceivable without improvements in technological and bureaucratic efficiency.

The commerce of sustenance is also founded upon a celebration of the human capacity to develop: to grow, to reproduce, and to nurture life. The ability of sustenance, however, rests on an awareness that humanity develops and grows always toward death. Sustainability emerges out of an awareness of death as the culmination, not the negation, of life. Sustainability emerges out of an awareness of the fact that birth and death embody the same principle of reproductivity: a principle embodied in all living communities of all mortal beings.

The ideal of sustainability, as I want to contest it, reveals development to be a craft before it is a technique. The craft of development is shared in a public world rather than taught to individuals. It is gained rather than achieved. It is celebrated rather than rewarded. Appreciating development as a craft, rather than as a technique, I have affirmed rather than rejected technology. The crucially important daily decisions I make about the technology choices open to me are not choices between good and bad technologies. The crucial choices at stake bear upon the nature of my relationships with those people and things that most enrich my life. The question to be asked of technology is not, "Is it productive?" but rather, "Does it bring me closer to that which I most want to embrace in my experience?"[24] Further, the very structure of this inquiry—its worldliness—resists attempts by the narratives of consumption to locate this embrace with the idea of private, subjective preference. This inquiry is mediated by forms of practical rationality that search out the symmetries that exist between our experience and the shared and necessarily political world of practice we inhabit.

Reconfiguring technology as our impulse to be intimately engaged with the reality around us, I can affirm that I enjoy greatly being a technological being. I enjoy the process of becoming skillfully resourceful. I greatly enjoy being crafty. I am enriched by learning the abilities of making, repairing, and growing things. I derive satisfaction from finding innovative uses for the "junk" that many of my neighbors throw away. Perhaps the greatest irony of our technological world is that the proliferation of devices is making it harder to be technological. Our interaction with technologies is increasingly being reduced to switch-flicking and button-pressing. But, in practices such as ecological gardening, for instance, I have found ample space to develop and hold together the skills of production with the skills of ecological care.

My garden is an ecological place. It hums with the activity of living things. But it is also an inherently technological place. It tells of my skill and of the manifold skill I still lack. Such skill as I have has not been applied to this small patch of land; it has been grown in me by the land. My garden is sustained in its present form only by my need to reproduce food within it. Left to itself, the sandy soil and the summer drought would soon exert their influence. In my garden, I move fluidly between the use of high-technology and age-old lore. My decisions about technology are shaped by my desire to walk the fine line in which food reproduction coexists with the flourishing of the biotic community in which it takes place. Thus, hanging from the plum tree last spring were two fruit-fly traps made out of old drink bottles. One contained the urine of a pregnant woman, this being in plentiful supply at the time and long known to be a powerful attractant for fruit flies. The other contained a pellet of laboratory-produced pheromone, the product of sophisticated science. These traps swung happily side by side in my garden (and, happily for Fremantle's feral fruit fly, neither were much good).

V THEORY AND PRACTICE

> How could we divorce ourselves completely and yet responsibly from the technologies and powers that are destroying our planet? The answer is not yet thinkable, and it will not be thinkable for some time—even though there are now groups and families and persons everywhere . . . who have begun the labor of thinking it.
>
> Wendell Berry[25]

The reign of technology in our lives should not be trivialized. Nor should it be totalized. Our experience of technology is always ambiguous; it is never conclusive. I do not consider contemporary life to be implacably unthinking. But, then, neither do I hold out hope of a planetary *Gestalt* to holistic consciousness; have faith in the existence of an otherwordly immortality; intend to grow old waiting for the descent of an unrealizable regime of equitable global governance; or proclaim the innocence of a previous age or another culture. I am, however, optimistic that gradual reform of the technological world will take place as the commerce of sustenance once again acquires a public, worldly character. Many are already engaging in the experiments out of which this reform will grow. I am confident that many more will join in these experiments as they experience the inability of the commerce of consumption to sustain richness in their lives. This reform will grow as the vague, widely shared feelings of unease elicited by the proliferation of technology crystallize into the realization that the technological world is itself a remarkable and ongoing experiment.

With this realization, we are rescued from the nihilistic conclusion that the disintegration of the ethics of sustainable development into empty rhetoric is entirely a consequence of sadistic (even masochistic) satisfaction or egoistic indifference or cynical pessimism on the part of latemoderns about the fate of their world. This disintegration speaks instead of the continued dominance of the narratives of technological security. It speaks of the fact that this project is not only a theoretical construction, it has become an actuality, it is our world.

We have built a remarkably insecure, risky world in which it seems, paradoxically, that we have no choice but to embrace all the more tightly the promise of technology. Latemoderns do care about their ecological and social reality. We care deeply about many things in and about our world and the yearning for sustainability is shared widely. However, in this technological world our desire for sustainability is readily subordinated to the imperatives of production. Enervated by the conditions of our latemodern world, our will to care is readily decentered within subjectivist moral discourses and dissipated within technocratic agendas.

The crucial task of reform is, of course, that of crafting apparently disparate experiments in the experience of sustenance together into new social structures capable of providing genuine alternatives to the imperatives of production. I am only too aware, for instance, that my own practices of home and garden are severely delimited by the instrumental institutions that reduce our care for the land to matters of capital and private ownership. And no doubt some will dismiss my embodied reflections in part III of this book as essentially private, essentially rarefied, and without relevance to the big, gritty questions of global economic reform with which I opened part I. To those readers I can do no more than remind them that the practices I have narrated—most notably practices of food production, neighborliness, gender differentiation, parenting and child rearing, home building and domestic maintenance, economic prudence, historical storytelling, communion with a more-than-human place, and mediation between cultural worlds—belong at the center of social experience. Our technological world has deformed these practices into technocratic structures that obscure their human essence. Any alternative to these structures appears somehow unrealistic, irrational, and idealistic. Yet they appear this way only in the terms of the explanation of our humanity that is encoded in the practices of our world. And, over time, over generations, I have no doubt these terms can be rewritten. Thus to reclaim even in small measure our practices from the promises of technological efficiency is to reclaim political possibilities for forms of practical decision making capable of honoring that which sustains us. It is to lift policy making off its precariously narrow pedestal of instrumentalism and found it on a more fully human footing. It is to join with Walt Whitman to celebrate the glorious fact that human wisdom is nourished, always, by a more-than-human world, for "something there is in the float of the sight of things provokes it out of the soul."[26]

The experimentation I champion will always be vulnerable to cooptation by the allure of commodification. Further and importantly, experimentation of this kind is always in part an affirmation of the freedoms made possible by the technological world. The optimistic stance of lived experimentation with technology I advocate is necessarily modest compared to the rampant techno-ethical optimism I cautioned against in my early chapters. I see only deceit and conceit in attempts to design and engineer the good life. My optimism takes root amidst my conviction that violence has been and will continue to be as perennial in human affairs as tolerance; devastation, ignorance, and fear as perennial as beauty, wisdom, and happiness. Death, impermanence in all its forms will persist, for it is as irrepressible and as likely as life itself. In whatever world it takes place, moral life is always contingent, struggling, perilous, and fraught. It is never sustainable; it can only ever and always be sustained within the continuous rhythm of human effort and insight in an everyday world of practice. The questions of sustainability do not promise answers that will liberate us from the ambiguity of our moral condition into a promised land of Sustainable Well-being.

There are no simple, universal, or transparent answers to the questions of sustainability. These questions are inherently normative, and as such they belong first and finally to practical reasoning. Any answers they may prompt will confound the logic of universalizing, instrumental reason and they will be directed always to limited, specific, and transient contexts. The crucial importance of questions about our ability to enter into the commerce of sustenance is that they prompt us to make questionable the kind of world we live in and the kind of goals this world establishes for our lives.

The ideal of sustainability gives rise to an agenda of good questions, practical questions that bear directly on our forms of life, drawing out and giving practical substance to our disquiet and to our hopes. Responses to these questions are essentially contestable, they demand of us not categorical certainty but the capacity to articulate what we feel to be most worthy of being sustained in our lives. These questions are valuable to us because they command our attention in an age of ecological crisis while simultaneously defying resolution and closure: they demand that we hold open for questioning our assumptions about what a resolution of this crisis might involve.

NOTES

PREFACE

1. Judith Wright, "Eve to Her Daughters," in *A Human Pattern: Selected Poems* (Sydney: ETT Imprint, 1996), 134–136, p. 135 (quotation from lines 18–42).

2. Stephen Bodian, "Simple in Means, Rich in Ends: An Interview with Arne Naess," in *Environmental Philosophy: From Animal Rights to Radical Ecology,* ed. M. Zimmerman (Englewood Cliffs, NJ: Prentice Hall, [1982] 1993), 182–192, p. 183.

INTRODUCTION

1. I avoid the terms *modern* and *postmodern* in describing contemporary affluent societies. *Modern* adequately covers the span from the late-sixteenth and early-seventeenth-century origins of science through to the flowering of industrialism, urbanization, and bureaucratic organization in the eighteenth, nineteenth, and early-twentieth centuries. The two world wars, however, mark a change to an unprecedented form of globalized and secularized society sufficiently different from the modern period to deserve another epithet. Yet declaring the late-twentieth and early-twenty-first centuries "postmodern," as is now common to do, is often the result of an overemphasis on the postwar proliferation of nonmodern theories. As will become evident as my discussion unfolds, I think that this overemphasis on new theories can serve to obscure and even to deny the ways in which our technological forms of life remain faithful to the modernist project of technological progress. I thus use *latemodern* throughout to refer to the historical span from the second world war through at least the early decades of this century, as well as to emphasize the present uncertain transition between the legacies of modernity and the appearance of truly postmodern possibilities.

2. Wendell Berry, *The Unsettling of America: Culture and Agriculture* (San Francisco: Sierra Club Books, 1977), p. 52.

3. I refer to sustainable development as a language, rather than as a concept, mindful of the "living use of language," as the philosopher Hans-Georg Gadamer put it, in which "words never exist in isolation. Their meaning is sustained and determined through the proximity and influence of neighbouring words." That this is the case for the burgeoning lexicon of sustainable development will become apparent in the first two chapters. "The Problem of Intelligence," in *The Enigma of Health: The Art of Healing in a Scientific Age*, trans. J. Gaiger and N. Walker (Cambridge: Polity, 1996), 45–60, p. 45.

4. Sharachchandra M. Lele, "Sustainable Development: A Critical Review," *World Development* 19, No. 6 (1991): 607–621, p. 607.

5. A great deal of up-to-date quantitative analysis of the symptoms of unsustainability is available in the following excellent annual publications: Lester R. Brown, et al. (eds), *State of the World: A Worldwatch Report on Progress Toward a Sustainable Society* (New York: W.W. Norton & Co., published annually since 1984); Lester Brown, et al. (eds), *Vital Signs: Trends that are Shaping our Future, Worldwatch Institute* (New York & London: W.W. Norton & Co., published annually since 1994); World Bank, *World Development Report* (New York & Oxford: Oxford University Press, published annually since 1978); and World Resources Institute, *World Resources: A Report by the World Resources Institute in collaboration with UNEP and UNDP* (New York & Oxford: Oxford University Press, published annually since 1988). A good historical discussion of the links between population growth, consumption levels, and environmental destruction can be found in Paul Harrison's *The Third Revolution: Population, Environment and a Sustainable World* (Harmondsworth, Middlesex: Penguin, [1992] 1993). Robert Chambers' *Rural Development: Putting the First Last* (New York: Longman, 1983) offers a blunt reminder to academics in affluent technological societies that our first task should be "critical self-examination" of our privilege, affluence, and bias.

6. See Paul Ekins, *A New World Order: Grassroots Movements for Global Change* (London & New York: Routledge, 1992), for a description of the four interlocking crises that give rise to the evidence of unsustainability. As Ekins describes them, these crises are (i) the military machine, (ii) the holocaust of poverty, (iii) the environmental crisis, and (iv) the denial of human rights. For a more recent, broader, and more hopeful overview, see Lester Brown's "Challenges of the New Century," in *State of the World 2000*, eds. L. R. Brown, C. Flavin, and H. French (New York & London: W.W. Norton & Co., 2000), 3–21.

7. Respectively: Brown et al., *Vital Signs 1994*, pp. 128–129, *Vital Signs 1998–1999*, pp. 68–69, *Vital Signs 1997*, pp. 130–131, *Vital Signs 1994*, pp. 72–73, and *Vital Signs 1998–1999*, pp. 74–75.

8. See Graeme Donald Snooks, *The Ephemeral Civilization: Exploding the Myth of Social Evolution* (London & New York: Routledge, 1997), ch. 7, pp. 176–205.

9. Ibid. See also Murray Bookchin, *The Limits of the City* (Montréal & New York: Black Rose Books, 1986), pp. 31–35; Michael Redclift, *Sustainable Development: Exploring the Contradictions* (London & New York: Routledge, 1987), pp. 108–110.

10. Bernal Diaz, cited in Redclift, *Sustainable Development*, pp. 108, 109.

11. Ibid., p. 110.

12. World Resources Institute, *World Resources 1992–93*, p. 88. See also UN Centre for Human Settlements (Habitat), *An Urbanizing World: Global Report on Human Settlements, 1996* (Oxford: Oxford University Press, 1996), pp. 145–146.

13. Brown et al., *Vital Signs 1998–1999*, p. 108; Herbert Girardet, *The Gaia Atlas of Cities: New Directions for Sustainable Urban Living* (London: Gaia Books, 1992), p. 108; UN Centre for Human Settlements (Habitat), *An Urbanizing World*, Table 4.3, pp. 52, 145.

14. Habitat, *An Urbanizing World*, p. 151.

15. Girardet, *The Gaia Atlas of Cities*, p. 90

16. Cited in Brown, et al., *State of the World 1995*, p. 179. For a detailed account of the nature of this failure see Gustavo Esteva and Madhu Suri Prakash, *Grassroots Post-Modernism: Remaking the Soil of Cultures* (London & New York: Zed Books, 1998).

17. Snooks, *The Ephemeral Civilization*, p. 178. The juxtaposition of Tenochtitlán and Mexico City also emphasizes the colonizing force of the Christian calendar, and I use the chronological notations B.C.E. (Before Christian Empire) and C.E. (Christian Empire) in a small effort to keep alive awareness of many temporal worlds that have been made invisible, if not obliterated, through Eurocentric hegemony.

18. As Guillermo Bonfil Batalla reminds us with his description of the ways in which mesoamerican civilization lives still in contemporary Mexico, despite its continued oppression, this "past" and this "present" do not stand discretely apart. See his *México Profundo: Reclaiming a Civilization*, trans. P. A. Dennis (Austin, TX: University of Texas Press, 1996).

CHAPTER ONE: AGENDA

1. IUCN, UNEP, WWF, *World Conservation Strategy: Living Resource Conservation for Sustainable Development* (Gland, Switzerland: IUCN, 1980), p. 1; World Commission on Environment and Development, *Our Common Future* (Oxford: Oxford University Press, 1987), p. 43; Nicholas A. Robinson (ed.), *Agenda 21 & the UNCED Proceedings, Vol. VI* (New York: Oceana Publications, 1993), para. 1.1; Commission on Global Governance, *Our Global Neighbourhood* (New York: Oxford University Press, 1995), p. 30.

2. See Michael Redclift's *Sustainable Development: Exploring the Contradictions* (London & New York: Routledge, 1987), pp. 15–36 and *Wasted: Counting the Costs of Global Consumption* (London: Earthscan, 1996), pp. 12–19.

3. IUCN, UNEP, WWF, *World Conservation Strategy.*

4. IUCN, UNEP, WWF, *Caring for the World: A Strategy for Sustainability* (Gland, Switzerland: IUCN, 1991). For a review of this term's evolution in the 1980s, see Linda Starke, *Signs of Hope: Working Towards Our Common Future* (Oxford: Oxford University Press, 1990), Ch. 3, pp. 39-61.

There is no wholly adequate nomenclature for describing the economic disparity between nations. The geographic metaphor adopted here is employed both for its neutrality with regard to the ideology of development and for its acceptance amongst "Southern" critics. *North* refers most directly to North America, EU, Japan, Australia, and New Zealand. *South* refers most directly to Africa, South and Central America, Mexico, the Indian subcontinent, East and Southeast Asia, and Pacific island states. Clearly many other nations and social groups occupy a problematic status in this division, such as Singapore in Southeast Asia, Kuwait in the Middle East, indigenous communities in Australia, and others.

5. After its Chair, and then Norwegian Prime Minister, Gro Harlem Brundtland.

6. WCED, *Our Common Future*, Ch. 2, pp. 43–65.

7. See Thomas G. Weiss, David P. Forsythe, and Roger A. Coate, *The United Nations and Changing World Politics* (Boulder, CO: Westview, 1994), pp. 207, 224.

8. Michael Jacobs, "Introduction: The New Politics of the Environment," in *Greening the Millennium? The New Politics of the Environment*, ed. M. Jacobs (Oxford: Blackwell, 1997), 1–17, p. 5.

9. Tarla Rai Peterson, *Sharing the Earth: The Rhetoric of Sustainable Development* (Columbia, SC: University of South Carolina Press, 1997), p. 6. I think Jacobs's suggestion in *Introduction: The New Politics of the Environment* that sustainable development has made "little impact on the public" (p. 6) is inaccurate; a fact I think is borne out in Robert Worcester's chapter in the same volume, "Public Opinion and the Environment," in *Greening The Millennium? The New Politics of the Environment*, ed. M. Jacobs (Oxford: Blackwell, 1997), 160–173, despite the fact that Worcester doesn't directly address the term *sustainable development*. I agree, however, that this term may have done little to increase public understanding of environmental problems.

10. I consider here only the postwar history of environmental concern, although, of course, such concern has an extensive modern history that includes the eighteenth-century political ecology of Thomas Malthus; eighteenth- and nineteenth-century English Romanticism; nineteenth-century German folk traditions; and the late-nineteenth- and early-twentieth-century land and wildlife conservation movements in North America that saw the establishment of the Sierra Club in 1892 and the National Audubon Society in 1905. See Anna Bramwell, *Ecology in the 20th Century: A History* (New Haven & London: Yale University Press, [1989] 1992); Kenn Kassman, *Envisioning Ecotopia: The U.S. Green Movement and the Politics of Radical Social Change* (Westport, CT & London: Praeger, 1997), and Donald Worster, *Nature's Economy: The Roots of Ecology* (San Francisco: Sierra Club Books, 1977). Even considering the postwar period, there are some, such as Mark Dowie, who place sustainable development in an eclectic third wave of environmental concern emerging in the 1990s, which in his terminology translates as the fourth wave of twentieth-century environmentalism, a distinction which seems to me somewhat arbitrary. *Losing Ground: American Environmentalism at the Close of the Twentieth Century* (Cambridge, MA & London: MIT Press, 1995), ch. 5, pp. 105–124, ch. 8, pp. 205–257. For a general discussion see Sharon Beder, *The Nature of Sustainable Development* (Newham, Victoria: Scribe, 1993), ch. 3, pp. 16–22.

11. Fairfield Osborn, *Our Plundered Planet* (Boston: Little, Brown & Co., 1948); William Vogt, *Road to Survival* (New York: William Sloane Associates, 1948); Aldo Leopold, *A Sand County Almanac: And Sketches Here and There*, intro. R. Finch, illust. C. W. Schwartz (New York & Oxford: Oxford University Press [1949] 1989); Rachel Carson, *Silent Spring* (Boston: Houghton Mifflin, 1962); Paul R. Ehrlich, *The Population Bomb* (New York: Ballantine, 1968); Theodore Roszak, *The Making of a Counter Culture: Reflections on Technocratic Society and its Youthful Opposition* (New York: Doubleday, 1969); Barry Commoner, *The Closing Circle: Nature, Man and Technology* (New York: Knopf, 1971); Donella H. Meadows, Dennis L. Meadows, Jørgen Randers, and William W. Behrens, *Limits to Growth* (New York: Universe Books, 1972); Herman Daly, *The Steady State Economy: Toward a Political Economy of Biophysical Equilibrium and Moral Growth* (San Francisco, CA: W. H. Freeman & Co., 1974).

12. Murray Bookchin, *Post-Scarcity Anarchism* (Palo Alto, CA: Ramparts Press, 1971); Arne Naess, "The Shallow and the Deep, Long Range Ecology Movement: A Summary," *Inquiry* 16 (1973): 95–100; Rosemary Radford Ruether, *New Woman New Earth: Sexist Ideologies and Human Liberation* (Minneapolis: The Seabury Press, 1975); Richard

Sylvan (Routley), "Is There a Need for a New, an Environmental Ethic?," in *Environmental Philosophy: From Animal Rights to Radical Ecology,* ed. M. E. Zimmerman, 2nd edn. (Upper Saddle River, NJ: Prentice Hall, [1971]1998), pp. 17–25.

13. Edward Goldsmith, Robert Allen, Michael Allaby, John Darvoll, and Sam Lawrence, *Blueprint for Survival* (Boston, MA: Houghton Mifflin, 1972), p. 15. Dennis Pirages's *The New Context for International Relations: Global Ecopolitics* (North Scituate, MA: Duxbury Press, 1978) provides a typical example of first-wave claims about an impending cultural paradigm shift.

14. Ivan D. Illich, *Celebration of Awareness: A Call for Institutional Revolution* (Harmondsworth, Middlesex: Penguin, 1973) and *Tools for Conviviality* (London: Calder & Boyars, 1973); Herbert Marcuse, *One Dimensional Man: Studies in the Ideology of Advanced Industrial Society* (London: Abacus, [1964] 1972); E. Fritz Schumacher, *Small is Beautiful: A Study of Economics as if People Mattered* (London: Abacus, 1973).

15. See, especially, Gordon Rattray Taylor's *The Doomsday Book* (London: Thames & Hudson, 1970). For a critical review of such accounts see Gregg Easterbrook, *A Moment on the Earth: The Coming Age of Environmental Optimism* (New York: Viking, 1995), pp. 79–88, 372–375.

16. Weiss et al., *The United Nations and Changing World Politics*, p. 177.

17. Garrett Hardin, "The Tragedy of the Commons," *Science* 162, December (1968): 1243–1248.

18. See Ramachandra Guha and Juan Martinez-Alier, *Varieties of Environmentalism: Essays North and South* (London: Earthscan, 1997), especially ch. 1–3, pp. 3–76.

19. Wilfred Beckerman, *In Defense of Economic Growth* (London: Jonathon Cape, 1974); John Maddox, *The Doomsday Syndrome* (London: MacMillan, 1972); Perry Pascarella, *Technology: Fire in a Dark World* (New York: Van Nostrand Reinhold, 1977); Julian Simon, *The Ultimate Resource* (Princeton: Princeton University Press, 1981).

20. Sharon Beder, *Global Spin: The Corporate Assault on Environmentalism* (Melbourne: Scribe, 1997), p. 16.

21. Beder, *The Nature of Sustainable Development*, p. 17. The *World Conservation Strategy* and other early interpretations of sustainable development such as that of Lester Brown's *Building a Sustainable Society* (New York: Norton, 1971) are examples of texts caught between the two waves of environmental concern I identify here.

22. Redclift, *Wasted*, p. 14.

23. An important contender was *ecodevelopment.* See Robert Riddell, *Ecodevelopment: Economics, Ecology and Development. An Alternative to Growth Imperative Models* (Westmead, UK: Gower, 1981).

24. Murray Bookchin thus claimed that out of a "movement that at least held the promise of challenging hierarchy and domination have [*sic*] emerged a form of *environmentalism* that is based more on tinkering with existing institutions, social relations, technologies and values than on changing them." "An Open Letter to the Ecological Movement," *Social Alternatives* 2, no. 3 (1982): 13–16, p. 14.

25. Analyses of ecological modernization as an emerging trend within policy making were pioneered by German social scientists, most notably Joseph Huber and Martin Jänicke in the early 1980s. The extent to which policy making reflects ecological

modernization in practice rather than in rhetoric is discussed by Marit Reitan in "Ecological Modernisation and 'realpolitik': Ideas, Interests and Institutions," *Environmental Politics* 7, no. 2 (1998): 1–26. For an overview of debates about ecological modernization, see John S. Dryzek, *The Politics of the Earth: Environmental Discourses* (Oxford & New York: Oxford University Press, 1997), pp. 141–143. Maarten A. Hajer in his seminal book on the subject, *The Politics of Environmental Discourse: Ecological Modernization and the Policy Process* (New York & Oxford: Clarendon, 1995), describes the Brundtland Report as a key document of ecological modernization but notes that this project "has a much sharper focus than does sustainable development on exactly what needs to be done with the capitalist political economy," p. 143. I suspect that in the new millennium this is largely no longer true. Many critiques of ecomodernism, including Hajer's, draw upon Ulrich Beck's analysis of "risk society." See, e.g., his *Ecological Politics in an Age of Risk,* trans. Amos Weisz (Cambridge: Polity Press, 1995).

26. Andrew Gouldson and Joseph Murphy, "Ecological Modernisation: Restructuring Industrial Economies," in *Greening the Millennium? The New Politics of the Environment*, ed. M. Jacobs (Oxford: Blackwell, 1997), 74–86. For a good example of how this synergy is working itself out in practice, and of how sustainable development is interwoven with policies for ecological modernization, see David Wallace, *Environmental Policy and Industrial Innovation: Strategies in Europe, the USA and Japan* (London: The Royal Institute of International Affairs/Earthscan, 1995). Marcel Wissenberg's impressive attempt to sketch out the features of a green liberal democratic theory provides insight into the theoretical underpinnings of ecomodernist agendas. He reveals some of the ways in which modernist theory can genuinely contest the ideal of sustainability, concluding that the prospects for a sustainable society rest with the "responsibility of each and every single individual" (most notably as consumers) to develop "sustainable preferences," thus guiding the free market toward environmental prudence. While I vigorously contest this vision of sustainability, Wissenberg's affirmation of modernity's quest toward a truly free society can only deepen and open the thinking of both proponents and critics of ecomoderism. *Green Liberalism: The Free and the Green Society* (London: UCL Press, 1998), p. 226.

27. See the chapters "Looming Tragedy: Survivalism" and "Growth Forever: The Promethean Response" in Dryzek, *The Politics of the Earth*, ch. 2, pp. 23–44, ch. 3, pp. 45–60.

28. Paul Ekins, *A New World Order: Grassroots Movements for Global Change* (London & New York: Routledge, 1992), ch. 2, pp. 14–39.

29. WCED, *Our Common Future*, p. 33.

30. The Secretary-General of the Brundtland Commission, Jim MacNeill, asserted that a "fivefold to tenfold increase in economic activity would be required over the next 50 years in order to meet the needs and aspirations of a burgeoning world population." "Strategies for Sustainable Economic Development," *Scientific American* 261, no. 3 (1989): 105–113, p. 106. This single-issue edition of *Scientific American*, entitled Managing Planet Earth, offers an excellent summary of nascent ecomodernism.

31. Redclift, *Wasted*, p. 20. This shift has given rise to what Anna Bramwell has called the "fading of the greens," that is, the weakening (but not disappearance by any means) of environmentalism as a discrete, identifiable, and transformative political movement. *The Fading of the Greens: The Decline of Environmental Politics in the West* (New Haven & London: Yale University Press, 1994).

32. C. J. Silas, CEO of Phillips Petroleum Co., 1990, cited in Beder, *Global Spin*, p. 23. See also Richard Welford (with contributions from Eloy Casagrande Jr., David Jones, Tarja Ketola, Nick Mayhew and Pall Rikhardsson), *Hijacking Environmentalism: Corporate Responses to Sustainable Development* (London: Earthscan, 1997).

33. David Helvarg, *The War Against the Greens: The "Wise Use" Movement, the New Right, and Anti-Environmental Violence* (San Francisco: Sierra Club Books, 1994), pp. 446, 459.

34. Easterbrook, *A Moment on the Earth*.

35. Wouter Van Dieren (ed.), *Taking Nature into Account: Toward a Sustainable National Income—A Report to the Club of Rome* (New York: Springer-Verlag, 1995), p. 97.

36. Ernst von Weizsäcker, Amory B. Lovins, and L. Hunter Lovins, *Factor Four: Doubling Wealth—Halving Resource Use, A New Report to the Club of Rome* (Sydney: Allen & Unwin, 1997), p. xviii.

37. WCED, *Our Common Future*, ch. 2, pp. 43–65.

38. Between the WCED and UNCED, the language of sustainable development was widely promoted by the Brundtland Commission's "public relations agency," the Centre for Our Common Future. See Starke, *Signs of Hope*. See also Michael Marien, "Environmental Problems and Sustainable Futures: Major Literature from WCED to UNCED," *Futures* 24, October (1993): 731–755, pp. 739–740.

39. In addition to 117 heads-of-state, the Earth Summit attracted observers from 1,400 Non-Government Organisations (NGOs) as well as 8,000 journalists. Over 18,000 NGO representatives took part in the '92 Global Forum, run concurrently with the Earth Summit. See Mark F. Imber, *Environment, Security and UN Reform* (New York: St. Martin's Press, 1994), ch. 5, pp. 84–113.

40. Tony Brenton, *The Greening of Machiavelli: The Evolution of International Politics* (London: Earthscan, 1994), ch. 9, pp. 207–221.

41. See Peter Doran, "The UN Commission on Sustainable Development," *Environmental Politics* 5, no. 1 (1996): 100–107.

42. UN Commission on Sustainable Development, *Documents of the First Session, June 14–25* (UN Document No.: E/CN.17/1993/8, 1993), p. 3.

43. See Reg Henry, "Adapting United Nations Agencies for Agenda 21: Programme Coordination and Organisational Reform," *Environmental Politics* 5, no. 1 (1996): 1–24, p. 6.

44. All nations are required by the UN, although not in a legally binding sense, to contribute at least 0.7 per cent of their Gross National Product (GNP) into the GEF. Ibid., p. 101. The GEF was formed in 1990 as part of the follow-up to the Brundtland Report and is administered by the World Bank, UNDP and UNEP. See Redclift, *Wasted*, pp. 28–31.

45. Such conferences include the International Conference on Population and Development (1994), the World Summit on Social Development (1995), the Fourth World Conference on Women (1995), and the Second UN Conference on Sustainable Human Settlements (1996).

46. Commission on Global Governance, *Our Global Neighbourhood*, p. 1. In contrast, the 1996 Independent Commission on Population and Quality of Life makes almost no use of the language of sustainable development. See *Caring for the Future: Making the*

Next Decades Provide a Life Worth Living (Oxford & New York: Oxford University Press, 1996).

47. UN Commission on Sustainable Development, *Overall Progress Achieved Since the United Nations Conference on Environment and Development: Report of the Secretary-General* (CSD, Fifth Session, April 7–25, UN Document No.: E/CN.17/1997/2, 1997). For a collection of essays that explore the lasting political impact, or otherwise, of *Agenda 21* see Tim O'Riordan and Heather Voisey (eds), *The Transition to Sustainability: The Politics of Agenda 21 in Europe* (London: Earthscan, 1998).

48. UN, *Assembly President Speaks of "Shame" that Commitments at Rio Conference of 1992 Remain Unfulfilled* (UN Press Release: GA/9261, ENV/DEV/426, 23 June 1997).

49. Earth Summit Watch, *What Have National Governments Done Since the 1992 Earth Summit in Rio?* (available at www.earthsummitwatch.org/implementation.html, accessed Jan. 26, 2000). See also Imber, *Environment, Security and UN Reform*, pp. 123–127.

50. UN Commission on Sustainable Development, *High-Level Advisory Group of Sustainable Development, Report of the 4th Session, 30 October 1995* (UN Document No.: E/CN.17/1996/2, 1996), p. 9.

51. Martin Khor, "Effects of Globalization on Sustainable Development After UNCED," *Third World Resurgence* 81/82 (1997): 5–11.

52. Weiss, et al., *The United Nations and Changing World Politics*, pp. 178–179.

53. World Trade Organization, *Trade and Environment Bulletin* 2, May 8 (1995), Appendix 1. For a wider discussion of the role of the WTO in the emerging global environmental politics, see Hilary French, "Coping with Ecological Globalization," in *State of the World 2000*, eds L. R. Brown, C. Flavin, and H. French (New York & London: W.W. Norton & Co., 2000), 184–202, pp. 189–194.

54. UN Commission on Sustainable Development, *Report on the Fourth Session of the CSD, 18 April - 3 May 1996* (UN Document No.: E/CN.17/1996/38, 1996), Decision 4.1.

55. Kenneth Piddington, "The Role of the World Bank," in *Green Planet Blues: Environmental Politics from Stockholm to Rio*, eds K. Conca, M. Alberthy, and G. Dablenko (Boulder: Westview, [1992] 1995), 199–203.

56. World Bank, *World Development Report 1992* (New York: Oxford University Press, 1992), p. 34.

57. Bruce Rich, *Mortgaging the Earth: The World Bank, Environmental Impoverishment and the Crisis of Development* (London: Earthscan, 1994).

58. World Bank, *Entering the 21st Century: World Development Report 1999/2000* (Oxford & New York: Oxford University Press, 2000).

59. Donald Brown and John Lemons, "Introduction," in *Sustainable Development: Science, Ethics and Public Policy*, eds. J. Lemons and D. A. Brown (Dordrecht: Kluwer Academic, 1995), 1–10, p. 7.

60. Commonwealth of Australia, *National Strategy for Ecologically Sustainable Development* (Canberra: Australian Government Printing Service, 1992). Another revealing national approach is that of the U.S. President's Council on Sustainable Development (PCSD) initiated in June 1993. See Molly Harris Olson, "Charting a Course for Sustainability," *Environment* 38, no. 4 (1996): 10–15, and the U.S. President's Council on Sustainable Development,

Sustainable America: A New Consensus (available at www.whitehouse.gov/PCSD, 1996, accessed March 20, 1997). For a critical discussion, see Dowie, *Losing Ground*, pp. 237–238.

61. Commonwealth of Australia, *Ecologically Sustainable Development: A Commonwealth Discussion Paper* (Canberra: Australian Government Printing Service, 1990).

62. Robinson, *Agenda 21 and the UNCED Proceedings, Vol. VI,* para. 23.2.

63. Stuart Harris, "Ecologically Sustainable Development: Implications for the Policy Process," *Canberra Journal of Public Administration* 69 (1992): 3–8, p. 8.

64. Ian Barns, "Value Frameworks in the Sustainable Development Debate," in *Ecopolitics V: Proceedings*, ed. R. Harding (Sydney: University of NSW, 1992), 199–208.

65. Bill Hare, *Ecologically Sustainable Development: Assessment of the ESD Working Groups* (Melbourne: Australian Conservation Foundation, Greenpeace and World Wide Fund for Nature Australia, 1991).

66. Australian Manufacturing Council, *The Environmental Challenge: Sustainable Businesses in the 1990s* (Melbourne: Australian Manufacturing Council, July 1991); Peter Colley, "Sustainable Development: The ACTU Perspective," *Canberra Journal of Public Administration* 69 (1992): 18–20.

67. For the government's account of these negotiations see Stuart Beil, "The Kyoto Protocol: Key Elements and Outcomes," *Climate Change Newsletter* 9, no. 4 (1998): 1–2.

68. Robinson, *Agenda 21 and the UNCED Proceedings, Vol. VI,* para. 28.2(a).

69. Redclift, *Wasted*, p. 32. There were, by the mid 1990s, over 1200 *Local Agenda 21's* in twenty-six countries. Doran, *The UN Commission on Sustainable Development,* p. 100. See also International Council for Local Environmental Initiatives, *The Local Agenda 21 Planning Guide: An Introduction to Sustainable Development Planning* (Toronto: ICLEI, 1996); Mary MacDonald, *Agendas for Sustainability: Environment and Development into the Twenty-first Century* (London & New York: Routledge, 1998).

70. See Stephen Young's discussion of this question in "Local Agenda 21: The Renewal of Local Democracy?," in *Greening the Millennium? The New Politics of the Environment*, ed. M. Jacobs (Oxford: Blackwell, 1997), 138–147.

71. Joke Waller-Hunter, "Sustainable Production: The Corporate Challenge," *UNEP Industry and Environment,* October-December (1995): 21–24, p. 24.

72. Stephen Schmidheiny, *Changing Course: A Global Business Perspective on Environment and Development* (Cambridge, MA: MIT Press, 1992), p. 14.

73. Ibid., p. 12.

74. Maurice Strong, cited in Pratap Chatterjee and Matthias Finger, *The Earth Brokers: Power, Politics and World Development* (London & New York: Routledge, 1994), p. 117.

75. Waller-Hunter, *Sustainable Production*, pp. 22–23. The Rotterdam Charter has been endorsed by over 2000 companies, including 200 of the world's largest. See Jean-Charles Rouher, "The ICC and UNEP: 23 Years of Close Collaboration," *UNEP Industry and Environment,* October-December (1995): 29–33, p. 29.

76. David Pearce, Anil Markandya, and Edward B. Barbier, *Blueprint for a Green Economy* (London: Earthscan, 1989); David Pearce, Edward B. Barbier, Anil Markandya, Scott Barrett, R. Kerry Turner, and Timothy Swanson, *Blueprint 2: Greening the World Economy* (London: Earthscan, 1991); David Pearce (CSERGE), *Blueprint 3: Measuring*

Sustainable Development (London: Earthscan, 1993); David Pearce, *Blueprint 4: Capturing Global Environmental Value* (London: Earthscan, 1995); David Maddison, David Pearce, Olaf Johansson, Edward Calthorp, Todd Litman and Eric Verhoef, *Blueprint 5: The True Costs of Road Transport* (London: Earthscan, 1996). For a discussion of how this theory is bound up in current institutional reforms see Timothy Swanson and Sam Johnston, *Global Environmental Problems and International Environmental Agreements: The Economics of International Institution Building* (Cheltenham, UK, & Northhampton, MA: Edward Elgar, 1999).

77. Pearce, et al., *Blueprint for a Green Economy*, Ch. 1, pp. 1–27.

78. Ibid., p. 2.

79. Pearce, *Blueprint 3*, p. 4.

80. Ibid., p. 8.

81. Massoud Karshenas reports that sustainability was a crucial element in the neoclassical definition of income as early as 1946. "Environment, Technology, and Employment: Towards a New Definition of Sustainable Development," *Development and Change* 25, no. 4 (1994): 723–756, p. 733.

82. UNDP, cited in Weiss, et al., *The United Nations and Changing World Politics*, p. 184 (emphasis added). See Walt W. Rostow, *The Stages of Economic Growth: A Non-Communist Manifesto* (New York: Cambridge University Press, 1960).

83. See Alan Holland's discussion of the problem of the substitutability of natural and human capital and the problematic connection between capital and social welfare in "Sustainability: Should we Start from Here?," in *Fairness and Futurity: Essays on Environmental Sustainability and Social Justice*, ed. A. Dobson (Oxford & New York: Oxford University Press, 1999), 46–68, pp. 48–57.

84. Pearce, *Blueprint 3*, pp. 15–19.

85. E.g., Wilfred Beckerman's "'Sustainable Development': Is it a Useful Concept?," *Environmental Values* 3 (1994): 191–209.

86. See Donald A. Brown, "Role of Economics in Sustainable Development and Environmental Protection," in *Sustainable Development: Science, Ethics and Public Policy*, eds. J. Lemons and D. A. Brown (Dordrecht: Kluwer Academic, 1995), 52–63. For a critique, see R. K. Blamey and Michael Common, "Sustainability and the Limits to Pseudo Market Valuation," in *Toward Sustainable Development: Concepts, Methods and Policy*, eds. J. C. J. M. van den Bergh and J. van der Straaten (Washington DC & Covelo, CA: Island Press, 1994), 165–205.

87. Pearce, *Blueprint 3*, pp. 24–5.

88. Pearce, *Blueprint 4*, p. 78.

89. Schmidheiny, *Changing Course*, pp. 16–18.

90. Elizabeth Dowdeswell, *Sharing Responsibilities in a Competitive World* (a speech to the Conference on Sustainable Industrial Development, Amsterdam, 22 February 1996), p. 5.

91. WCED, *Our Common Future*, p. 8.

92. Ibid., p. 217.

93. M. Taghi Farvar and John P. Milton, *The Careless Technology: Ecology and International Development* (New York: The Natural History Press, 1972), p. xv.

94. Ernest Braun, *Wayward Technology* (Westport, CT: Greenwood Press, 1984), p. 217. Interestingly, his later book, *Futile Progress: Technology's Empty Promise* (London: Earthscan, 1995), takes a much less optimistic view, offering trenchant criticism of *Agenda 21*. See, especially, pp. 162–167.

95. A view given lavish treatment by Lewis M. Branscomb in *Confessions of a Technophile: A Grand Tour of High Technology and Guide for the Future* (New York: American Institute of Physics Press, 1995), pp. 165–219. See also Scientific American (ed), *Key Technologies for the 21st Century* (New York: W. H. Freeman, [1995] 1996).

96. George Heaton, Robert Repetto and Rodney Sobin, *Transforming Technology: An Agenda for Environmentally Sustainable Growth in the 21st Century* (Washington DC: World Resources Institute, 1991), pp. ix, 7.

97. WCED, *Our Common Future*, p. 60.

98. Ibid., pp. 217–218.

99. Ibid., p. 219.

100. Ibid., p. 4.

101. Easterbrook, *A Moment on the Earth*, p. 668.

102. Chatterjee and Finger, *The Earth Brokers*, p. 59.

103. Robinson, *Agenda 21 and the UNCED Proceedings, Vol. VI*, para. 34.1.

104. Kazuo Matsushita, "Officer in Charge of Technology Transfer at UNCED Secretariat," *Network '92* 9 (August 1991); Nicholas A. Robinson (ed), *Agenda 21 & the UNCED Proceedings, Vol. 1* (New York: Oceana Publications, 1992), p. 199. For examples of the large-scale conventional development projects funded through *Agenda 21* under the cover of this relativism (if not sophism), see World Bank, *Making Development Sustainable: The World Bank Group and the Environment* (Washington DC: World Bank, 1994), Annex E, pp. 238–242.

105. Robinson, *Agenda 21 and the UNCED Proceedings, Vol. VI*, para. 34.5.

106. Claude Fussler, "The Cure for Marketing Myopia: Meeting Emerging Market Needs for Eco-efficiency Products," *UNEP Environment and Industry*, October-December (1995): 34–37, p. 35.

107. Heaton et al., *Transforming Technology*, p. vii.

108. Council of Academies of Engineering and Technological Sciences, *The Role of Technology in Environmentally Sustainable Development: A Declaration of the Council of Academies of Engineering and Technological Sciences* (Kiruna, Sweden: CAETS, 21 June 1995), p. 14.

109. Sharon Baker, "Sustainable Development and Sustainable Consumption: The Ambiguities—The Oslo Ministerial Roundtable Conference on Sustainable Production and Consumption, Oslo, 6–10 February 1995," *Environmental Politics* 5, no. 10 (1996): 93–99.

110. Adapted from Waller-Hunter, *Sustainable Production*, p. 22.

111. There are, in addition, a raft of secondary terms including the UN's cumbersome *environmentally sound technology* and the favorite of marketeers, *sustainable technology*. Other adjectives commonly used to distinguish technology for sustainable development include *green, clean, cleaner, environmentally superior,* and *environmentally advanced*. More recently *sustainable technology* and *sustainable development technology* have gained prominence.

112. That ideas of ecological technology are closely related to those about ecological economics is evident in the continuing influence of Fritz Schumacher's *Small is Beautiful.* For a commentary, see Kelvin Willoughby, *Technology Choice: A Critique of the Appropriate Technology Movement* (London: Westview Press/IT Publications, 1990). In addition to appropriate (or intermediate) technology movements, cultural visions of ecological technology are central to many related social movements such as bioregionalism, biodynamic farming and permaculture, and can be found in the writings of activist theorists such as Wendell Berry, Jerry Mander, Masanobu Fukuoka, Wes Jackson, and Bill Mollison. For a textbook on ecological technology, see William J. Mitsch and Sven E. Jørgensen, *Ecological Engineering: An Introduction to Ecotechnology* (New York: John Wiley & Sons, 1989). Three of the best of the many recent discussions by ecological technology practitioners are Robert L. Thayer, *Grey World, Green Heart: Technology, Nature, and the Sustainable Landscape* (New York: John Wiley & Sons, 1995); Nancy J. Todd and John Todd, *From Eco-Cities to Living Machines: Principles of Ecological Design* (Berkeley, CA: North Atlantic Books, 1994); and Sim van der Ryn and Stuart Cowan, *Ecological Design* (Washington DC; Covelo, CA: Island Press, 1996).

113. Jesse H. Ausubel, "Directions for Environmental Technologies," *Technology in Society* 16 (1994): 139–154.

114. Robert Socolow, "Six Perspectives from Industrial Ecology," in *Industrial Ecology and Global Change*, eds. R. Socolow, C. Andrews, F. Berkhout, and V. Thomas (Cambridge: Cambridge University Press, 1994), 3–16, p. 3.

115. Ausubel, *Directions for Environmental Technologies*, p. 140.

116. See Robert U. Ayers and Udo E. Simonis (eds*), Industrial Metabolism: Restructuring for Sustainable Development* (Tokyo: United Nations University Press, 1994).

117. Thomas Graedel, "Industrial Ecology: Definition and Implementation," in *Industrial Ecology and Global Change*, eds. R. Socolow, C. Andrews, F. Berkhout, and V. Thomas (Cambridge: Cambridge University Press, 1994), 23–41, p. 40.

118. As Daniel B. Botkin reveals, the trend to displace machine-based metaphors of nature with nature-based metaphors of machines reflects the widespread erosion of longstanding mechanistic explanations of both ecosystems and technosystems. See his *Discordant Harmonies: A New Ecology for the Twenty-first Century* (New York & Oxford: Oxford University Press, 1990).

119. C. Andrews, "Policies to Encourage Clean Technology," in *Industrial Ecology and Global Change*, eds. R. Socolow, C. Andrews, F. Berkhout, and V. Thomas (Cambridge: Cambridge University Press, 1994), 405–422, p. 418.

120. Ausubel, *Directions for Environmental Technologies*, p. 140

121. For general discussions of decarbonisation and dematerialisation, see, respectively, Christopher Flavin and Seth Dunn, "Reinventing the Energy System," and Gary Gardner and Payal Sampat, "Forging a Sustainable Materials Economy," in *State of the World 1999*, eds. L. R. Brown and C. Flavin (London: Earthscan/Worldwatch Institute, 1999), 22–40; 41–59.

122. Heaton et al., *Transforming Technology*, p. 9.

123. Bertram Wolfe, "Why Environmentalists Should Promote Nuclear Energy," *Issues in Science and Technology,* Summer (1996): 55–60, p. 60.

124. Socolow, *Six Perspectives From Industrial Ecology*, p. 14. See, e.g., Wallace, *Environmental Policy and Industrial Innovation*, ch. 10, 177–218.

125. Ausubel, *Directions for Environmental Technologies*, p. 148.

126. E. Talero and P. Gaudette, *Harnessing Information for Development: A Proposal for a World Bank Group Strategy,* Finance and Private Sector Vice Presidency, Industry and Energy Dept., Telecommunications and Informatics Division (available at www.worldbank.org/html/fpd/harnessing, 1994, accessed October 11, 1999).

127. Graedel, *Industrial Ecology*, p. 40.

128. See, e.g., the neglect of agricultural, fishery, and forestry issues in the reports of the National Commission on the Environment, *Choosing a Sustainable Future: The Report of the National Commission on the Environment* (Washington DC & Covelo, CA: Island Press, 1993) and the U.S. National Science and Technology Council, *Technology for a Sustainable Future: A Framework for Action* (Washington DC: Government Printing Office, 1994).

129. WCED, *Our Common Future*, p. 218. On biotechnology, see Donald L. Plunknett and Donald L. Winkelmann, "Technology for Sustainable Agriculture," in *Key Technologies for the 21st Century,* ed. Scientific American (New York: W. H. Freeman, [1995] 1996), 133–138, and Bernhard Zechendorf, "Sustainable Development: How Can Biotechnology Contribute?" *Trends in Biotechnology* 17 (1999): 218–225.

130. Stuart R. Taylor, "Green Management: The Next Competitive Weapon," *Futures* 24, no. 7 (1992): 669–680, p. 669.

131. Iain Anderson, "Manufacturing a Sustainable Future," *Science and Public Affairs,* Spring (1998): 24–27.

132. Ernst U. von Weizsäcker, *Earth Politics* (London & Atlantic Highlands, NJ: Zed Books, 1994), p. 177.

133. Marcello Colitti, "Economic Stagnation and Sustainable Development," in *Sustainable Development and the Energy Industries: Implementation and Impacts of Environmental Legislation,* ed. N. Steen (London: Earthscan, 1994), 35–43, p. 35.

134. Ibid., p. 36.

135. Ibid., p. 35.

136. Frank J. Comes and Christopher Power, "21st Century Capitalism," *Business Week,* December 12, 1994: 16–17, p. 16.

137. U.S. National Science and Technology Council, *Technologies for a Sustainable Future*, p. 2.

138. Abdus Salam and Azim Kidwai, "A Blueprint for Science and Technology in the Developing World," *Technology in Society* 13 (1991): 389–404, p. 389.

139. Jean-Francois Rischard, *Revolutions in Technology for Development: Technet Working Paper Series* (available at www.worldbank.org/html/fpd/technet/rischard.htm, 1995, accessed October 11, 1999).

140. Harry Truman, cited in Otto Ullrich, "Technology," in *Development Dictionary: A Guide to Knowledge as Power*, ed. W. Sachs (London & Atlantic Highlands, NJ: Zed Books, 1992), 275–287, p. 275.

141. J. Ronald Engel, "Introduction: The Ethics of Sustainable Development," in *Ethics of Environment and Development: Global Challenge, International Response*, eds. J. R. Engel and J. G. Engel (London: Belhaven Press, 1990), 1–23, pp. 2–5.

142. IUCN, UNEP, WWF, *World Conservation Strategy,* para. 13.1.

143. Engel, *Introduction: The Ethics of Sustainable Development*, p. 3.

144. An example of how expressions of this ethic have developed since 1980 is offered by Martin Holgate, former Director General of the IUCN, in *From Care to Action: Making a Sustainable World* (London: Earthscan, 1996), especially ch. 5, pp. 120–145.

145. Dowdeswell, *Sharing Responsibilities in a Competitive World*, p. 8.

146. WCED, *Our Common Future*, p. 1.

147. Ibid., p. 308 (emphasis added).

148. Ibid.

149. Ibid., p. 43.

150. UN Secretariat, cited in Aidan Davison and Ian Barns, "The Earth Summit and the Ethics of Sustainable Development," *Current Affairs Bulletin* 69, no. 1 (1992): 4–13, p. 7.

151. Ibid., pp. 7–11.

152. Howard Glasser, Paul P. Craig, and Willett Kempton, "Ethics and Values in Environmental Policy: The Said and the Unced," in *Toward Sustainable Development: Concepts, Methods, and Policy*, eds. J. C. van den Bergh and J. van den Straaten (Washington DC & Covelo, CA: Island Press, 1994), 83–107, pp. 87–88.

153. Brenton, *The Greening of Machiavelli*, p. 214. See the critique of Chatterjee and Finger, *The Earth Brokers*, pp. 49–53.

154. R. Kerry Turner and David Pearce, *Ethical Foundations of Sustainable Economic Development, LEEC Paper 90–01* (London: International Institute for Economic Development/LEES, 1990), p. 2.

155. Michael Walzer, "The Idea of Civil Society," in *Toward a Global Civil Society*, ed. M. Walzer (Providence: Berghahn Books, 1994), 7–27, p. 7. See also Jenny Pearce, "Civil Society—Trick or Treat?," *Third World Resurgence* 77/78 (1997): 48–50.

156. Ronnie D. Lipschutz with Judith Mayer, *Global Civil Society and Global Environmental Governance: The Politics of Nature from Place to Planet* (Albany, NY: State University of New York Press, 1996), p. 233. This book provides an excellent countervailing argument to ecomodernist claims about the need for global management regimes.

157. Frank Popoff, cited in Schmidheiny, *Changing Course*, p. 87.

158. Weizsäcker et al., *Factor Four*, p. 299.

159. Even the recent and puzzling book on global equity by Michael Carley and Philippe Spapens of the Sustainable Europe Campaign of Friends of the Earth assumes, without explicit reason, that technological increases in ecoefficiency, while not sufficient on their own to produce global justice, are entirely compatible with "changing values which recognise that, beyond meeting basic needs, quality of life derives from a sufficiency, rather than an excess of consumption." Explaining overconsumption as arising out of "genetic programming to secure life-supporting basic needs," these authors largely

ignore the iniquitous techno-economic structures that are being reinforced by the pursuit of ecoefficiency. *Sharing the World: Sustainable Living and Global Equity in the 21st Century* (London: Earthscan, 1998), pp. 107, 133.

160. Commission on Global Governance, *Our Global Neighbourhood*, p. 46.

161. Ibid., p. 55.

162. UN Development Programme, *Human Development Report 1994* (New York & Oxford: Oxford University Press/UNDP, 1994), p. 17.

163. Ibid., p. 4.

164. Ibid., p. 19.

165. Welford, *Hijacking Environmentalism,* pp. 30–31.

CHAPTER TWO: POLITICS

1. Ashok Khosla, "Foreword," in *A Sustainable World: Defining and Measuring Sustainable Development*, ed. T. C. Trzyna (London: IUCN/ICEP/Earthscan, 1995), 7–9, p. 8.

2. Sharachchandra M. Lélé, "Sustainable Development: A Critical Review," *World Development* 19, no. 6 (1991): 607–21, p. 613.

3. Langdon Winner, *The Whale and the Reactor: A Search for Limits in an Age of High Technology* (Chicago: University of Chicago Press, 1986), p. 54.

4. Lélé, *Sustainable Development,* p. 607.

5. Johan Holmberg and Richard Sandbrook, "Sustainable Development: What is to be Done?," in *Policies for a Small Planet: From the International Institute for Environment and Development*, eds. J. Holmberg and R. Sandbrook (London: Earthscan, 1992), 19–38, p. 20.

6. Richard B. Norgaard, "Sustainable Development: A Co-evolutionary View," *Futures* 20, December (1988): 606–620, p. 607.

7. Timothy O'Riordan, "Democracy and the Sustainability Transition," in *Democracy and the Environment: Problems and Prospects*, eds. W. M. Lafferty and J. Meadowcroft (Cheltenham, UK & Brookfield, MA: Edward Elgar, 1996), 140–156, p. 144.

8. Holmberg and Sandbrook, *Sustainable Development*, pp. 23, 37.

9. Donald Worster, *The Wealth of Nature: Environmental History and the Ecological Imagination* (New York & Oxford: Oxford University Press, 1993), p. 142.

10. See Michael Jacobs, "Sustainable Development as a Contested Concept," in *Fairness and Futurity: Essays on Environmental Sustainability and Social Justice*, ed. A. Dobson (Oxford & New York: Oxford University Press, 1999), 21–45, p. 24.

11. Ivan D. Illich, *Tools for Conviviality* (London: Calder and Boyars, 1973), p. 9.

12. Donella H. Meadows, Dennis L. Meadows, Jørgen Randers, and William W. Behrens, *Limits to Growth* (New York: Universe Books, 1972), p. 24.

13. These were the *Intergovernmental Conference of Experts on the Biosphere*, initiated by the UN Educational, Scientific, and Cultural Organization as a result of scientific concern about the industrial transformation of the biosphere, and the Washington-based *Ecological Aspects of International Development*. See Thomas G. Weiss, David P. Forsythe,

and Roger A. Coate, *The United Nations and Changing World Politics* (Boulder, CO: Westview, 1994), pp. 195–196. For an example of the thinking about sustainability that emerged at this latter meeting, see Robert Dasmann, John P. Milton, and Peter H. Freeman, *Ecological Principles for Economic Development* (London: John Wiley & Sons, 1973), p. 22.

14. See the advertisements reproduced in Jed Greer and Kenny Bruno, *Greenwash: The Reality Behind Corporate Environmentalism* (Penang, Malaysia: Third World Network, 1996).

15. Jacobs, *Sustainable Development as a Contested Concept*, p. 279 (ff.1).

16. Ernest J. Yanarella and Richard S. Levine, "Does Sustainable Development Lead to Sustainability?," *Futures* 24 (1992): 759–776, p. 760.

17. Arne Naess, "Sustainable Development and Deep Ecology," in *Ethics of Environment and Development: Global Challenge, International Response*, eds. J. R. Engel and J. G. Engel (London: Belhaven, 1990), 87–96, p. 96. A view he reiterated in a 1994 conference paper, see George Myerson and Yvonne Rydin, *The Language of Environment: A New Rhetoric* (London: UCL Press, 1996), pp. 122–123. In my view, Naess's commitment to a "methodological vagueness" capable of embracing substantial diversity of opinion about terms like *sustainability*—see Howard Glasser, "Demystifying the Critiques of Deep Ecology," in *Environmental Philosophy: From Animal Rights to Radical Ecology*, eds M. Zimmerman et al., 2nd edn. (Upper Saddle River, NJ: Prentice Hall, 1998), 212–226, pp. 217–218—may come unstuck here. While I admire and share Naess's commitment to antidogmatism, what is required is not loose, embracing use of the ideal of sustainability, but, rather, highly focussed, specific elaborations that can resist the colonizing force of ecomodernity yet remain reflectively self-aware of their own essential contestability (see section III of this chapter).

18. Tarla Rai Peterson, *Sharing the Earth: The Rhetoric of Sustainable Development* (Columbia, SC: University of South Carolina Press, 1997), p. 8.

19. In *Sustainable Development as a Contested Concept,* Jacobs argues that we recognize in sustainable development discourses two levels of meaning: one unitary but essentially vague, the other politically meaningful but essentially contested. My suggestion, however, is that these two layers be actively held apart through resisting the attempt to conflate the technocratic agenda of sustainable development and the ideal of sustainability.

20. Myerson and Rydin, *The Language of Environment*, p. 100.

21. Ibid., pp. 119–126.

22. Ozay Mehmet, *Westernizing the Third World: The Eurocentricity of Economic Development Theories* (London & New York: Routledge, 1995), p. 124.

23. Ibid., p. 149.

24. Jacobs, *Sustainable Development as a Contested Concept*, p. 22.

25. Andrew Dobson, "Environmental Sustainability: A View from the Terraces," *Environmental Politics* 6, no. 3 (1997): 176–179, p. 176.

26. Michael Redclift, *Wasted: Counting the Costs of Global Consumption* (London: Earthscan, 1996), p. 57. For an extended version of this argument see Alan Fricker, "Measuring up to Sustainability," *Futures* 30, no. 4 (1998): 367–375.

27. For a discussion of indicators that attempts to respect the plurality and autonomy of local communities, see J. Gary Lawrence, "Getting the Future that you Want: The Role of Sustainability Indicators," in *Community and Sustainable Development: Participation in the Future*, ed. D. Warburton (London: Earthscan, 1998), 68–80.

28. See the use of soft-hard typology by Marshall, cited in Redclift, *Wasted*, p. 58; minimal-maximal typology in Robyn Eckersley, "Sustainable Development and the Politics of Language," *Canberra Journal of Public Administration* 69 (1992): 36–41; shallow-deep typology in Imogen Zethoven, "Sustainable Development: A Critique of Perspectives," in *Immigration, Population and Sustainable Environments: The Limits to Australia's Growth*, ed. J. W. Smith (Adelaide: The Flinders Press, 1991), 295–320; weak-strong typology in Jacobs, *Sustainable Development as a Contested Concept*, pp. 31–32; very weak-very strong typology in O'Riordan, *Democracy and the Sustainability Transition*, Table 8.1, p. 145. See the more rewarding approach to typing sustainabilities in Kjell Dahle, "Toward Governance for Future Generations: How do we Change Course?," *Futures* 30, no. 4 (1998): 277–292.

29. See my discussion of Pearce in chapter 1.

30. It is in this way, then, that I suggest that we interpret Michael Redclift's claim that debates about sustainability draw upon "two frequently opposed intellectual traditions"—*Sustainable Development: Exploring the Contradictions* (London & New York: Routledge, 1987), p. 199—or Vandana Shiva's assertion that there are "two different meanings of 'sustainability'"—"Rediscovering the Real Meaning of Sustainability," in *The Environment in Question*, eds. D. E. Cooper and J. A. Palmer (London & New York: Routledge, 1992), 187–193, p. 192.

31. David Orr distinguishes between technological sustainability and ecological sustainability in his essay, "Two Meanings of Sustainability," in *Ecological Literacy: Education and the Transition to a Postmodern World* (Albany, NY: State University of New York Press, 1992), ch. 2, pp. 23–40.

32. Redclift, *Sustainable Development*, p. 56. In *Wasted*, he tells of his unease at the use of the notion of 'sustainable development' at the Earth Summit, p. 1.

33. Gerald Schmitz, "Democratization and Demystification: Deconstructing 'Governance' as Development Paradigm," in *Debating Development Discourse: Institutional and Popular Perspectives*, eds. D. B. Moore and G. J. Schmitz (London & New York: MacMillan/St. Martins Press, 1995), 54–90, p. 56.

34. Maarten A. Hajer, *The Politics of Environmental Discourse: Ecological Modernization and the Policy Process* (New York & Oxford: Clarendon, 1995), pp. 8–15.

35. Dennis Pirages, *The New Context for International Relations: Global Ecopolitics* (North Scituate, MA: Duxbury Press, 1978), p. 5.

36. Wolfgang Sachs, "Global Ecology and the Shadow of 'Development'," in *Global Ecology: A New Arena of Political Conflict*, ed. W. Sachs (London & Atlantic Highlands, NJ/Halifax: Zed Books/Fernwood Publishers, 1993), 3–21, p. 3.

37. Ibid., p. 5.

38. Wolfgang Sachs, "Introduction," in *Global Ecology: A New Arena of Political Conflict*, ed. W. Sachs (London & Atlantic Highlands, NJ/Halifax: Zed Books/Fernwood Publishers, 1993), xv–xvii, p. xv.

39. Redclift, *Wasted*, pp. 15–19, 25–28.

40. Tellingly, the South's answer to the Brundtland Commission, the South Commission, talked not about sustainable development but about sustained development. Pratap Chatterjee and Matthias Finger, *The Earth Brokers: Power, Politics and World Development* (London & New York: Routledge, 1994), p. 32.

41. Nicholas A. Robinson (ed.), *Agenda 21 & the UNCED Proceedings, Vol. VI* (New York: Oceana Publications, 1993), p. xxx.

42. Nicholas Hildyard, "Foxes in Charge of the Chickens," in *Global Ecology: A New Arena of Political Conflict*, ed. W. Sachs (London & Atlantic Highlands, NJ/Halifax: Zed Books/Fernwood Publishers, 1993), 22–35, p. 31.

43. Shiva, *Rediscovering the Real Meaning of Sustainability*, p. 188.

44. Worster, *The Wealth of Nature,* p. 146.

45. Global Forum, *Alternative Treaty: Ethical Commitments to Global Ecological Posture and Behaviour* (available at www.igc.apc.org/habitat/treaties/ethics.html, 1992, accessed January 26, 2000).

46. These issues are well discussed in the collections of Majid Rahnema and Victoria Bawtree (eds), *The Post-Development Reader* (London & Atlantic Highlands, NJ: Zed Books, 1997) and Wolfgang Sachs (ed), *The Development Dictionary: A Guide to Knowledge as Power* (London & Atlantic Highlands, NJ: Zed Books, 1992). Theo Coleborn, D. Dumanoski, and John P. Myers' *Our Stolen Future: Are we Threatening our Fertility, Intelligence, and Survival?—A Scientific Detective Story* (London: Little, Brown & Co., 1996) is essential reading on the subject of the systemic risks to human and ecological health posed by contemporary technosystems.

47. O'Riordan, *Democracy and the Sustainability Transition*, pp. 151–152.

48. On the failure of the biosafety conference, see Chee Yoke Ling, "US Behind Collapse of Cartagena Biosafety talks," *Third World Resurgence* 104–105 (1999): 5–7.

49. For discussion of these matters, see, respectively, Redclift, *Wasted*, p. 24; Alain Lipietz, *Green Hopes: The Future of Political Ecology*, trans. M. Slater (Cambridge: Polity Press, 1995), pp. 102–103; Michael McCoy and Patrick McCully, *The Road from Rio: An NGO Action Guide to Environment and Development* (Amsterdam: International Books/WISE, 1993), p. 9; Chatterjee and Finger, *The Earth Brokers*, p. 58.

50. Chatterjee and Finger, *The Earth Brokers*, p. 129. For a critical discussion of the public relations firm employed by the Business Council for Sustainable Development, see Greer and Bruno, *Greenwash*, pp. 28–30.

51. Chatterjee and Finger, *The Earth Brokers*, pp. 114–117.

52. Mark F. Imber, *Environment, Security and UN Reform* (New York: St. Martin's Press, 1994), pp. 144–145.

53. Redclift, *Wasted*, pp. 72–81.

54. See Patricia Adams, *Odious Debts: Loose Lending, Corruption and the Third World's Environmental Legacy* (London: Earthscan/Probe International, 1991).

55. Imber, *Environment, Security and UN Reform*, p. 118.

56. Reg Henry, "Adapting United Nations Agencies for Agenda 21: Programme Coordination and Organisational Reform," *Environmental Politics* 5, no. 1 (1996): 1–24, p. 19.

57. Richard J. Barnet and John Cavanaugh, *Global Dreams: Imperial Corporations and the New World Order* (New York: Touchstone, 1995), p. 427.

58. Chatterjee and Finger, *The Earth Brokers*, p. 111.

59. Marian A. L. Miller, *The Third World in Global Environmental Politics* (Buckingham, UK: Open University Press, 1995), pp. 143–144.

60. Lipietz, *Green Hopes*, p. 99.

61. See, e.g., Ramachandra Guha and J. Martinez-Alier, *Varieties of Environmentalism: Essays North and South* (London: Earthscan, 1997).

62. Redclift, *Sustainable Development*, pp. 73–78.

63. Chakravarthi Raghavan, "What is Globalization?," *Third World Resurgence* 74 (1996): 11–14, p. 13.

64. See the special issue of *Third World Resurgence* entitled "Multilateral Agreement on Investment: A Charter for Corporations, a New Threat to the South," 90/91 (1998).

65. Martin Khor, "Globalization: Implications for Development Policy," *Third World Resurgence* 74 (1996): 15–21; Michael Tanzer, "Globalizing the Economy: The Role of the IMF and the World Bank," *Third World Resurgence* 74 (1996): 22–26.

66. See Gustavo Esteva and Madhu Suri Prakash, *Grassroots Postmodernism: Remaking the Soil of Cultures* (London & New York: Zed Books, 1998), pp. 110-151.

67. See, e.g., Greer and Bruno, *Greenwash*.

68. Maria Mies, "The Need for a New Vision: The Subsistence Perspective," in *Ecofeminism*, M. Mies and V. Shiva (Melbourne: Spinifex, 1993), 297–324, p. 297.

69. Chatterjee and Finger, *The Earth Brokers*, p. 103. See also pp. 79–91.

70. A discussion of the everyday reality of this exclusion is provided in Aidan Davison and Ian Barns, "The Earth Summit and the Ethics of Sustainable Development," *Current Affairs Bulletin* 69, no. 1 (1992): 4–13.

71. See Ian Tellam, "Introduction," in *The Road to Rio: An NGO Action Guide to Environment and Development*, eds. M. McCoy and P. McCully (Amsterdam: International Books/WISE, 1993), 5–7.

72. Barnet and Cavanaugh, *Global Dreams*, p. 429. Although Redclift claims that "divisions within the ranks of NGOs became more apparent, as their public profile grew." *Wasted*, p. 23.

73. Molly O'Meara's recent discussion about "networking for sustainable development" provides interesting examples of the integration of information technology within environmental activism, but I suspect that for each of these examples we can find thousands of instances where the centrifugal forces of information technology emerging from the corporate/managerial core of global technological society fragment and disperse fragile civic webs. "Harnessing Information Technologies for the Environment," in *State of the World 2000*, eds. L. R. Brown, C. Flavin, and H. French (New York & London: W.W. Norton & Co., 2000), 121–141, pp. 136–140.

74. Lipietz, *Green Hopes*, p. 96.

75. Chatterjee and Finger, *The Earth Brokers*, p. 104.

76. Maurice Strong, "Foreword," in *Grassroots Environmental Action: People's Participation in Sustainable Development,* eds. D. Ghai and J. Vivian (London & New York: Routledge, 1992), xiii–xiv, p. xiv.

77. Schmitz, *Democratization and Demystification*, p. 74.

78. Majid Rahnema, "Participation," in *Development Dictionary: A Guide to Knowledge as Power,* ed. W. Sachs (London & New Jersey: Zed Books, 1992), 116–131.

79. Sachs, *Global Ecology and the Shadow of 'Development'*, p. 19.

80. Hildyard, *Foxes in Charge of the Chickens*, p. 23.

81. Esteva and Prakash, *Grassroots Postmodernism,* pp. 19–20.

82. World Commission on Environment and Development, *Our Common Future* (Oxford: Oxford University Press, 1987), p. 1.

83. Yaakov J. Garb, "The Use and Misuse of the Whole Earth Image," *Whole Earth Review,* March (1985), 266–267.

84. Wolfgang Sachs, "The Blue Planet: An Ambiguous Modern Icon, *The Ecologist* 24 (1994): 170–175, p. 171.

85. Vandana Shiva, "The Greening of the Global Reach," in *Global Ecology: A New Arena of Political Conflict*, ed. W. Sachs (London & Atlantic Highlands, NJ/Halifax: Zed Books/Fernwood Publishing, 1993), 149–156, pp. 149–150.

86. Chatterjee and Finger, *The Earth Brokers*, p. 73.

87. Ibid., p. 136.

88. Miller, *The Third World in Global Environmental Politics,* p. 144.

89. Ibid., p. 151.

90. Ibid., p. 149.

91. Robin Broad and John Cavanaugh, *Plundering Paradise: The Struggle for the Environment in the Philippines* (Berkeley, Los Angeles & Oxford: University of California Press, 1993), p. 140.

92. Sachs, *Global Ecology and the Shadow of 'Development'*, p. 12.

93. Esteva and Prakash, *Grassroots Post-Modernism*, p. 36.

94. J. Ronald Engel, "Introduction: The Ethics of Sustainable Development," in *Ethics of Environment and Development: Global Challenge, International Response*, eds. J. R. Engel and. J. G. Engel (London: Belhaven, 1990), 1–23, p. 1.

95. Peterson, *Sharing the Earth,* p. 179.

96. Ibid., p. 85. Peterson does recognize that Earth Summit organisers "began with a terministic screen constructed out of neoclassical economics vocabulary turned to market economists for a revision and ended up pretty much where they began," p. 179.

97. Ibid., p. 52.

98. A lack mirrored in the lack of any reference to radical, Southern critics of development in her study of the Earth Summit.

99. Redclift, *Wasted*, p. 139.

100. Ibid., p. 136.

101. Robert Chambers, *Sustainable Rural Livelihoods: A Strategy for People, Environment, and Development* (Brighton: University of Sussex, 1987). See also the case studies related in Dahram Ghai (ed), *Development and Environment: Sustaining People and Nature* (Cambridge, MA: Blackwell/UNRISD, 1994).

102. Arturo Escobar, "Reflections on 'Development': Grassroots Approaches and Alternative Politics in the Third World," *Futures* 24 (1992): 411–435, p. 412.

103. Esteva and Prakash, *Grassroots Post-Modernism*, p. 199.

104. Wendell Berry, *Sex, Economy, Freedom, and Community: Eight Essays* (New York: Pantheon, 1993), p. 20.

105. Richard B. Norgaard, *Development Betrayed: The End of Progress and a Coevolutionary Revisioning of the Future* (London & New York: Routledge, 1994).

106. Ibid., p. 23.

107. Vandana Shiva, *Staying Alive: Women, Ecology and Development* (London: Zed Books, 1989), ch. 4, pp. 55–95.

108. Shiva, *Rediscovering the Real Meaning of Sustainability*.

109. Ibid., p. 188.

110. David Goodman and Michael Redclift, *Refashioning Nature: Food, Ecology and Culture* (London & New York: Routledge, 1991).

111. Redclift, *Sustainable Development,* p. 171.

112. Ibid., p. 51. Helena Norberg-Hodge illustrates this point wonderfully in her historical account of the technocratic 'development' of Ladakh, in the Western Himalayas, *Ancient Futures: Learning from Ladakh* (London: Rider, 1991).

113. John Peet, *Energy and the Ecological Economics of Sustainability* (Washington DC & Covelo, CA: Island Press, 1992), p. 252.

114. Richard J. Smith, "Sustainability and the Rationalisation of the Environment," *Environmental Politics* 5 (1996): 25–47, p. 42.

115. The idea that moral and political concepts are 'essentially contested' was first laid out by W. B. Gallie in the 1950s and elaborated upon most fully by William E. Connolly in *The Terms of Political Discourse*, 2nd edn. (Oxford: Martin Robertson, 1983), ch. 2, pp. 10–44. A number of environmental theorists have drawn this analysis into the orbit of sustainability discourses including Eckersley, *Sustainable Development and the Politics of Language*; Jacobs, *Sustainable Development as a Contested Concept*; and David Moore, "Development Discourse as Hegemony: Towards an Ideological History 1945–1995," in *Debating Development Discourse: Institutional and Popular Perspectives*, eds. D. B Moore and G. J. Schmitz (London & New York: MacMillan/St. Martins Press, 1995), 1–53, p. 3.

116. Connolly, *The Terms of Political Discourse*, pp. 27–28.

117. Ibid., p. 40.

118. Ibid., pp. 40–41.

119. Michael Zimmerman, *Contesting Earth's Future: Radical Ecology and Postmodernity* (Berkeley, Los Angeles & London: University of California Press, 1994), see pp. 16–17.

120. John Barry, "Sustainability, Political Judgement and Citizenship: Connecting Green Politics and Democracy," in *Democracy and Green Political Thought: Sustainability,*

Rights, and Citizenship, eds. B. Doherty and M. de Geus (London & New York: Routledge, 1996), 115–131, p. 116.

121. See Phil Macnaghten and John Urry's excellent discussion of the way in which narrations of "nature," including that of "modern science," cannot be unbound from embodied, historical socio-cultural practices in *Contested Natures* (London, Thousand Oaks, CA & New Delhi: Sage, 1998).

122. Engel, *Introduction: The Ethics of Sustainable Development*, p. 1.

123. I agree with Richard Norgaard that "Sustainable Development cannot be defined operationally." *Development Betrayed*, p. 20.

124. Orr, *Ecological Literacy*, p. 1.

CHAPTER THREE: METAPHYSICS

1. Carl Mitcham, "The Sustainability Question," in *The Ecological Community: Environmental Challenges for Philosophy, Politics, and Morality*, ed. R. S. Gottlieb (New York & London: Routledge, 1997), 359–379, p. 375.

2. A definition I draw largely from Tarla Rai Peterson, *Sharing the Earth: The Rhetoric of Sustainable Development* (Columbia, SC: University of South Carolina Press, 1997), p. 26.

3. Herbert Marcuse, *One Dimensional Man: Studies in the Ideology of Advanced Industrial Society* (London: Abacus, [1964] 1972), p. 14.

4. Marcuse noted that "when technics becomes the universal form of material production, it circumscribes an entire culture; it projects a historical totality—a 'world'." Ibid., p. 127.

5. See *The Oxford English Dictionary*, prepared by J. A. Simpson and E. S. C. Weiner, 2nd edn. (Oxford: Clarendon Press, 1989), Vol. XVII, p. 327.

6. Donald Worster, *The Wealth of Nature: Environmental History and the Ecological Imagination* (New York & Oxford: Oxford University Press, 1993), p. 146.

7. Gifford Pinchot cited in Derek Wall, *Green History: A Reader in Environmental Literature, Philosophy, and Politics* (London & New York: Routledge, 1994), p. 136.

8. *The Oxford English Dictionary*, Vol. XVII, p. 327.

9. Bill McKibben, *The End of Nature* (Harmondsworth, Middlesex: Penguin, 1990).

10. Julian Simon, "Scarcity or Abundance?" in *The Business of Consumption: Environmental Ethics and the Global Economy*, eds. L. Westra and P. H. Werhane (Lanham, Maryland: Rowman & Littlefield, [1994] 1998), 237–245.

11. World Commission on Environment and Development, *Our Common Future* (Oxford: Oxford University Press, 1987), p. 8.

12. Indeed, some such as John Gowdy persist in the myopic belief that "the notion of progress is all but dead among intellectuals. See his "Progress and Environmental Sustainability," *Environmental Ethics* 16 (1994): 41–55, p. 42.

13. William Ophuls, *Ecology and the Politics of Scarcity* (San Francisco: W. H. Freeman, 1977), p. 2.

14. Ibid., p. 3. Warren Johnson, writing around the same time, similarly claimed that "[s]carcity is the mechanism that is inexorably diverting industrial society from the path of sustained growth that has characterized the modern era." *Muddling Toward Frugality* (Boulder, CO: Shambhala, [1978], 1979), p. 12.

15. For a recent discussion of this literature, see Robert U. Ayers, *Turning Point: An End to the Growth Paradigm* (London: Earthscan, 1998).

16. Pieter Tijmes and Reginald Luijf, "The Sustainability of Our Common Future: An Inquiry into the Foundations of an Ideology," *Technology in Society* 17 (1995): 327–336, p. 330.

17. Ibid., p. 331.

18. Marcuse, *One Dimensional Man*, p. 28.

19. Frederick E. Dessauer, *Stability* (New York: MacMillan, 1949), p. 9.

20. Ibid., p. 269.

21. See Morris Berman, *The Reenchantment of the World* (London: Cornell University Press, 1981), pp. 27–37.

22. René Descartes, *Discourse on Method and the Meditations*, trans. by F. E. Sutcliffe (Harmondsworth, Middlesex: Penguin Classics, [1637] 1968).

23. Francis Bacon, *The New Organon and Related Writings*, ed. F. H. Anderson (Indianapolis & New York: Bobbs-Merrill Co., [1620] 1960); *The Advancement of Learning and New Atlantis*, intro. T. Case, The World's Classics Vol. 93 (London: Oxford University Press, [1627] 1974).

24. Neil Postman, *Technopoly: The Surrender of Culture to Technology* (New York: Alford A. Knopf, 1992), p. 35.

25. See Berman, *The Reenchantment of the World*, pp. 16–17 and Carolyn Merchant, *The Death of Nature* (San Francisco: Harper & Row, 1980), p. 223.

26. Dessauer, *Stability*, p. 18.

27. Mitcham, *The Sustainability Question*, p. 360.

28. St. Augustine and Roger Bacon cited in David F. Noble, *The Religion of Technology: The Divinity of Man and the Spirit of Invention* (Harmondsworth, Middlesex: Penguin [1997] 1999), pp. 12, 28. Noble goes on to argue that the medieval turn toward technology as transcendence is completed and transferred into the "emergent mentality of modernity" largely "through the enormous and enduring influence of Francis Bacon," p. 53.

29. Erigena, cited in Ibid., p. 17. This court philosopher is credited with introducing the term *mechanical arts*, a precursor to the modern term *technology*, into European languages, p. 15.

30. Lewis Mumford, *Technics and Civilization* (London: Routledge & Kegan Paul, 1934), p. 17. For a more recent discussion of the changes initiated by mechanical timekeeping, see Arnold Pacey, *The Maze of Ingenuity: Ideas and Idealism in the Development of Technology*, 2nd edn. (Cambridge, MA: MIT Press, 1992), pp. 63–86.

31. Tijmes and Luijf, *The Sustainability of Our Common Future*, p. 328

32. John Maynard Keynes, cited in Hans Achterhuis, "Scarcity and Sustainability," in *Global Ecology: A New Arena of Political Conflict*, ed. W. Sachs (London & Atlantic Highlands, NJ/Halifax: Zed Books/Fernwood Publishers, 1993), 104–116, p. 109.

33. Freya Mathews, *The Ecological Self* (London: Routledge, 1991), p. 25.

34. See, e.g., the chapter on Malthus and Ricardo in Robert L. Heilbroner, *The Worldly Philosophers: The Lives, Times and Ideas of the Great Economic Thinkers*, 4th edn. (New York: Simon & Schuster, 1972), ch. IV, pp. 73–101. William Ophuls provides a discussion of Edmund Burke in *Ecology and the Politics of Scarcity*, pp. 233–234.

35. John Stuart Mill, cited in Wall, *Green History*, p. 121.

36. Gowdy, *Progress and Environmental Sustainability*, p. 41.

37. Mitcham, *The Sustainability Question*, pp. 361–362.

38. See Herman Daly's essay "The Steady-State Economy: Alternative to Growthmania," in his *Steady-State Economics*, 2nd edn. (Washington DC & Covelo, CA: Island Press, 1991), pp. 180–194.

39. Achterhuis, *Scarcity and Sustainability*, p. 106.

40. Murray Bookchin, *Post-Scarcity Anarchism* (Palo Alto, CA: Ramparts Press, 1971), p. 11.

41. Nicholas Rescher, *Unpopular Essays on Technological Progress* (Pittsburgh: University of Pittsburgh Press, 1980), p. 12.

42. See, especially, David Bohm's *Wholeness and the Implicate Order* (London & Boston: Routledge & Kegan Paul, 1980). For a lay person's overview of this literature, see John Gribben and Paul Davies, *The Matter Myth: Beyond Chaos and Complexity* (Harmondsworth, Middlesex: Penguin, 1992).

43. Fritjof Capra, *The Tao of Physics* (London: Wildwood House, 1975); Gary Zukav, *The Dancing Wu Li Masters: An Overview of the New Physics* (New York: Bantam, 1979). See also the fascinating interplay of science and mysticism in J. Krishnamurti and David Bohm, *The Ending of Time: Thirteen Dialogues* (London: Victor Gollancz, 1988).

44. Fritjof Capra, *The Turning Point: Science, Society, and the Rising Culture* (London: Fontana, 1983), p. xviii.

45. Ibid, pp. 230–231.

46. Thomas Kuhn, *The Structure of Scientific Revolutions* (Chicago: University of Chicago Press, 1970).

47. See, e.g., Jacob Norman, "Towards a Theory of Sustainability," *The Trumpeter* 6 (1989): 93–97, and Ingrid L. Stefanovic, "Evolving Sustainability: A Re-thinking of Ontological Foundations," *The Trumpeter* 8 (1991): 194–200. For a more practical account of the application of holism to thinking about sustainability, see Alan Savory, *Holistic Resource Management* (Covelo, CA: Island Press, 1988).

48. See Michael Zimmerman's interesting discussion of these movements and their ambiguous debt to modernism, and thus their vulnerability to ecomodernism, in *Contesting Earth's Future: Radical Ecology and Postmodernity* (Berkeley, Los Angeles & London: University of California Press, 1994), ch. 2, pp. 57–90.

49. See, for instance, the approach of the International Institute for Applied Systems Analysis (IIASA) in William C. Clark and R. Edward Munn, *Sustainable Development of the Biosphere* (New York: Cambridge University Press/IIASA, 1986).

50. McKibben, *The End of Nature*, p. 54.

51. See John Allen, *Biosphere 2: The Human Experiment* (New York: Penguin, 1991). Citing Lewis Mumford as an inspiration, Allen claims that "[b]ringing technology in the service of life has been one of the most revolutionary aspects of the making of Biosphere 2," p. 71. However, this rhetoric falls away at the end of this book to reveal the guiding vision of cosmic colonization: "This ability of humans to move off Earth and establish themselves in permanent communities would assure the evolutionary expansion of Earth's biospheric life as well as satisfy our insatiable instinct to explore," p. 151. This project has now been abandoned as a failure, and the complexity of ecological systems ensures that such biospheric technology is a long way, if ever, from being realized. See the editorial, "The Origins of Biosphere 2," *The Ecologist* 25, no. 4 (1995): 158–162.

52. Allen, *Biosphere 2*, p. 151.

53. Guy Beney, "'Gaia': The Globalitarian Temptation," in *Global Ecology: A New Arena of Political Conflict,* ed. W. Sachs (London & Atlantic Highlands, NJ/Halifax: Zed Books/Fernwood Publishing, [1991] 1993), 179–193, p. 188.

54. Ivan D. Illich, *Tools for Conviviality* (London: Calder & Boyars, 1973), pp. 100–101.

55. For the technoutopian Buckminster Fuller, "[t]he universe is a comprehensive system of technology. Humanity is discovering and beginning to employ it." "Technology and the Human Environment," in *The Ecological Conscience*, ed. R. Disch (Englewood Cliffs, NJ: Prentice Hall, 1970), 174–180, p. 178.

56. Beney, *'Gaia'*, p. 188.

57. This phrase forms a chapter title in James Lovelock, *Gaia: The Practical Science of Planetary Medicine* (Sydney: Allen & Unwin, 1991), ch. 8, pp. 153–171. More recently, it can be found as a chapter title in David Shearman and Gary Sauer-Thompson, *Green or Gone* (Kent Town, South Australia: Wakefield Press, 1997) and as the theme of Van B. Weigel's *Earth Cancer* (Westport, CT: Preager, 1995). Characteristically, these authors speak of a generic "humanity" and draw heavily upon unitary, biologically determinist accounts of human evolution in doing so, where there is often a crucial need to acknowledge the historic hegemony of Eurocentrism. The parallels between Lovelock's Gaian argument and Garrett Hardin's earlier account of Lifeboat ethics is clear. See Hardin's "Living on a Lifeboat," *BioScience* 24 (1974): 561–568.

58. Lovelock, *The Ages of Gaia*, pp. 210–211.

59. Beney, *'Gaia'*, p. 187.

60. Lovelock does not himself advocate that humanity attempt to become managers of Spaceship Earth. His reasoning seems less based on the moral claim that humanity has no right to this position, however, and more on a cynical, and perhaps even misanthropic, pessimism: thus, "I would sooner expect a goat to succeed as a gardener than expect humans to become responsible stewards of the Earth," *Gaia*, p. 186. For further evidence of Lovelock's neglect of the force of technological imperatives which will shift geophysiology toward global managerialism see *The Ages of Gaia*, ch. 7, pp. 152–182.

61. Peter Russel, cited in Beney, *Gaia*, p. 190. For an unrestrainedly technoutopian account of the "cyborgisation" of people and planet, see the futurist Jerome C. Glenn's book *Future Mind: Artificial Intelligence, Merging the Mystical and the Technological in the 21st Century* (Washington DC: Acropolis Books, 1989). Indeed, scientific Gaians enjoy speculating about the prospects for humanity, the nervous system of the cosmos, to give

birth to a new Gaian system on Mars. See, e.g., Lynn Margulis's essay "Gaia and the Colonization of Mars" in Lynn Margulis and Dorian Sagan, *Slanted Truths: Essays on Gaia, Symbiosis, and Evolution* (New York: Springer-Verlag, 1997), pp. 221–234.

62. Murray Bookchin, "An Open Letter to the Ecological Movement," *Social Alternatives* 2, no. 3 (1982): 13–16, p. 14.

63. Three of the many recent anthologies in environmental philosophy are Anthony Weston (ed), *An Invitation to Environmental Philosophy* (Oxford & New York: Oxford University Press, 1999); Joseph DesJardins, *Environmental Ethics: Concepts, Policy, Theory* (Mountain View, CA: Mayfield Pub. Co., 1999); Michael E. Zimmerman (general ed), *Environmental Philosophy: From Animal Rights to Radical Ecology,* 2nd edn. (Englewood Cliffs, NJ: Prentice Hall, 1998).

64. The language of sustainable development emerged with little recognition from radical ecophilosophers that environmentalism was being coopted within a modernist agenda. A revealing justification of the early isolation of environmental ethics from everyday environmentalism can be found in Eugene Hargrove, "On Reading Environmental Ethics," *Environmental Ethics* 6, no. 4 (1984): 291–292. A recent paper in the same journal that shows the benefits of drawing environmental philosophy into sustainability contests is that of Christopher B. Barrett and Ray Grizzle, "A Holistic Approach to Sustainability Based on Pluralism Stewardship," *Environmental Ethics* 21, no. 1 (1999): 22–42.

65. Zimmerman, *Contesting Earth's Future,* p. 375; Carolyn Merchant, *Radical Ecology: The Search for a Livable World* (London & New York: Routledge, 1992), p. 212.

66. Marcuse, *One Dimensional Man*; Lewis Mumford, *The Myth of the Machine*: Vol. 1, *Technics and Human Development*, Vol. 2, *The Pentagon of Power* (New York: Harcourt Brace Jovanovich, 1967, 1970); William Leiss, *The Domination of Nature* (New York: George Braziller, 1972); Merchant, *The Death of Nature*; Berman, *The Reenchantment of the World*; Murray Bookchin, *The Ecology of Freedom: The Emergence and Dissolution of Hierarchy* (Palo Alto, CA: Cheshire Books, 1982); Shiva, *Staying Alive*; Val Plumwood, *Feminism and the Mastery of Nature* (London & New York: Routledge, 1993); Noble, *The Religion of Technology*.

67. Plumwood, *Feminism and the Mastery of Nature*, p. 31.

68. Ibid., p. 2.

69. Examples of this confusion abound. See, for instance, the preface and first chapter in John R. Smythies and John Beloff (eds), *The Case for Dualism* (Charlottesville: University Press of Virginia, 1989).

70. Plumwood, *Feminism and the Mastery of Nature*, p. 47.

71. See Richard Warner and Tadeusz Szubka (eds), *The Mind-Body Problem: A Guide to the Current Debate* (Oxford & Cambridge, MA: Blackwell, 1994).

72. See, e.g., Drew Leder, *The Absent Body* (Chicago: University of Chicago Press, 1990).

73. Richard Rorty, *Philosophy and the Mirror of Nature* (Princeton: Princeton University Press, 1979), p. 357.

74. See, e.g., Lynda M. Glennon, *Women and Dualism: A Sociology of Knowledge Analysis* (New York: Longman, 1979).

75. I note that there are those, such as Gordon P. Baker and Katherine J. Morris in *Descartes' Dualism* (London & New York: Routledge, 1996), who suggest that Descartes did not advocate what has become known as Cartesian dualism: an argument that returns us with unease to Descartes' own plea for "future generations never to believe that the things people tell them come from me, unless I myself have published them." *Discourse on Method and the Meditations*, p. 84. Nonetheless, even if there is some historical accuracy in Baker's and Morris's claim—and interpretations of historical philosophical texts should always be structured so as to self-reflectively display their interpretive character—dualism, as described by Plumwood, seems to me an accurate account of the founding logic of modernist epistemology.

76. Plumwood, *Feminism and the Mastery of Nature*, ch. 3, pp. 69–103. Her attempt to trace dualism back to the Greek roots of philosophy is not without precedent. See, e.g, Albert Borgmann's claim that Plato was working open the tiny rupture between culture and nature established in classical thought by Thales in "The Nature of Reality and the Reality of Nature," in *Reinventing Nature? Responses to Postmodern Deconstruction*, eds. M. E. Soulé and G. Lease (Washington DC & Covelo, CA: Island Press, 1995), 31–45, p. 32; Hannah Arendt's discussion of the dualism between contemplation and action in Socratic philosophy in *The Human Condition* (Chicago & London: University of Chicago Press, 1958); and Berman's claim that Plato helped establish subject/object dualism in *The Reenchantment of the World*, pp. 78–79.

77. Descartes, *Discourse on Method and the Meditations,* Discourse 6, p. 78. This is founded on his claim in the previous dialogue that "there is nothing which leads feeble minds more readily astray from the straight path of virtue than to imagine that the soul of animals is of the same nature as our own . . . when one knows how much they differ, one can understand much better the reasons which prove that our soul is of a nature entirely independent of the body," Discourse 5, p. 76.

78. Plumwood, *Feminism and the Mastery of Nature*, pp. 45–46.

79. Ibid., p. 5.

80. Ibid., p. 22.

81. Ibid., p. 48.

82. Here Plumwood cites Aristotle's defense of slavery in *Politics* (Book 1), Ibid., p. 46.

83. Ibid., p. 140.

84. Plumwood's criticisms of deep ecology are overstated, however, and again reflect the penchant of radical ecophilosophers to denigrate their own rather than concentrate more purposefully on challenging the project of ecomodernization. For instance, one of the most philosophically sophisticated elaborations of the ontological therapy implied by deep ecology, provided by Freya Mathews, includes a detailed framework for dealing with individuation within a monist totality. The task of holistic thinking, in Mathews's view, is not to absorb difference within some metaphysical principle of the 'Whole', but to develop forms of human culture that embody the reality of our multilayered, that is, in my shorthand, our ideocentric, homocentric, biocentric, and cosmocentric, selfhood. *The Ecological Self*, ch. 4, 117–163. That Mathews's argument translates to a rich communitarian politics is evident in her "Community and the Ecological Self," in *Ecology and Democracy*, ed. F. Mathews (London & Portland, OR: Frank Cass, 1996), 66–100.

85. Plumwood, *Feminism and the Mastery of Nature*, p. 128.

86. Ibid., p. 126.

87. Ibid., p. 193.

88. Ibid., p. 195.

89. Ibid., p. 194.

90. Plumwood here builds on the efforts of those such as Carol Gilligan who have pioneered a feminist critique of modern moral philosophy. See her "Remapping the Moral Domain: New Images of Self in Relationship," in *Mapping the Moral Domain,* eds. C. Gilligan et al. (Cambridge, MA: Harvard University Press, 1988), 3–20.

91. Plumwood, *Feminism and the Mastery of Nature*, p. 134.

92. Ibid., p. 160.

93. Ibid., p. 154.

94. Ibid., p. 146.

95. Ibid., p. 186.

96. See, for instance, John Carroll's discussion of shopping as therapy and the mythology of computers, cars, and celebrities such Princess Diana in *Ego and Soul: The Modern West in Search of Meaning* (Sydney: Harper Collins, 1998).

97. Although Plumwood provides a list of over twenty dualist pairs, which admittedly she doesn't claim to be exhaustive, none of these dualisms figure here or at any point in her discussion. *Feminism and the Mastery of Nature*, p. 43.

98. Anthony Weston's essay "Before Environmental Ethics," *Environmental Ethics* 14, Winter (1992): 321–338, represents by far the most thoughtful attempt to make this point in environmental philosophy that I know of. Weston's critique of many attempts to move beyond anthropocentrism can be directly transposed to the theme of dualism I develop here. I agree that environmental philosophers will benefit from giving up pretensions to building novel and self-contained theoretical structures, and in their stead taking up the more modest, but eminently practical, task of articulating new values in the context of practical possibilities for their cultural constitution, in this way participating in the actual, historical coevolution of moral ideas and moral experience.

99. Fritjof Capra, absorbed as he is with Eastern accounts of mystical enlightenment, suggests that Descartes's philosophical dualism arose "after several hours of intense concentration, during which he [Descartes] reviewed systematically all the knowledge he had accumulated, he perceived, in a sudden flash of intuition, the 'foundations of a marvellous science' which promised the unification of all knowledge." *The Turning Point*, p. 42.

100. Arendt, *The Human Condition*, p. 273.

101. See Don Ihde, *Technology and the Lifeworld: From Garden to Earth* (Bloomington & Indianapolis: Indiana University Press, 1990), pp. 48–58.

102. Lynn White, Jr., *Medieval Technology and Social Change* (Oxford: Oxford University Press, 1962), argued that medieval historians have significantly underestimated the extent of diffusion of technologies from the technologically fecund Orient to the Occident in the Middle Ages.

103. Arendt describes the combination of the defeat of distance by global exploration, the innerworldy focus of the Reformation and the Archimedean point revealed by Galileo's telescope as the decisive events in the establishment of the world alienation characteristic of dualism. *The Human Condition,* ch. VI, 248–325.

104. For a tangential discussion of Descartes' scepticism about vision, see Dalia Judovitz, "Vision, Representation, and Technology in Descartes," in *Modernity and the Hegemony of Vision*, ed. D. M. Levin (Berkeley, Los Angeles & London: University of California Press, 1993), 63–86.

105. Freya Mathews, "The Soul of Things," in *Sense of Place* (Proceedings, Hawkesbury Colloquium, New South Wales, December 1996), 133–142, p. 134. In this regard, Arne Naess's response to David Rothenberg's book, *Hand's End: Technology and the Limits of Nature* (Berkeley: University of California Press, 1993) is also welcome. See *The Influence of Technology Upon How We Think and Feel About Nature: Reflections Upon Reading a Book on Technology* (Intramural Note, unpublished, March, 1994).

106. See Langdon Winner's fine essay "The State of Nature Revisited," in *The Whale and the Reactor: A Search for Limits in an Age of High Technology* (Chicago & London: University of Chicago Press, 1986), ch. 7, pp. 121–137. See also the essays in George Robertson, Melinda Mash, Lisa Tickner, Jon Bird, Barry Curtis, and Tim Putnam (eds), *FutureNatural: Nature, Science, Culture* (London & New York: Routledge, 1996).

CHAPTER FOUR: BUILDING A DEFORMED WORLD

1. David W. Orr, *Ecological Literacy: Education and the Transition to a Postmodern World* (Albany, NY: State University of New York Press, 1992), p. 15

2. César Cuello Nieto and Paul T. Durbin, "Sustainable Development and Philosophies of Technology," in *Technology and Ecology: The Proceedings of the VII International Conference of the Society for Philosophy and Technology*, eds. L. A. Hickman and E. F. Porter (Carbondale, IL: STP, Southern Illinois University, 1993), 215–239, p. 234.

3. David Strong, *Crazy Mountains: Learning from Wilderness to Weigh Technology* (Albany, NY: State University of New York Press, 1995), p. 80.

4. Langdon Winner, *The Whale and the Reactor: A Search for Limits in an Age of High Technology* (Chicago & London: University of Chicago Press, 1986), p. 3.

5. Ibid., p. 4.

6. A summary of the history of philosophy of technology as a discrete subdiscipline, originating in the U.S. with a series of conferences in the mid-1960s, can be found in Carl Mitcham, "What is the Philosophy of Technology?," *International Philosophical Quarterly* 25 (1985): 73–88. Two landmarks in the emergence of this subdiscipline are (i) the anthology, Carl Mitcham and Robert Mackay (eds), *Philosophy and Technology: Readings in the Philosophical Problems of Technology* (New York: The Free Press, 1972) and (ii) the inaugural volume of the International Society for Philosophy and Technology's annual journal *Research in Philosophy and Technology* (Greenwich: JAI Press, 1978) which contains introductions to the philosophy of technology by Albert Borgmann, Paul Durbin, and Robert McGinn. The founding texts of the philosophy of technology include Albert Borgmann, *Technology and the Character of Contemporary Life: A*

Philosophical Inquiry (Chicago: University of Chicago Press, 1984); James K. Feibleman, *Technology and Reality* (den Hague: Martinus Nijhoff, 1982); Don Ihde, *Technics and Praxis: A Philosophy of Technology* (Dordrecht: D. Reidel, 1979); Frederick Rapp, *Analytical Philosophy of Technology* (Dordrecht: D. Reidel, 1981); Langdon Winner, *Autonomous Technology: Technics Out-of-Control as a Theme in Political Thought* (Massachusetts: MIT Press, 1977). Of more recent note are (i) the introductory texts, Frederick Ferré, *Philosophy of Technology* (Englewood Cliffs, NJ: Prentice Hall, 1988); Don Ihde, *Philosophy of Technology: An Introduction* (New York: Paragon, 1993); Carl Mitcham, *Thinking Through Technology: The Path Between Engineering and Philosophy* (Chicago: University of Chicago Press, 1994); Andrew Feenberg, *Questioning Technology* (London & New York: Routledge, 1999); and (ii) the Series in the Philosophy of Technology, edited by Don Ihde and published by Indiana University Press since 1990. This subdiscipline has, to date, been dominated by North American males. For a European perspective, see Wim Zweers and Jan J. Boersema (eds), *Ecology, Technology and Culture* (Cambridge, UK: The White Horse Press, 1994). For a collection by English philosophers, and one that opens by questioning Ihde's definition of the philosophy of technology, see Roger Fellows (ed), *Philosophy and Technology* (Cambridge: Cambridge University Press, 1995). For evidence that the gap between feminist critiques of technology and the philosophy of technology is closing see Joan Rothschild (guest ed) and Fred Ferré (series ed), "Technology and Feminism," *Research in Philosophy and Technology* 13 (Greenwich, CT: JAI Press, 1993) and Patrick D. Hopkins (ed), *Sex/Machine: Readings in Culture, Gender, and Technology* (Bloomington: Indiana University Press, 1998).

7. Since Marx re-established *praxis* as a primary philosophical subject, philosophers as diverse as Max Weber, John Dewey, Nicholas Berdayev, Ortega y Gasset, George Grant, Maurice Merleau-Ponty, Theodore Adorno, Martin Heidegger, Jacques Ellul, Hannah Arendt, Herbert Marcuse, Hans-Georg Gadamer, and Jürgen Habermas have laid, often unwittingly, the foundations for the philosophy of technology. For an overview, see Ihde, *Philosophy of Technology*, pp. 32–38. For detailed studies of some of those in this list, see Ian H. Angus, *George Grant's Rejoinder to Heidegger: Contemporary Political Philosophy and the Question of Technology* (New York: Edwin Mellen, 1987); Andrew Feenberg, *A Critical Theory of Technology* (Oxford: Oxford University Press, 1991); Larry Hickman, *John Dewey's Pragmatic Technology* (Bloomington & Indianapolis: Indiana University Press, 1990). For an acknowledgment of the much older philosophical concern with technology in the German and to lesser extent French languages, see Mitcham, *What is the Philosophy of Technology?*

8. A fact that testifies to the lingering power of the dualism of theory and practice in Academia. Ihde, *Philosophy of Technology*, pp. 17–22.

9. During the 1990s, however, the philosophy of technology did become increasingly aligned with the literature of radical ecophilosophy. See, e.g., Alan Drengson, *The Practice of Technology: Exploring Technology, Ecophilosophy, and Spiritual Disciplines for Vital Links* (Albany, NY: State University of New York Press, 1995); David Rothenberg, *Hand's End: Technology and the Limits of Nature* (Berkeley: University of California Press, 1993); Strong, *Crazy Mountains*. This convergence is also evident in the anthology of Kristin Schrader-Frechette and Laura Westra (eds), *Technology and Values* (Lanham, MD: Rowman & Littlefield Pub., 1997) and the edited collection of David Macauley (ed), *Minding Nature: The Philosophers of Ecology* (New York & London: The Guilford Press,

1996). But we can be a little sceptical of Drengson's claim that "[a]ccording to ecophilosophy, the lack of a comprehensive understanding of technology is a major problem," *The Practice of Technology*, p. 45. While it is certainly true that ecophilosophy has deplored our technological alienation from nature, and while this remark can certainly be applied to Drengson's own work—his writings on the philosophy of technology extend back over two decades—direct philosophical considerations of technology have until very recently been sparse and superficial in this literature. Further, Drengson's adoption of the work of a number of technology critics, notably, Fritz Schumacher, Langdon Winner, and Ivan Illich, into the deep ecology literature is somewhat problematic as these writers do not necessarily seek to defend the ecocentric metaphysics Drengson has in mind. See, for instance, Winner's review of deep ecology in his essay "The State of Nature Revisited," in *The Whale and the Reactor*, ch. 7, pp. 121–137. The fact that the ecophilosopher with the most sustained philosophical interest in technology, Murray Bookchin, has been one of deep ecology's most virulent critics should also not be missed.

10. Anthony Giddens, *Modernity and Self Identity: Self and Society in the Late Modern Age* (Stanford: Stanford University Press, 1991), p. 27.

11. Sean Cubitt, "Supernatural Futures: Theses on Digital Aesthetics," in *FutureNatural: Nature, Science, Culture*, eds. G. Robertson, M. Mash, L. Tickner, J. Bird, B. Curtis, and T. Putnam (London & New York: Routledge, 1996), 237–255, p. 237.

12. Giddens, *Modernity and Self-Identity*, p. 27.

13. Indeed, the distinction between the modern and premodern is frequently drawn too sharply: it remains complex and problematic. Three recent books that make this point by uncovering the inherent coherence between the latest information high-tech and premodern Christian yearnings are David F. Noble, *The Religion of Technology: The Divinity of Man and the Spirit of Invention* (Harmondsworth, Middlesex: Penguin [1997] 1999); William A. Stahl, *God and the Chip: Religion and the Culture of Technology* (Waterloo, Ontario: Wilfred Laurier University Press, 1999); Margaret Wertheim, *The Pearly Gates of Cyberspace: A History of Space from Dante to the Internet* (Sydney, New York & London: Doubleday, 1999).

14. Hans Jonas, "Toward a Philosophy of Technology," in *Technology as a Human Affair*, ed. L. Hickman (New York: McGraw-Hill, [1979] 1990), 40–72, p. 41.

15. My understanding of correlational practice is indebted to David Strong's description of the "need for an account of technology as correlational environment" in *Crazy Mountains*, p. 87.

16. Winner, *The Whale and the Reactor*, pp. 5–10.

17. Carl Mitcham, "Types of Technology," in *Research in Philosophy and Technology* 1 (Greenwich, CT: JAI Press, 1978): 229–294, pp. 234–256. Modified in Mitcham, *Thinking Through Technology*.

18. Ibid.

19. See Mitcham's diagrammatic representation of these modes in Figure 1, *Thinking Through Technology*, p. 160.

20. My constitutive analysis of technology is indebted to many rich conversations with Ian Barns. See, for instance, his "Technology and Citizenship," in *Poststructuralism,*

Citizenship, and Social Policy, eds. A. Peterson, I. Barns, J. Dudley, and P. Harris (London: Routledge, 1999), 154–198.

21. *The Oxford English Dictionary, Vol.* XVII (Oxford: Clarendon Press, 1989). In fact, prior to the eighteenth century, the English term *technology* (derived from the Greek *technologia* in the sixteenth century) referred to "a discourse or treatise on an art or arts." In the now prevalent everyday use of this term, however, the distinction between the scientific study of the "practical arts" and the knowledge, actions, and artefacts associated with these arts is blurred. Both French and German retain greater clarity through the use of two distinct terms; one to denote systematic scientific study (*technologie* in both) and the other to indicate the object of this study (French *technique*; German *technik*). The classical (especially Aristotelian) philosophical interest in *techne* (the Greek root of the term *technology*) was matched by almost total neglect in the modern era until recently. For instance, William Fleming's *The Vocabulary of Philosophy* (London: Charles Griffin & Co., 1876) and the postwar *The Encyclopedia of Philosophy*, ed.-in-chief P. Edwards, Vols. 1–8 (New York: MacMillan & The Free Press, 1967) make no mention whatsoever of topics such as *techne*, practice (and the related *poiesis*, *praxis*, *phronesis*) or technology, while even recent dictionaries such as Thomas Mautner's *A Dictionary of Philosophy* (Oxford & Cambridge: Blackwell, 1996) make only the most perfunctory of remarks about *techne*.

22. Albert Borgmann tells us that "[i]nformation is about to overflow and suffocate reality." *Holding on to Reality: The Nature of Information at the Turn of the Millennium* (Chicago: University of Chicago Press, 1999), p. 213. See also pp. 1–6.

23. Winner, *Autonomous Technology*, p. 10.

24. See, e.g., Ulrich Beck, *Ecological Politics in an Age of Risk*, trans. Amos Weisz (Cambridge: Polity Press, 1995).

25. Hugh MacKay, *Turning Point: Australians Choosing their Future* (Sydney: MacMillan, 1999), p. 237.

26. George Orwell, *The Road to Wigin Pier: The Complete Works of George Orwell*, Vol. 5 (London: Secker & Warburg, [1937] 1986), p. 191.

27. Winner, *The Whale and the Reactor*, p. 10.

28. Orwell, *The Road to Wigin Pier*, p. 189.

29. For a broad analysis of technological determinism in political thought, see Winner, *Autonomous Technology*. Bookchin, in his 1965 essay, "Towards a Liberatory Technology," *in Post-Scarcity Anarchism* (London: Wildwood House [1971] 1974), 85–139, argued that while a positive determinism was dominant in the 1920s and 1930s, in the 1950s and early 1960s, "[t]o an ever-growing extent, technology is viewed as a demon, imbued with a sinister life of its own," p. 86.

30. See, for instance, Arthur C. Clarke's *Profiles of the Future: Enquiry into the Limits of the Possible* (London: Victor Gollancz, 1962).

31. For an overview, see William Kuhns, *The Post-Industrial Prophets: Interpretations of Technology* (New York: Weybright & Talley, 1971). Marshall MacLuhan would seem to have occupied an interesting (and influential) place somewhere between these extremes. Mumford and Marcuse also displayed both a dominant negative determinism and a marginal positive determinism. See Mumford's "Authoritarian and Democratic Tech-

nics," *Technology and Culture* 5 (1964): 1–8, and Marcuse's *An Essay on Liberation* (Boston: Beacon Press, 1969).

32. One of the most persistent critics of this literature has been the engineer Samuel Florman. See *Blaming Technology: The Irrational Search for Scapegoats* (New York: St. Martin's Press, 1981). Many leftist radicals, such as Bookchin, have also had harsh words for those that denied the liberatory potential of technology; see *Towards a Liberatory Technology*, p. 86.

33. Kirkpatrick Sale, *Rebels Against the Future: The Luddites and their War on the Industrial Revolution, Lessons for the Computer Age* (Reading, MA, Menlo Park, CA & New York: Addison-Wesley, 1995). Neo-Luddite discussions that seek to recover the true spirit of reflective scepticism and political activism of the Luddites continue to flourish (as reflected in organizations such as the Jacques Ellul Society). See the dialogues (that include Wendell Berry, Jerry Mander, Vandana Shiva, and Winner) in Stephanie Mills (ed), *Turning Away from Technology: A New Vision for the 21st Century* (San Francisco: Sierra Books, 1997).

34. The essence of Ellul's argument can be seen in his claim that "[w]hat is established is not the subordination of man to technology . . . but, far more deeply, a new totality." *The Technological System*, trans. J. Neugroschel (New York: Continuum, 1980), p. 203. As David Lovekin demonstrates in *Technique, Discourse, and Consciousness: An Introduction to the Philosophy of Jacques Ellul* (London & Toronto: Associated University Presses, 1991), there is for Ellul no dualism between the external reality of artefacts and the internal reality of human consciousness. Technology does not represent for him an external force acting on an essentially human domain. Rather, modern technology constitutes this domain from within in a way that suppresses humanity's dialectical awareness of the tension between consciousness and object, meaning and symbol, word and image, God and humanity, heaven and earth. That is, modern technology is nondialectical: it points only to itself. Lovekin discusses the idea of *lethotechny* on p. 98.

35. As Ellul tells his readers in the preface to *The Technological Society* (New York: Vintage Books, 1964), his purpose in describing *la technique* was to "call the sleeper to awake," p. xxxiii.

36. For a collection of the work of many leading figures in constructivist STS studies, see Sheila Jasanoff, Gerald E. Markle, James C. Peterson, and Trevor Pinch (eds), *Handbook of Science and Technology Studies* (Thousand Oaks, CA: Sage, 1995). The longstanding journal *Technology and Culture* has been the site of much constructivist debate. See Terry S. Reynolds and Stephen H. Cutcliffe (eds), *Technology and the West: A Historical Anthology from Technology and Culture* (Chicago & London: University of Chicago Press, 1997).

37. See the early essays in Joan Rothschild (ed), *Machina ex Dea: Feminist Perspectives on Technology* (New York: Permagon, 1983) and the more recent contributions in Keith Grint and Rosalind Gill (eds), *The Gender-Technology Relation: Contemporary Theory and Research* (London: Taylor & Francis, 1995).

38. Arnold Pacey, *Meaning in Technology* (Cambridge, MA & London: MIT Press, 1999).

39. Langdon Winner, "Technology Today: Utopia or Dystopia?," *Social Research* 64, no. 3 (1997): 989–1017, p. 998.

40. Ibid., pp. 998–1000.

41. I have modified Mitcham's model, in *Thinking Through Technology,* Figure 1, p. 160, so as to express the shared dualistic parentage of instrumental and determinist explanations:

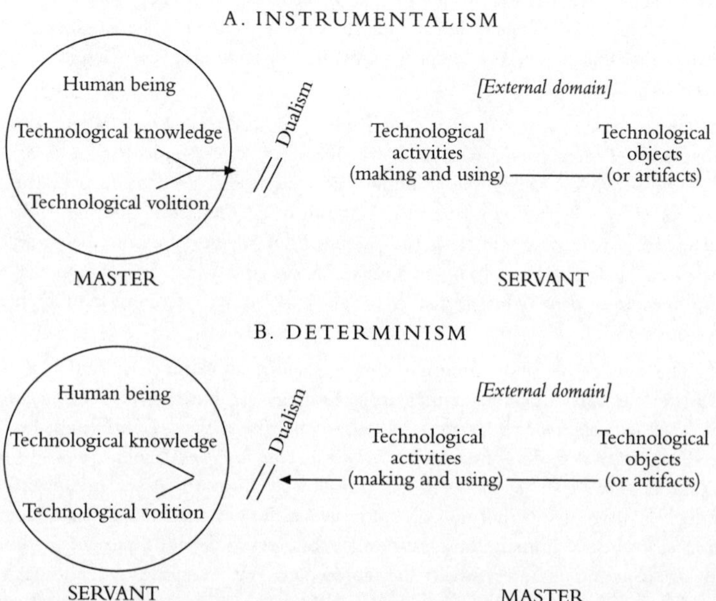

42. And, let us not forget that to be liberated from the prior constraints of paraplegia by the mobility of virtual identities is to be placed in a fundamentally and irreconcilably different position with respect to technological culture than one whose children is being poisoned by the waste created by a microchip factory.

43. Robert L. Thayer offers an excellent discussion of this ambivalence in *Grey World, Green Heart: Technology, Nature, and the Sustainable Landscape* (New York: John Wiley & Sons, 1995).

44. It is true, as MacKay's 1998 domestic survey of the public attitudes to technology found, that "technology continues to polarize the Australian community," *Turning Point,* p. 238. Yet it is often overlooked that this polarization occurs as much within different aspects of a single individual's experience as it does between different individuals. In making this claim, I draw particularly on my involvement in focus group research exploring public attitudes to genetic medicine. See Ian Barns, Renato Schibeci, Aidan G. Davison, and Robyn Shaw, "'What Do You Think About Gene Therapy?' Facilitating Sociable Public Discourse on Developments in Genetic Medicine," *Science, Technology, and Human Values* 25, no. 3 (2000): 283–308. For a discussion of difficulties encountered in exploring public attitudes to technology (in this case, biotechnology) see Aidan Davison, Ian Barns, and Renato Schibeci, "Problematic Publics: A Critical Review of Surveys of Public Attitudes to Biotechnology," *Science, Technology, and Human Values* 22, no. 3 (1997): 317–348.

45. The lure of mastering technology is influential in much radical as well as much conservative thought. In *The Practice of Technology*, for instance, the supporter of deep

ecology Drengson hopes, with what I find to be a somewhat Cartesian cadence, for an actualization of "human potentials to master the whole technological process so as to subordinate it to higher values that transcend the present transient and ephemeral ones of industrial culture," p. 114.

46. Winner, *Technology Today*, p. 992.

47. Wertheim, *The Pearly Gates of Cyberspace*, p. 228.

48. Sherry Turkle, *Life on the Screen: Identity in the Age of the Internet* (New York: Simon & Schuster, 1995), pp. 233–254.

49. George Kateb, "Technology and Philosophy," *Social Research* 64, no. 3 (1997): 1225–1246, p. 1226.

50. See Giddens's chapter on "The Trajectory of the Self," in *Modernity and Self Identity*, ch. 3, pp. 70–108.

51. Borgmann, *Technology and the Character of Contemporary Life*, p. 3.

52. Albert Borgmann, "Theory, Practice, Reality," *Inquiry* 38 (1995): 143–156, p. 148.

53. Consider the sentiment of Michael E. Zimmerman's claim, in *Eclipse of the Self: The Development of Heidegger's Concept of Authenticity*, revised edn. (Athens: Ohio University Press, 1986), that we "often imagine that the moment of enlightenment must occur in circumstances befitting a movie epic. In fact, however, every moment of everyday life, every cup of coffee we drink, every word we offer others, involves the revelation of Being," p. 251. I agree, but, equally, it is worth stressing that we often fail to appreciate that the moment of vicious oppression need not take on the form of an epic, blood-spattered event: it too takes place in every cup of coffee we drink. See my "Coffee: The Aroma of Freshly Brewed Globalization," in *Ambivalence and Hope*, eds. M. Booth and T. Hogan (Perth, WA: Murdoch University/ISTP, 1997), 169-173.

54. Winner, *The Whale and the Reactor*, p. xi.

55. Ibid., p. 11.

56. Ibid., pp. 28–29.

57. Ibid., p. 9.

58. Ibid., ch. 1, pp. 3–18.

59. Richard Ford cited in MacKay, *Turning Point*, p. 250.

60. See Neil Postman, *Technopoly: The Surrender of Culture to Technology* (New York: Alford A. Knopf, 1992), p. 18.

61. Wendell Berry, *The Unsettling of America: Culture and Agriculture* (San Francisco: Sierra Club Books, 1977), p. 18.

62. Lewis Mumford, *Technics and Civilization* (London: Routledge & Kegan Paul, 1934), p. 3. These three paradigms are the wood-water complex of the Ecotechnic world of the tenth to the eighteenth centuries, the coal-iron complex of the Paleotechnic world of the eighteenth to the early-twentieth centuries, and the electricity-alloy complex of the Neotechnic world of the early twentieth century. See also *The Myth of the Machine: Vol .1, Technics and Human Development, Vol. 2, The Pentagon of Power* (New York: Harcourt Brace Jovanovich, 1967, 1970).

63. See, e.g., George Basalla, *The Evolution of Technology* (Cambridge: Cambridge University Press, 1988). For a countervailing record of the historical plurality of technological paradigms see Helaine Selin (ed), *Encyclopedia of the History of Science, Technology, and Medicine in Non-Western Cultures* (Dordrecht & Boston: Kluwer Academic, 1997) and the recent *Technology and Social Agency: Outlining a Practice Framework for Archeology* (Oxford & Malden, MA: Blackwell, 2000) by archeologist Marica-Anne Dobres.

64. Ivan D. Illich, *Tools for Conviviality* (London: Calder & Boyars, 1973), pp. 79–82.

65. Paul Graves-Brown's interesting study of the convergence of car and cyber practices is relevant here. "From Highway to Superhighway: The Sustainability, Symbolism, and Situated Practices of Car Culture," *Social Analysis* 41, no. 1, March (1997): 64–75.

66. See Peter Newman and Jeff Kenworthy, *Sustainability and Cities: Overcoming Automobile Dependence* (Covelo, CA: Island Press, 1998).

67. Bent Flyvbjerg's account of the Aalborg Project in *Rationality and Power: Democracy in Practice*, trans. S. Sampson (Chicago & London: University of Chicago Press, [1991] 1998), is instructive. His study, influenced by the analyses of power developed by Nietzsche and Foucault, illuminates the intensely cultural, political substance of planning practices that cannot be tackled within an instrumentalist framework, but that require a different epistemological basis for talking about practical (in contrast to technical) matters such as transport.

68. Winner, *Autonomous Technology*, p. 229. See also his "Do Artifacts Have Politics?," "Techné and Politeia" and "On Not Hitting the Tar-Baby" in *The Whale and the Reactor*, pp. 19–39, 40–58, 138–154.

69. Beck, *Ecological Politics in an Age of Risk*, ch. 7, pp. 158–184. As Postman points out, latemoderns typically wait for scientists, physical and social, to feed back to us as fact what our experience could have made obvious. *Technopoly*, pp. 144–146.

70. Brian J. Fleay, *The Decline of the Age of Oil* (Sydney: Pluto Press Australia, 1995), ch. 2, pp. 13–30.

71. It is remarkable that some appear to be neverendingly surprised, even affronted, by the unexpected and unplanned results of technological change. See, e.g., Edward Tenner, *Why Things Bite Back: Technology and the Revenge of Unintended Consequences* (New York: Alfred A. Knopf, 1996).

72. Written and produced by Jamie Uys, CAT Films, 1980 (Twentieth Century Fox Release).

73. In general terms, latemodern technological trajectories are being set by the direct manipulation of (i) molecules, particularly genes (such as synthetic, nano-manipulative, and biological technologies), (ii) the electromagnetic radiation spectrum (such as fibre optic, laser, microwave, radiowave, and solar energy technology), and (iii) digitized information (information and multimedia technologies of all kinds).

74. Stephen Hill, *The Tragedy of Technology: Human Liberation Versus Domination in the Late Twentieth Century* (UK: Pluto Press, 1988), p. 28.

75. Borgmann, *Technology and the Character of Contemporary Life*, p. 35.

76. Ibid., p. 34.

77. Ibid., p. 10.

78. Ibid., p. 39.

79. Ibid., p. 44.

80. Ibid., p. 76.

81. Strong, *Crazy Mountains*, p. 86.

82. Frank Dexter, "An Interview with Satan," in *FutureNatural: Nature, Science, Culture,* eds. G. Robertson, M. Mash, L. Tickner, J. Bird, B. Curtis, and T. Putnam (London & New York: Routledge, 1996), 293–302, p. 302.

83. Borgmann, *Technology and the Character of Contemporary Life*, p. 161.

84. Ibid., p. 158.

85. Ibid., p. 47.

86. Albert Borgmann, "The Nature of Reality and the Reality of Nature," in *Reinventing Nature? Responses to Postmodern Deconstruction*, eds. M. E. Soulé and G. Lease (Washington DC & Covelo, CA: Island Press, 1995), 31–45.

87. Donna Haraway, "A Manifesto for Cyborgs: Science, Technology, and Socialist Feminism in the 1980s," *Australian Feminist Studies* 4 (1987): 1–42. As Barns says, "the trouble with much of this literature is that despite its scornful dismissal of an abstract modernism, its own abstractness and rhetorical overstatement also mystifies and distorts the complex ways in which contemporary technologies are involved in forming ordinary people in the ordinariness of everyday life." *Technology and Citizenship*, p. 158.

88. Borgmann, *Technology and the Character of Contemporary Life*, p. 225.

89. An argument Borgmann develops with rich detail in his description of postmodern realism in *Crossing the Postmodern Divide* (Chicago: University of Chicago Press, 1992), ch. 5, pp. 110–147.

90. This fact is diffusely evident throughout his texts in the nature of the essentially conservative examples Borgmann employs (examples that hark back to the Eurocentric, Christian, patriarchal, pioneer, and middle-class ideals of family and communal life that were likely to have defined life in the late 19th and early 20th centuries and that have come under increasing challenge since the second world war). This fact is also sometimes startlingly evident in remarks like this, from his recent study of information technology: "The recent burst of information technology has further, and fortunately, silenced the voices of overt misery, of disease, poverty, and violence, both here and around the globe. There is still unspeakable suffering in many parts of the world. But information technology is both the channel and the energy that is carrying the free-market economy and its blessings to every corner on earth," *Holding on to Reality*, p. 232. That Borgmann's interest in the prospects of the world's poor is genuine, but nonetheless sanguine and somewhat abstract, is nowhere clearer than in his recent article, "A Scarcity of Focal Things: Reply to Pieter Tijmes," *Technology in Society* 21 (1999): 191–199. In general, Borgmann appears confident that the device paradigm will be able to realize its promise of security. As part I ought to have made clear, I do not, in the broad, agree with him that "[t]echnology has the conceptual resources and thus the physical and social ones as well to deal with its crises," nor that "the global ecological situation is more promising [that is, in 1984] socially and physically than most critics had dared to hope a decade ago," *Technology and the Character of Contemporary Life*, pp. 145, 147.

CHAPTER FIVE: REVEALING AN INHOSPITABLE REALITY

1. The publication of Martin Heidegger's *Sein und Zeit* (*Being and Time*) in 1927 is the pivot on which we can delineate distinct early (pre-1930) and late (post-1935) philosophies of technology, a shift catalyzed by his study of Nietzsche. For a discussion of Heidegger's early philosophy of technology, see Don Ihde, *Technology and the Life-world: From Garden to Earth* (Bloomington & Indianapolis: Indiana University Press, 1990), pp. 31–34. For a comparison of his earlier and later approaches to technology, see Hubert Dreyfus, "Heidegger's History of the Being of Equipment," in *Heidegger: A Critical Reader*, eds. H. Dreyfus and H. Hall (Oxford & Cambridge, MA: Blackwell, 1993), 173–185. Despite the centrality of technology in Heidegger's thought, his philosophy of technology was largely neglected until well after his death in 1976. See Albert Borgmann, with the assistance of Carl Mitcham, "The Question of Heidegger and Technology: A Critical Review of the Literature," *Philosophy Today* 31, no. 2 (1987): 97–194. Among the works of Heidegger I consider here in English translation are, in chronological order, the 1935 lecture/1950 essay "The Origin of the Work of Art" (originally published in *Holzwege*); 1936 essay "Hölderlin and the Essence of Poetry" (originally published as *Hölderlin und das Wesen der Dichtung*); 1938 lecture/1950 essay "The Age of the World Picture" (originally published in *Holzwege*); 1949–1950 lectures/1954 essays "The Question Concerning Technology," "The Turning," and "The Thing" (originally published in *Vorträge und Aufsätze*); letter to Jean Beaufret "Letter on Humanism" (originally published in 1947 as *Über den Humanismus*); 1951 lecture/1954 essay "Building, Dwelling, Thinking" (originally published in *Vorträge und Aufsätze*); 1955 public address/1959 essay "Memorial Address" (originally published as *Gelassenheit*); 1955–1956 lecture course/1957 book "The Principle of Reason" (originally published as *Der Satz vom Grund*); and the 1966 interview/published posthumously in 1976 "*Der Spiegel* Interview."

2. Don Ihde, *Technics and Praxis: A Philosophy of Technology* (Dordrecht: D. Reidel, 1979), p. 103.

3. Martin Heidegger, "The Question Concerning Technology," in The *Question Concerning Technology and Other Essays,* trans. and ed. W. Lovitt (New York: Harper & Row, 1977), 3–35, p. 4.

4. As he did so often, Heidegger appropriated such terms in an extraordinary way in an attempt to unleash their original, ancient force. That conventional philosophical use of these terms diverges from Heidegger's can be seen by comparison of his texts with the definitions provided in *The Cambridge Dictionary of Philosophy,* general ed. A. Robert (Cambridge: Cambridge University Press, 1995) and *The Oxford Dictionary of Philosophy*, general ed. S. Blackburn (Oxford & New York: Oxford University Press, 1994). In general, I agree with Leslie Paul Thiele, *Timely Meditations: Martin Heidegger and Postmodern Politics* (Princeton: Princeton University Press, 1995), that we need to see Heidegger's obscure language as "resonant ontological speech," p. 131. Nonetheless, many slavishly adopt Heideggerese (often perfecting it well beyond Heidegger's own use) thereby wittingly or unwittingly insulating themselves from broader discourses.

5. William Lovitt, "Introduction," in *The Question Concerning Technology and Other Essays,* by M. Heidegger, trans. and ed. W. Lovitt (New York: Harper & Row, 1977), xiii–xxxix, p. xiii. See also William Lovitt and Harriet Brundage Lovitt, *Modern Technol-*

ogy in the Heideggerian Perspective (Lewiston, Queenstown & Lampeter: Edwin Mellen, 1995), Vol. I, ch. V, pp. 171–222.

6. Joan Stambaugh, *Thoughts on Heidegger* (Washington DC: Centre for Advanced Research in Phenomenology/University Press of America, 1991), p. 5.

7. Lovitt, *Introduction,* p. xxxiv. See Heidegger's "On The Essence of Truth," in *Existence and Being,* ed. W. Brock, trans. R.F.C. Hull and A. Crick, 3rd edn. (Chicago: Henry Regnery, 1968), 319–351, p. 338.

8. One of the antecedents of Heidegger's understanding of truth can be seen in Friederich Hölderlin's 1799 aphorism entitled 'The Root of all Evil': "Being at one is god-like and good, but human, too human, the mania/Which insists there is only One, one country, one truth, one way." *Friedrich Hölderlin: Poems and Fragments,* trans. M. Hamburger (London: Routledge & Kegan Paul, 1966), p. 71.

9. See Lovitt and Lovitt, *Modern Technology in the Heideggerian Perspective,* Vol. I, ch. VIII, pp. 293–328.

10. An idea Heidegger expands on in his "Letter on Humanism," in *Basic Writings,* trans. D. F. Krell, revised and expanded edn. (San Francisco: Harper & Row, 1993), 217–265, pp. 221–223. See also Michael Zimmerman, *Heidegger's Confrontation with Modernity: Technology, Politics and Art* (Bloomington & Indianapolis: Indiana University Press, 1990), pp. 191–192.

11. Lovitt and Lovitt, *Modern Technology in the Heideggerian Perspective,* Vol. I, pp. 19–21. "It is as if modern technology were a lens that allows us to see the world as only one special set of offerings and possibilities," says David Rothenberg. "We frame the world within the shape of these techniques, and we perceive the world as fuel for human opportunity," *Hand's End: Technology and the Limits of Nature* (Berkeley: University of California Press, 1993), p. 81.

12. Heidegger, *The Question Concerning Technology,* p. 14.

13. I adopt the term *technological ontology* in an attempt to bypass the convolutions involved in Heidegger's idiosyncratic application of the term *metaphysics.* Zimmerman claims that, in fact, Heidegger traces the earliest expressions of technological ontology to Heraclitus and Parmenides. *Heidegger's Confrontation with Modernity,* p. 168.

14. Martin Heidegger, *Plato's Doctrine of Truth,* cited in Ian H. Angus, *George Grant's Rejoinder to Heidegger: Contemporary Political Philosophy and the Question of Technology* (New York: The Edwin Mellen Press, 1987), p. 47.

15. Heidegger, *The Question Concerning Technology,* p. 13.

16. See, e.g., *The Cambridge Dictionary of Philosophy.*

17. Lovitt and Lovitt, *Modern Technology in the Heideggerian Perspective,* Vol. I, p. 394. Heidegger's derivation of the term *techné* diverges from its conventional location in *teks* (to fabricate or weave) and turns instead to *tek* (to bear or beget, also closely related to *techo* which related to the procreating and raising of children), p. 346.

18. Heidegger, *The Question Concerning Technology,* pp. 6-12.

19. In *Letter on Humanism,* Heidegger claims that "thinking comes to an end by slipping out of its element . . . [and] replaces this loss by procuring a validity for itself as

techné, as an instrument of education. . . . One no longer thinks; one occupies oneself with philosophy," p. 221.

20. Lovitt, *Introduction*, p. xxv.

21. Just as Val Plumwood describes three steps in the establishment of dualism in philosophy, the first two taken by Plato, the last by Descartes (see my discussion in chapter 2), Zimmerman describes three steps in the emergence of technological ontology: (1) the shift from presencing to enduring presence metaphysical ground, (2) the shift from truth as unconcealment to certainty being taken by Plato, and (3) the elevation of humanity to self-certain subject. *Heidegger's Confrontation with Modernity*, p. 182.

22. In Descartes, Lovitt and Lovitt contend, we witness the "self-exaltation . . . [of] self-sufficient subjectness." *Modern Technology in Heideggerian Perspective*, Vol. II, p. 473.

23. See Martin Heidegger, "The Age of the World Picture," in *The Question Concerning Technology and Other Essays*, trans. and ed. W. Lovitt (New York: Harper & Row, 1977), 115–154, p. 130.

24. Martin Heidegger, *The Principle of Reason*, trans. R. Lilly (Bloomington & Indianapolis: Indiana University Press, 1991), p. 120.

25. Ibid., p. 128.

26. In this context, Albert Borgmann's suggestion that "philosophy is the child of cosmopolitanism" deserves special consideration. "Cosmopolitanism and Provincialism: On Heidegger's Errors and Insights," *Philosophy Today* 36 (1992): 131–145, p. 131. Murray Bookchin's *The Limits of the City* (Montréal & New York: Black Rose Books, 1986) is also instructive on the convergence of philosophy and urban life. Equally important, as David Abram shows us so wonderfully in *The Spell of the Sensuous: Perception and Language in a More-Than-Human World* (New York, Vintage Books, 1997), is the slow historical movement from oral to formal (and especially phonetic) written forms of storytelling, the emergence of the technology of writing, in Occidental traditions. I consider that Heidegger himself does not make his historical transformation of forms of life sufficiently clear. In particular, his talk of the sleep of the principle of reason in the "extraordinarily long incubation period" between Plato and Descartes, seems inadequate in reflecting the dramatically changing world of Europe during this time. *The Principle of Reason*, p. 118.

27. Heidegger, *The Principle of Reason*, p. 121.

28. Martin Heidegger, (*Overcoming Metaphysics*) *Vorträge und Aufsätze*, translated by and cited in Joanna Hodge, *Heidegger and Ethics* (London & New York: Routledge, 1995). As Hodge explains, for Heidegger modern technological forms of life "are the culmination of the history of philosophy, for they make actual the abstract relations which in metaphysical enquiry are only hypothesised," p. 20.

29. Ibid., pp. 35–36.

30. Michael Zimmerman, *Eclipse of the Self: The Development of Heidegger's Concept of Authenticity*, revised edn. (Athens: Ohio University Press, 1986), p. 226.

31. Ibid., p. 227.

32. Martin Heidegger, "What are Poets For?" in *Poetry, Language, Thought*, trans. A. Hofstadter (New York: Harper & Row, 1971), 91–142, p. 117.

33. Philosophy, Heidegger tells us, dissolves into cybernetics. "The Spiegel Interview," in *Martin Heidegger and National Socialism: Questions and Answers,* eds. G. Neske and E. Kettering (New York: Paragon, 1990), 41–66, pp. 58–59.

34. Martin Heidegger, "The Turning," in *The Question Concerning Technology and Other Essays,* trans. and ed. W. Lovitt (New York: Harper & Row, 1977), 36–49, p. 47.

35. Ibid., p. 37. He says that "[w]hen we look into the ambiguous essence of technology, we behold the constellation, the stellar course of the mystery." *The Question Concerning Technology*, p. 33.

36. Borgmann, *The Question of Heidegger and Technology*, p. 138.

37. Heidegger, *The Question Concerning Technology*, p. 18.

38. I note but cannot here contribute to those debates concerning Heidegger's place within the historical determinism of German idealism, see Zimmerman, *Heidegger's Confrontation with Modernity*, pp. 250–255; his indebtedness to both Eastern religion and the Christian mystic Meister Eckhart, see Thiele, *Timely Meditations*, pp. 227–233 and John Caputo, *The Mystical Element in Heidegger's Thought* (Athens: Ohio University Press, 1978); his perception of a foundational Greek-German axis in philosophy, see Larry Hickman, *John Dewey's Pragmatic Technology* (Bloomington & Indianapolis: Indiana University Press, 1990), p. 199; and the Christian residues latent in his thought, see John Macquarie, *Heidegger and Christianity: The Hensley Henson Lectures 1993–94* (New York: Continuum, 1994).

39. Martin Heidegger, "Hölderlin and the Essence of Poetry," in *Existence and Being*, ed. W. Brock, trans. D. Scott, 3rd edn. (Chicago: Henry Regnery, 1968), 291–315, p. 310.

40. Heidegger, *The Question Concerning Technology*, p. 34. "Within and beyond the looming presence of modern technology," Lovitt explains, "there dawns the possibility of a fuller relationship between man and Being—and hence between man and all that is—than there has ever been," *Introduction*, p. xxxvii.

41. Heidegger, *The Turning*, p. 37. "Heidegger made it very clear that there was nothing people could 'do' to inaugurate a new age." Zimmerman, *Heidegger's Confrontation with Modernity*, p. 236.

42. Heidegger, *The Spiegel Interview*, p. 57.

43. Heidegger, *Der Spiegel Interview*, cited in and translated by Thiele, *Timely Meditations*, p. 206.

44. Heidegger, *The Turning*, p. 42.

45. Thiele, *Timely Meditations*, p. 206.

46. Ibid., pp. 39–40.

47. Martin Heidegger, "Memorial Address," in *Discourse on Thinking*, trans. J. M. Anderson and E. H. Freund (New York: Harper & Row, 1966), 43–57, pp. 54–55.

48. Heidegger, *Hölderlin and the Essence of Poetry*, p. 282.

49. Art is, Heidegger asserts, "a realm that is, on the one hand, akin to the essence of technology and, on the other, fundamentally different from it." *The Question Concerning Technology*, p. 35. See also his critique of "the literature industry" in ". . . Poetically Man Dwells . . . ," in *Poetry, Language, Thought*, trans. and introduced by A. Hofstadter (New

York: Harper & Row, 1971), 213–229. As Lovitt and Lovitt interpret it, genuine art illuminates "human life from beyond itself," *Modern Technology in the Heideggerian Perspective*, Vol. II, p. 634.

50. Heidegger, *The Question Concerning Technology*, p. 14.

51. Ibid., p. 5.

52. Ibid., p. 14. See also Heidegger's *Discourse on Thinking*, p. 50.

53. Windmills emerged into Europe from Persia in the twelfth century. See, e.g., Lynn White, Jr., *Medieval Technology and Social Change* (Oxford: Oxford University Press, 1962).

54. In *Being and Time* Heidegger says that "[t]he wood is a forest of timber, the mountain a quarry of rock; the river is water-power; the wind is wind 'in the sails'," trans. J. Macquarrie and E. Robinson (Oxford: Basil Blackwell, 1962), p. 100.

55. Martin Heidegger, "The Origin of the Work of Art," in *Poetry, Language, Thought*, trans. and introduced by A. Hofstadter (New York: Harper & Row, 1971), 17–87, p. 34.

56. Charles Taylor suggests that the techniques of dimensional perspective, which first emerged in Italian renaissance painting, encouraged the artist to see themselves "as standing over against the object. . . . The artist is looking at what he depicts from a determinate point of view." Taylor rightly asks whether these practices didn't "prepare the ground for the more radical break, in which the subject frees himself decisively by objectifying the world?" *Sources of the Self: The Making of the Modern Identity* (Cambridge, MA: Harvard University Press, 1989), pp. 201, 202.

57. Heidegger, *The Origin of the Work of Art*, p. 34. It should not go unnoted here that we once again find a man romanticizing the hardship, poverty, and deprivations of a woman's lot (his mother's lot in nineteenth century rural Germany?).

58. There is no record of Van Gogh ever painting the shoes (boots) of a peasant woman. Lovitt and Lovitt, *Modern Technology in the Heideggerian Perspective,* Vol. II, p. 688.

59. Heidegger, *The Origin of the Work of Art*, p. 34.

60. See, e.g., James Klagge's "The Good Old Days: Age-Specific Perceptions of Progress," in *Technological Transformation: Contextual and Conceptual Implications*, Philosophy and Technology Vol. 5, eds. E. F. Byrne and J. C. Pitt (Dordrecht: Kluwer Academic, 1989), 93–104. I am not arguing here that judgments about technology can be reduced to expressions of historical relativism. I am arguing, however, that such judgments are always contextualized by the forms of life within which we first come to know our world.

61. Hölderlin, *Friedrich Hölderlin*, p. 82.

62. Plautus, cited in Daniel J. Boorstin, *The Discoverers: A History of Man's Search to Know his World and Himself* (London: J. M. Dent & Sons, 1983), p. 28.

63. Consider the dislocation the poet Matthew Arnold felt living through the first decades of industrialization: "Wandering between two worlds, one dead/The other powerless to be born,/With nowhere yet to lay my head,/Like them, on earth I wait forlorn," cited in John Dewey, "Poetry and Philosophy," in *The Early Works 1882–1898,* Vol. 3 1882–1892 (Carbondale & Edwardsville: Southern Illinois University Press, [1891] 1969), 110–124, p. 114.

64. In *The Age of the World Picture* Heidegger asserts that "[m]achine technology remains up to now the most visible outgrowth of the essence of modern technology, which is identical with the essence of modern metaphysics," p. 116.

65. See, e.g., John Merson, *Roads to Zanadu: East and West in the Making of the Modern World* (Adelaide, Australia: Wervin Weldon/ABC, 1988).

66. See Martin Heidegger, "Building, Dwelling, Thinking," in *Poetry, Language, Thought*, trans. and introduced by A. Hofstadter (New York: Harper & Row, 1971), 145–161, pp. 152–154.

67. See Zimmerman, *Heidegger's Confrontation with Modernity*, p. 216.

68. Heidegger, *The Turning*, p. 48.

69. Heidegger, *The Principle of Reason*, p. 124.

70. Martin Heidegger, "The Thing," in *Poetry, Language, Thought*, trans. and introduced by A. Hofstadter (New York: Harper & Row, 1971), 165–186, p. 165.

71. This abstractness is well discussed by Abram in his chapter "Time, Space, and the Eclipse of the Earth," in *The Spell of the Sensuous*, pp. 181–201.

72. Heidegger, *Gesamtausgabe Vol. 55*, translated by and cited in Zimmerman, *Heidegger's Confrontation with Modernity*, p. 209.

73. Heidegger, *The Question Concerning Technology,* p. 33.

74. Thiele, *Timely Meditations*, p. 214.

75. In his television interview with Richard Wisser, Heidegger responds to the question of whether or not philosophy has a social mission by saying that "today's society is only modern subjectivity made absolute. A philosophy that has overcome a position of subjectivity therefore has no say in the matter [of social change]." "The Television Interview," in *Martin Heidegger and National Socialism: Questions and Answers,* eds. G. Neske and E. Kettering (New York: Paragon, 1990), 81–88, p. 82.

76. Ibid., p. 84.

77. Ibid., p. 85.

78. A responsibility compounded by his lack of acknowledgment that his postwar philosophy of technology provides the basis for a critique of the instrumentalist orientation of *Being and Time.* See Dreyfus, *Heidegger's History of the Being of Equipment*; Drew Leder, "Modes of Totalization: Heidegger on Modern Technology and Science," *Philosophy Today* Fall (1985): 245–256; Zimmerman, *Heidegger's Confrontation with Modernity*, ch. 10, pp. 150–165.

79. Hodge, *Heidegger and Ethics*, pp. 32–33, 55. According to Hodge, "the question of ethics is the definitive, if unstated problem of his [Heidegger's] thinking", p. 1.

80. Heidegger, *The Turning*, p. 40.

81. Hubert L. Dreyfus, "Heidegger on the Connection between Nihilism, Art, Technology, and Politics," in *The Cambridge Companion to Heidegger*, ed. C. Guignon (Cambridge: Cambridge University Press, 1993), 289–316, p. 303.

82. Ibid., p. 305. In *Identity and Difference*, Heidegger describes how ensuring the peaceful use of atomic energy is likely only to extend metaphysical domination; see Hodge, *Heidegger and Ethics*, p. 60.

83. Compare my claim here with Borgmann's hopeful conviction that: "Professors, by and large, lead a commendable life, one that deserves praise and admiration. It is commendable also in deserving to be commended to others. Extensive air travel excepted, it is a relatively simple life and could be widely shared without driving the global environment to destruction." "Theory, Practice, Reality," *Inquiry* 38 (1995): 143–156, p. 155.

84. Dreyfus, *Heidegger on the Connection between Nihilism, Art, Technology, and Politics*, p. 307.

85. Hubert L. Dreyfus, "Heidegger on Gaining a Free Relation to Technology," in *Technology and the Politics of Knowledge*, eds. A. Feenberg and A. Hannay (Bloomington & Indianapolis: Indiana University Press, 1995), 97–107, pp. 101–102. Dreyfus appears to hold out hope of an artistic *Gestalt* whereby a work of art will refocus background practices and launch us thereby into a new paradigm, a new world. So, for instance, he says that the Woodstock music festival came close to stabilizing a "new understanding of being."

86. I agree with Rothenberg: "Those of us caught in the thick of the struggle to develop real solutions to immediate problems might find the ephemeral answer of poetry quite far from the difficulties we face." *Hand's End*, p. 83.

87. In my case, I retreat from artificial light, clocks, telephones, and computers to my hammock in the garden, where I frequently look up from *The Question Concerning Technology* to admire the busy activity of insects amongst the vegetables. Heidegger himself chose to do much of his thinking in the costume of the Swabian peasant in a remote, austere mountain hut. Yet in latemodernity the very possibility of nonhybrid vegetable seed or of remote huts in mountainous wilderness is only kept open by vigorous, worldly, political activism.

88. Dreyfus, *Heidegger on the Connection between Nihilism, Art, Technology, and Politics*, p. 305.

89. Heidegger, *Memorial Address*, p. 54.

90. Heidegger, *The Spiegel Interview*, pp. 60–61.

91. Heidegger, *Letter on Humanism*, p. 259.

92. Ibid., p. 262.

93. Thiele, *Timely Meditations*, p. 164.

94. In *Building, Dwelling, Thinking,* Heidegger describes for us the nondualistic character of old, poetic forms of life that make possible what he calls *dwelling*. Everyday actions within these forms of life were characterized by the bringing-forth of *Techne*. Everyday thoughts led always to the essential ground of meditative thinking. Building and thinking "belong to dwelling," pp. 160–161.

95. As can be seen in the exemplars of focal practice such as the hearth, wilderness, and the culture of the table that Albert Borgmann offers in *Technology and the Character of Contemporary Life: A Philosophical Inquiry* (Chicago: University of Chicago Press, 1984).

96. Dreyfus, *Heidegger on the Connection Between Nihilism, Art, Technology, and Politics*, p. 311.

97. Heidegger, *The Turning*, p. 40. See also his *Letter on Humanism*, p. 217.

98. Richard Bernstein, *The New Constellation: The Ethical-Political Horizons of Modernity/Postmodernity* (Cambridge: Polity, 1991), p. 101. As Zimmerman observes, "Heideg-

ger denies the efficacy of human praxis in bringing about the fully human world. The highest form of action is thinking—opening ourselves to a new manifestation of Being." *Eclipse of the Self*, p. 252.

99. Heidegger, *Memorial Address*, p. 53.

100. Heidegger's tendency to create such bald distinctions may extend back to *Being and Time* in which he suggests, as Zimmerman summarizes it, "that the only two alternatives in everyday life were either compulsive working or detached observing." *Heidegger's Confrontation with Modernity*, p. 153.

101. Bernstein, *The New Constellation*, p. 126. I agree with Hodge that we witness a movement in Heidegger's works from a privileging of *praxis* over *poiesis* to a privileging of *poiesis* over *praxis*, *Heidegger and Ethics*, pp. 3–4.

102. There is considerable evidence that Heidegger was at best a passive observer of the horrors perpetrated by Nazism. His role as a public intellectual figurehead for National Socialism in the early 1930s is also undoubted. Speculation surrounding his actions during the war remain inconclusive. Heidegger himself never apologized for his involvement in National Socialism, and he never admitted to collusion with the Nazis. For an overview of this debate, in addition to many of the works already cited, see the collections edited by Alan Milchman and Alan Rosenberg (eds), *Martin Heidegger and the Holocaust* (New Jersey: Humanities Press, 1996) and Richard Wolin (ed), *The Heidegger Controversy* (Cambridge, MA: MIT Press, 1993). For a focused analysis of the relation between Heidegger's philosophy of technology and Nazism, see Tom Rockmore, "Heidegger on Technology, Nazism, and the Thought of Being," *Research in Philosophy and Technology* 13 (Greenwich, CT: JAI Press, 1993): 265–282. The danger of this debate drifting off into philosophical gamesmanship is typified for me by Richard Rorty's "Another Possible World," in *Martin Heidegger: Politics, Art, and Technology*, eds. K. Harries and C. Jamme (New York & London: Holmes & Meier, 1994), 34–40.

103. See Bernstein, *The New Constellation*, p. 81.

104. Given this silence, we should perhaps not be surprised that the 195-page (!) index to Lovitt and Lovitt's monumental *Modern Technology in the Heideggerian Perspective*, a work that is arguably more Heideggerian in tone than Heidegger himself, contains no entries for *praxis*, practice, or action.

105. Jacques Derrida, "Heidegger's Silence (excerpts from a talk given on 5 February 1988)," in *Martin Heidegger and National Socialism: Questions and Answers*, eds. G. Neske and E. Kettering (New York: Paragon, 1990), 145–148, p. 147.

106. From the original transcript, translated by and cited in Bernstein, *The New Constellation*, p. 130.

107. Martin Heidegger, "The Rectorate 1933/4: Facts and Thoughts," in *Martin Heidegger and National Socialism: Questions and Answers,* eds. G. Neske and E. Kettering (New York: Paragon, 1990), 15–32, p. 18.

108. Ibid., p. 19.

109. Ibid., p. 29.

110. Bernstein, *The New Constellation*, p. 136.

111. I am sympathetic to Bernstein's proposition that Heidegger's postwar texts, written around the time of his public condemnation, exile from university life, and de-Nazification, can be "read as an *apologia*." Ibid. p. 130.

112. Bernstein says that the "entire rhetorical construction of 'The Question Concerning Technology' seduces us into thinking that the only alternative to the threatening danger of . . . [the essence of technology] is *poiésis*. It excludes and conceals the possible response of *phronésis* and *praxis*." Ibid., p. 122. This point I shall develop further in chapter 7.

113. Ibid., pp. 84–85.

114. Michael Zimmerman, "The Role of Spiritual Discipline in Learning to Dwell on Earth," in *Dwelling, Place, and Environment: Toward a Phenomenology of Person and Planet*, eds. D. Seamon and R. Mugerauer (Dordrecht: Martinus Nijhoff, 1986), 247–256, p. 248.

CHAPTER SIX: DISORIENTING MORAL LIFE

1. Ian Barns, "Technology and Citizenship," in *Poststructuralism, Citizenship, and Social Policy*, eds. A. Peterson, I Barns, J. Dudley, and P. Harris (London: Routledge, 1999), 154–198, p. 160.

2. See Wendell Berry's *The Unsettling of America: Culture and Agriculture* (San Francisco: Sierra Club Books, 1977) and his delightful essay "The Pleasures of Eating," in *What Are People For?* (San Francisco: North Point Press, 1990), 145–152.

3. Joanna Hodge, *Heidegger and Ethics* (London & New York: Routledge, 1995), p. 28.

4. Langdon Winner, "Citizen Virtues in a Technological Order," *Inquiry* 35, nos. 3/4 (1992): 341–361, p. 341.

5. John Hart, *Ethics and Technology: Innovation and Transformation in Community Contexts* (Cleveland: The Pilgrim Press, 1997), p. 141.

6. Hans Jonas, *Philosophical Essays: From Ancient Creed to Technological Man* (Chicago: University of Chicago Press, 1974), p. 18.

7. James Gouinlock, *Rediscovering the Moral Life: Philosophy and Human Practice* (New York: Prometheus, 1993), p. 11.

8. Ibid., p. 15.

9. "A student of Ethics seeks," claimed Henry Sidgwick, "systematic and precise general knowledge of what ought to be, and in this sense his aims and methods may properly be termed 'scientific'." *The Methods of Ethics*, 7th edn. (Indianapolis & Cambridge: Hackett Pub. Co., [1907] 1981), p. 1. Reflecting this view, *The Oxford English Dictionary*, prepared by J. A. Simpson and E.S.C. Weiner, 2nd edn. (Oxford: Clarendon Press, 1989), Vol. V, defines ethics as the "science of morals" and an ethic as "a scheme of moral science" (pp. 421–422). It is worth remarking that the forum for learned moral debate at Cambridge University in the early 1900s (the famed era of the "apostles" and the Bloomsbury set), which included Ludwig Wittgenstein, George Moore, Bertrand Russell, and Maynard Keynes, was called the "moral science club." For a summary of the epistemological footings of moral science, see William K. Frankena, *Ethics*, 2nd edn. (Englewood Cliffs, NJ: Prentice-Hall, 1973), pp. 12–60. For a critique, see Charles Taylor, *Sources of the Self: The Making of the Modern Identity* (Cambridge, MA: Harvard University Press, 1989), pp. 75–90.

10. Jean Piaget, *The Moral Judgement of the Child* (London: Routledge and Kegan Paul, 1932), p. 404.

11. For a recent discussion of these debates, see Stephen Darwall, Allan Gibbard, and Peter Railton, "Toward *Fin de siécle* Ethics: Some Trends," in *Moral Discourse and Practice: Some Philosophical Approaches*, eds S. Darwall, A. Gibbard, and P. Railton (New York & Oxford: Oxford University Press, 1997), 3–47.

12. Sidgwick, *The Methods of Ethics*, p. 382. For a definitive critique of Archimedean objectivity, see Hannah Arendt, *The Human Condition* (New York: Doubleday, 1958), pp. 263–265. See also Bernard Williams's chapter entitled "The Archimedean Point," in *Ethics and the Limits of Philosophy* (Cambridge, MA: Harvard University Press, 1985), ch. 2, pp. 22–29, and his "The Point of View of the Universe: Sidgwick and the Ambitions of Ethics," in *Making Sense of Humanity and Other Philosophical Papers, 1982–1993* (Cambridge: Cambridge University Press, 1995), pp. 153–171.

13. Gouinlock, *Rediscovering the Moral Life*, pp. 73–75. For a recent example of this ambition, see Richard Boyd, "How to be a Moral Realist," in *Moral Discourse and Practice: Some Philosophical Approaches*, eds. S. Darwall, A. Gibbard, and P. Railton (New York & Oxford: Oxford University Press, [1988] 1997), 105–135.

14. For a seminal discussion of the crippling hold of absolutism and relativism on wider modern culture, see Richard Bernstein's *Beyond Objectivism and Relativism: Science, Hermeneutics, and Praxis* (Oxford: Basil Blackwell, 1983).

15. Gouinlock, *Rediscovering the Moral Life*, pp. 287–288.

16. Alasdair MacIntyre, *After Virtue: A Study in Moral Theory,* 2nd edn. (Notre Dame, IN: University of Notre Dame Press, 1984), pp. 14–15. For an introduction to the emotivist theory of Ayer and Stevenson, see Theodore C. Denise, Sheldon P. Peterfreund, and Nicholas P. White, *Great Traditions in Ethics,* 8th edn. (Belmont, CA: Wadsworth Publishing Co., 1996), pp. 345–365.

17. The latemodern resurgence of Archimedean "moral geometry" is evident in the widespread influence of Rawls's neo-Kantian constructivist assertion that although there are no moral facts, "moral objectivity is to be understood in terms of a suitably constructed social point of view that all can accept." Rawls, cited in Darwall, Gibbard, and Railton, *Toward Fin de siécle Ethics*, p. 13.

18. MacIntyre, *After Virtue*, p. 12.

19. Ibid., p. 39.

20. Anne Manne, "In Freedom's Shadow," *The Australian's Review of Books*, July 1998, 12–14, p. 14.

21. MacIntyre, *After Virtue*, ch. 4, pp. 36–50.

22. Ibid., pp. 158, 185.

23. Michael J. Sandel, "The Procedural Republic and the Unencumbered Self," *Political Theory* 12, no. 1 (1984): 81–96, p. 82.

24. MacIntyre, *After Virtue*, p. 118.

25. Alasdair MacIntyre, "The Spectre of Communitarianism," *Radical Philosophy* 70 (1995): 34–35, p. 35.

26. Charles Taylor, "A Most Peculiar Institution," in *World, Mind, and Ethics: Essays on the Ethical Philosophy of Bernard Williams*, eds. J.E.J. Altham and R. Harrison (Cambridge: Cambridge University Press, 1995), 132–155, p. 134.

27. Marilyn Friedman, *What are Friends For? Feminist Perspectives on Personal Relationships and Moral Theory* (Ithaca & London: Cornell University Press, 1993).

28. Taylor continues that we can no longer argue about these moral reactions "once we assume a neutral stance and try to describe the facts as they are independent of these reactions, as we have done in natural science since the seventeenth century." *Sources of the Self*, p. 8.

29. Ibid., part I, pp. 3–107. For a brief description of narratology, see Bent Flyvbjerg, *Rationality and Power: Democracy in Practice*, trans. S. Sampson (Chicago & London: University of Chicago Press, [1991] 1998), pp. 7–8.

30. Taylor, *Sources of the Self*, p. 93.

31. Ibid., pp. 27–28.

32. Gouinlock, *Rediscovering the Moral Life*, p. 79.

33. Taylor, *Sources of the Self*, p. 10.

34. Richard M. Hare, *Moral Thinking: Its Levels, Method, and Point* (Oxford: Clarendon Press, 1981), p. 6.

35. Taylor, *A Most Peculiar Institution*, p. 151.

36. Taylor, *Sources of the Self*, ch. 1–2, pp. 3–52.

37. Barns, *Technology and Citizenship*, p. 160.

38. Ibid., p. 178.

39. The place of *praxis* and practical reason (*phronesis)* in the thought of these two thinkers is sympathetically reviewed in Joseph Dunne, *Back to the Rough Ground: 'Phronesis' and 'Techne' in Modern Philosophy and in Aristotle* (Notre Dame, IN: University of Notre Dame Press, 1993). See also Bernstein, *Beyond Objectivism and Relativism.*

40. Hans-Georg Gadamer, "Theory, Technology, Praxis," in *The Enigma of Health: The Art of Healing in a Scientific Age*, trans. J. Gaiger and N. Walker (Cambridge: Polity Press, [1972] 1996), 1–30. Gadamer claimed that "the concept of 'praxis' which was developed in the last two centuries is an awful deformation of what practice really is," cited in Bernstein, *Beyond Objectivism and Relativism*, p. 39.

41. In *The Human Condition* (Chicago & London: University of Chicago Press, 1958), Hannah Arendt argued that "the action of the scientists, since it acts into nature from the standpoint of the universe and not into the web of human relationships, lacks the revelatory character of action as well as the ability to produce stories and become historical, which together form the very source from which meaningfulness springs into and illuminates human existence," p. 324.

42. Consider, for instance, Descartes' claim "that it is possible to arrive at knowledge which is most useful in life, and that, instead of the speculative philosophy taught in the Schools, a practical philosophy can be found by which, knowing the power and the effects of fire, water, air, the stars, the heavens, and all the other bodies which surround us, as distinctly as we know the various trades of our craftsmen." *Discourse on Method and*

the Meditations, trans. F. E. Sutcliffe (Harmondsworth, Middlesex: Penguin Classics, [1637] 1968), Discourse 6, p. 78.

43. Gadamer, *The Enigma of Health*, p. 17.

44. Williams, *Ethics and the Limits of Philosophy*, p. 18. Nonetheless, Williams continues, this requirement remains in the technological world "an influential ideal and, by a reversal of the order of causes, it can look as if it were the result of applying to the public world an independent ideal of rationality." Or, alternatively, we can say that it is social features of the modern world which create the environment in which the instrumental reason can dominate, and not the other way around.

45. Ibid., p. 198.

46. Immanuel Kant, *Critique of Practical Reason*, trans. L. W. Beck (Indianapolis & New York: Liberal Arts Press, [1788] 1956), pp. 167–168.

47. Lewis White Beck, "Translator's Introduction," in *Critique of Practical Reason*, I. Kant, trans. L. W. Beck (Indianapolis & New York: Liberal Arts Press, 1956), vii–xix, p. xii.

48. Kant, *A Critique of Practical Reason*, p. 18.

49. Immanuel Kant, *On the Old Saw: That Might be Right in Theory But it Won't Work in Practice*, trans. E. B. Ashton (Philadelphia: University of Pennsylvania Press, 1974), p. 55.

50. Ibid., pp. 80–81.

51. Taylor, *Sources of the Self*, pp. 86–89. See also his essay "Overcoming Epistemology" in *Philosophical Arguments* (Cambridge, MA, & London: Harvard University Press, 1995), 1–19. Procedural explanations continue to dominate in moral philosophy and can be loosely grouped into neo-Hobbesean and neo-Kantian forms. These descriptions are united, for all their differences, in the claim that ethics can aspire to objectivity because of "universal demands imposed within an agent's practical reasoning." Neo-Hobbeseans such as Kurt Baier argue that the particular interests and aims of the (atomistic and egoistic) moral agent are the touchstone of practical reasons and that "morality as a system of practical reasoning is in each person's interests; each gains by using it since this is necessary for mutually advantageous cooperation." For neo-Kantians such as Thomas Nagel, "practical reasoning is subject to a formal constraint which effectively requires that any genuine reason to act be agent-neutral [that is, universally categorical]." Darwall, Gibbard, and Railton, *Toward Fin de siécle Ethics: Some Trends*, pp. 9, 10, 11.

52. Kant, *On The Old Saw*, pp. 41, 43.

53. Charles Taylor, "Explanation and Practical Reason," in *The Quality of Life*, eds. M. Nussbaum and A. Sen (Oxford: Clarendon Press, 1993), 208–231, p. 208.

54. Frederick L. Will, "The Rational Governance of Practice," in *Pragmatism and Realism*, by F. Will, ed. K. R. Westphal (London: Rowman & Littlefield, [1981] 1997), 63–83, pp. 65–66.

55. Bernstein, *Beyond Objectivism and Relativism*, p. 18.

56. Charles Taylor, "Engaged Agency and Background in Heidegger," in *The Cambridge Companion to Heidegger*, ed. C. Guignon (Cambridge: Cambridge University Press, 1993), 317–336, pp. 320–325. Taylor argues that the crucial and fateful step in the Cartesian model of agency is its "ontologizing, that is, the reading of the ideal method into the very constitution of the mind," p. 321.

57. Taylor, *Sources of the Self*, p. 151.

58. Jürgen Habermas, *Theory and Practice*, trans. J. Viertel from 4th ed. (London: Heinemann, [1971] 1974), p. 42.

59. Hannah Arendt, "The Concept of History: Ancient and Modern," in *Between Past and Future: Eight Exercises in Political Thought* (New York: The Viking Press, [1954] 1968), 41–90, p. 57.

60. Berry, *The Unsettling of America*, p. 57.

61. Habermas, *Theory and Practice*, p. 43.

62. Ibid.

63. Ibid., p. 255.

64. Charles Taylor, *The Ethics of Authenticity* (Cambridge, MA, & London: Harvard University Press, 1991), p. 105.

65. Taylor, *Sources of the Self*, p. 7.

66. As Taylor says in "Overcoming Epistemology," "the epistemological tradition is connected with some of the most important moral and spiritual ideas of our civilization—and also with some of the most controversial and questionable. To challenge them is sooner or later to run up against the force of this tradition." *Philosophical Arguments*, p. 8.

67. As is still the case amongst the world's poor, the immediate impact of industrialization was not to create shared abundance, but was, rather, to undermine the local subsistence economies of relatively isolated (relatively bioregional) rural communities (eroding their traditions of skilled craft) and to create vast amounts of (life-eroding) waste. See, e.g., Kirkpatrick Sale, *Rebels Against the Future: The Luddites and their War on the Industrial Revolution, Lessons for the Computer Age* (Reading, MA, Menlo Park, CA, & New York: Addison-Wesley, 1995).

68. Taylor, *The Ethics of Authenticity*, pp. 25–29.

69. Ibid., p. 28.

70. Ibid., p. 29.

71. Ibid.

72. Ibid., p. 65.

73. A theme pioneered by Richard Rorty in *Philosophy and the Mirror of Nature* (Princeton: Princeton University Press, 1979), part I, pp. 15–127.

74. Taylor's claim is that it was Locke who first took ontological disengagement to the point where it holds out the prospect of self-remaking. *Sources of the Self*, ch. 9, pp. 159–176.

75. Hans-Georg Gadamer, cited in Richard J. Bernstein, *The New Constellation* (Cambridge: Polity Press, 1991), p. 126.

76. Albert Borgmann, *Crossing the Postmodern Divide* (Chicago: University of Chicago Press, 1992), p. 112.

CHAPTER SEVEN: RECOVERING PRACTICAL POSSIBILITIES

1. While there is much to distinguish the work of these philosophers, I understand them as belonging to an emerging latemodern discussion about moral *praxis*. A

good overview of this emerging conversation in fields such as hermeneutics, phenomenology, existentialism, poststructuralism, pragmatism, communitarianism, and feminism is provided in Richard Bernstein, *Beyond Objectivism and Relativism: Science, Hermeneutics, and Praxis* (Oxford: Basil Blackwell, 1983) and in Hubert L. Dreyfus and Stuart E. Dreyfus, "What is Morality? A Phenomenological Account of the Development of Ethical Expertise," in *Universalism vs. Communitarianism: Contemporary Debates in Ethics*, ed. D. Rasmussen (Cambridge, MA: The MIT Press, 1990), 237–263. Recent interpretations of Aristotle's practical philosophy can be found in Joseph Dunne, *Back to the Rough Ground: 'Phronesis' and 'Techne' in Modern Philosophy and in Aristotle* (Notre Dame, IN: University of Notre Dame Press, 1993); Eugene Garver, *Aristotle's Rhetoric: An Art of Character* (Chicago: University of Chicago Press, 1994); and James Gouinlock, *Rediscovering the Moral Life: Philosophy and Human Practice* (New York: Prometheus, 1993).

2. Lorraine Code reminds us that, for Aristotle, "slaves, children, and women were not persons in any meaningful sense." She argues that the androcentricity of Aristotle's account does not exist simply as a historical overlay on his explanation of practical reason, but systematically informs this account. *Rhetorical Spaces: Essays on Gendered Locations* (New York & London: Routledge, 1995), p. 99. See also her discussion pp. 215–217.

3. I am generally convinced by attempts, such as Carol Gilligan's, to demonstrate the need for case-by-case judgment about the applicability of maxims such as telling the truth, keeping promises, and respecting the sanctity of life. See her *In a Different Voice: Psychological Theory and Women's Development* (Cambridge, MA & London: Harvard University Press, 1982), pp. 25–31. But, nonetheless, the crucial and vexing problem of whether or not I should steal a life-saving drug not otherwise available to my dying friend, or, alternatively, of whether I should administer (illegally) the lethal injection they beg of me, takes place upon a background world of institutional artefacts and relations; they occur in a world of built structures, and my focus, in this discussion at least, rests on the building and transformation of these structures.

4. See the Preface of Richard Bernstein's *Praxis and Action* (London: Duckworth, [1971] 1972). Significantly, the majority of dictionaries of philosophy do not even include a reference for the term *practice*, apparently considering this meaning of this term so low, biological, and banal as to have no philosophical content at all. See, e.g., *The Oxford Dictionary of Philosophy*, S. Blackburn (general ed.) (Oxford & New York: Oxford University Press, 1994); Thomas Mautner, *A Dictionary of Philosophy* (Oxford & Cambridge: Blackwell, 1996).

5. *Phronesis* is commonly translated as "prudence" and has been variously explained in terms of notions of "practical reason," "practical knowledge," and "practical wisdom." *Phronesis* is discussed by Aristotle, in relation to the other intellectual virtues, in Book VI of his *Nicomachean Ethics.* According to Aristotle, *phronesis* is a virtuous intellectual state, "a true state, reasoned and capable of action in the sphere of human goods." Although this concept is central to his practical philosophy, it is absent from the majority of his corpus. *Ethics: The Nicomachean Ethics*, trans. J.A.K. Thomson, revised H. Tredennick, intro. J. Barnes, revised edn. (Harmondsworth, Middlesex: Penguin Classics, 1976), p. 210 [1140b12–33]. For contemporary interpretations of *phronesis,* see Bernstein, *Beyond Objectivism and Relativism*, pp. 1–49; Dunne, *Back to the Rough Ground*; Albert Jonsen and Stephen Toulmin, *The Abuse of Casuistry: A History of Moral Reasoning* (Berkeley & Los Angeles: University of California Press, 1988), pp. 25–28; Alasdair MacIntyre, *After Virtue: A Study in Moral Theory*, 2nd edn. (London: Duckworth, 1985), pp. 159–162.

6. Bernstein, *Praxis and Action*, pp. ix–x.

7. This is a fall from awareness that can be traced back before Aristotle to Socrates' use of the term *episteme* (theoretical knowledge) to describe *phronesis* (practical knowledge). Aristotle, *Ethics*, Book VI (xiii), p. 224, n. 3. In more recent times, this is a fall from awareness due in no small measure to Marx. Reflecting his trust in the Enlightenment promise of technological emancipation (albeit a trust beset by notable lapses), Marx conflated doing and making. He largely interpreted the substantive character of production as the essence of social action itself. The appropriation of *techne* and *poiesis* is the emancipatory *praxis* of the proletariat. See Bernstein, *Praxis and Action*, Part I, pp. 11–83, and also Andrew Feenberg, *A Critical Theory of Technology* (Oxford: Oxford University Press, 1991).

8. Bernstein, *Praxis and Action*, p. xi.

9. This remark comes from John Dewey's wonderfully sensible early essay "Moral Theory and Practice," published in 1891; an essay that could, had it been heeded, have saved moral philosophy from many decades wandering in the positivist wasteland of abstractness. In *The Early Works 1882–1898,* Vol. 3 1882–1892 (Carbondale & Edwardsville: Southern Illinois University Press, [1981] 1969), 93–109, p. 106. A discussion of this essay can be found in Larry Hickman's *John Dewey's Pragmatic Technology* (Bloomington & Indianapolis: Indiana University Press, 1990), ch. 5, pp. 107–139.

10. Nel Noddings was one of the first to point out the obvious that the "hand that steadied us as we learned to ride our first bicycle did not provide propositional knowledge, it guided and supported us all the same, and we finished up 'knowing how'." *Caring: A Feminine Approach to Ethics and Moral Education* (Berkeley: University of California Press, 1984), p. 3. A book that takes up the theme of mothering in an insightful way is Sara Ruddick, *Maternal Thinking: Toward a Politics of Peace* (Boston: Beacon, [1989] 1995). For an account of the relation between health, illness, and caring practices by nursing educators strongly influenced by the work of philosophers like Taylor and Dreyfus, see Patricia Benner and Judith Wrubel, *The Primacy of Caring: Stress and Coping in Health and Illness* (Menlo Park, CA: Addison-Wesley, 1989). For a recent study that covers the caring literature on mothering and nursing as well as friendship and citizenship, see Peta Bowden, *Caring: Gender-Sensitive Ethics* (London and New York: Routledge, 1997). I will take up some of the issues raised in this literature indirectly in a discussion of my experience of fathering in part III.

11. Chapter 10 of Bernard Williams, *Ethics and the Limits of Philosophy* (Cambridge, MA: Harvard University Press, 1985) is aptly entitled "Morality, the Peculiar Institution."

12. Hans-Georg Gadamer, *Truth and Method* (New York: The Seabury Press, 1975), p. 288.

13. Raimond Gaita, *A Common Humanity: Thinking about Love, Truth and Justice* (Melbourne: Text Publishers, 1999), p. 14.

14. Charles Taylor, *Sources of the Self: The Making of the Modern Identity* (Cambridge, MA: Harvard University Press, 1989), p. 72.

15. Ibid., p. 86.

16. And, as Aristotle observed in his discussion of *techne*, *episteme,* and *phronesis*, while the development of technical and scientific knowledge does not by itself bring with it

practical wisdom, the possession of practical wisdom "will carry with it the possession of them all." *Ethics*, p. 225 [1144b33–1145a11].

17. Albert Borgmann, *Technology and the Character of Contemporary Life: A Philosophical Inquiry* (Chicago: University of Chicago Press, 1984), p. 25.

18. Dreyfus and Dreyfus, *What is Morality? A Phenomenological Account of the Development of Ethical Expertise*, p. 258.

19. As Borgmann observes, it is the subject upon which our thinking is focused, rather than its procedures, that defines the character of practical rationality. *Technology and the Character of Contemporary Life*, p. 72.

20. Bernstein, *Praxis and Action*, p. 216.

21. Jim Cheney and Anthony Weston, "Environmental Ethics as Environmental Etiquette: Toward an Ethics-Based Epistemology," *Environmental Ethics* 21, Summer (1999): 115–134, p. 128.

22. Ibid., p. 117.

23. Ibid., p. 118.

24. See, for instance, Hubert Dreyfus's description of ethical expertise in Bent Flyvbjerg, "Sustaining Non-Rationalized Practices: Body-mind, Power, and Situational Ethics, An Interview with Hubert and Stuart Dreyfus," *Praxis International* 11, no. 1 (1991): 93–112.

25. See Gadamer, *Truth and Method*, pp. 283–288.

26. Dewey suggested that wherever "there is anything which deserves the name of conduct, there is an idea, a 'theory', at least as large as the action." *The Early Works 1882–1898,* Vol. 3 1882–1892, pp. 95–96. Practical theory, for Dewey, is neither abstract nor speculative, but involves the application of intelligence to a specific situation in the busy hustle of everyday life so as to cut a "cross-section" through it, thereby momentarily arresting its movement: "Theory is [thus] the cross-section of the given state of action in order to know the conduct that should be; practice is the realization of the idea thus gained: it is theory in action," p. 109.

27. MacIntyre, *After Virtue,* p. 184.

28. Ibid., p. 187.

29. Taylor, *Sources of the Self*, p. 93.

30. See Williams, *Ethics and the Limits of Philosophy*, pp. 34–36. For Aristotle's description of *eudaimonia,* see, *Ethics*, Book I, pp. 63–90 [1094a1–1103a10].

31. "Moral goodness," Aristotle proclaimed, "is the result of habit [custom], from which it actually got its name, being a slight modification of the word *ethos*." *Ethics*, Book II (i), p. 91 [1103a14–b1]. And, he insists, well-being refers not to moments or experiences but can only be applied to a "complete lifetime. One swallow does not make a summer; neither does one day. Similarly neither can one day, or a brief space of time, make a man blessed and happy." Book I (vii), p. 76 [1098a8–27].

32. MacIntyre, *After Virtue*, pp. 187–188.

33. Ibid., p. 191.

34. Patricia Benner, "The Role of Articulation in Understanding Practice and Experience as Sources of Knowledge in Clinical Nursing," in *Philosophy in an Age of Pluralism:*

The Philosophy of Charles Taylor in Question, ed. J. Tully, assisted by D. M. Weinstock (Cambridge & New York: Cambridge University Press, 1994), 136–155, p. 144.

35. Aristotle, *Ethics,* Book I (vii), pp. 73–74 [1097a15–b21].

36. For Aristotle, claims MacIntyre, in the acquisition and exercise of the virtues in *phronesis,* "the relationship of means to end is internal and not external." *After Virtue*, p. 184.

37. Taylor, *Sources of the Self*, p. 86.

38. Borgmann, *Technology and the Character of Contemporary Life*, p. 7

39. Ibid., p. 41.

40. Taylor, *Sources of the Self,* p. 86.

41. Ibid., p. 20.

42. As Charles Taylor makes clear in "Engaged Agency and Background in Heidegger," in *The Cambridge Companion to Heidegger*, ed. C. Guignon (Cambridge: Cambridge University Press, 1993), 317–336, pp. 320–325, our background of practices can never be fully explicated because "explicating itself presupposes a background. The very fashion in which we operate as engaged agents with such a background makes the prospect of total explication incoherent. The background cannot in this sense be thought of quantitatively at all," p. 329.

43. I am here appropriating David Bohm's rich processual language in a fresh context. Bohm noted that the Cartesian "way of thinking about things . . . is convenient and useful mainly in the domain of practical, technical, and functional activities. . . . However, when this mode of thought is applied more broadly to man's notion of himself and the whole world in which he lives (i.e., to his self-world view), then man ceases to regard the resulting divisions as merely useful or convenient and begins to see and experience himself and his world as actually constituted of separately existent fragments." *Wholeness and the Implicate Order* (London & Boston: Routledge & Kegan Paul, 1980), p. 2.

44. Robert Browning, cited by John Dewey in "Poetry and Philosophy," *The Early Works 1882–1898,* Vol. 3 1882–1892 (Carbondale & Edwardsville: Southern Illinois University Press, [1981] 1969), 110–124, p. 120.

45. David Abram, *The Spell of the Sensuous: Perception and Language in a More-Than-Human World* (New York, Vintage Books, 1997), p. 201.

46. See, for instance, MacIntyre's description of the "interlocking set of social relationships" in premodern, traditional societies in which "I am brother, cousin, grandson, member of this household, that village, this tribe." *After Virtue*, pp. 33–34.

47. Mary Oliver, "Home," *Aperture* 150, Winter (1998): 22, 25, p. 25.

48. Abram, *The Spell of the Sensuous,* p. 69.

49. Bernstein tells us that "for Aristotle, *theoria* is a form of life that involves strenuous disciplined activity. It is not entirely accurate [however] to call *theoria* and *praxis* ways or forms of life, for according to Aristotle they emerge as two dimensions of the truly human and free life." *Praxis and Action*, p. x. I agree with Williams that the "world to which ethical thought now applies is irreversibly different, not only from the ancient world but from any world in which human beings have tried to live and have used ethical concepts." *Ethics and the Limits of Philosophy*, p. 197.

50. Aristotle, *Ethics*, Book VI (v), p. 209 [1140a24–b12].

51. Albert Borgmann, *Crossing the Postmodern Divide* (Chicago: University of Chicago Press, 1992), p. 110.

52. Borgmann, *Technology and the Character of Contemporary Life*, p. 173.

53. Borgmann, *Crossing the Postmodern Divide*, pp. 96–97.

54. Ibid., p. 110. Borgmann asserts that "what shows itself in the vacuity or arbitrariness of most private moral discourse is neither ethical pluralism nor ethical chaos but complicity with technology." *Technology and the Character of Contemporary Life*, pp. 172–173.

55. Charles Taylor, "Social Theory as Practice" in *Philosophy and the Human Sciences: Philosophical Papers,* Vol. 2 (Cambridge & New York: Cambridge University Press, 1985), 91–115, p. 92. See the parallels here with Bent Flyvbjerg's description of the social sciences as a phronetic rather than an epistemic discipline, in "Aristotle, Foucault, and Progressive Phronesis: Outline of an Applied Ethics for Sustainable Development," in *Applied Ethics: A Reader*, eds. E. R. Winkler and J. R. Coombs (Oxford & Cambridge: Basil Blackwell, 1993), 11–27, pp. 15–18, and Pierre Bourdieu's influential sociological account of *habitus* in *Outline of a Theory of Practice* (Cambridge: Cambridge University Press, 1977).

56. Taylor, *Social Theory as Practice*, p. 94.

57. Ibid., p. 104.

58. Ibid., p. 111.

59. Taylor, *Sources of the Self*, p. 204.

60. Ibid. MacIntyre's discussion of practice is more rarefied still. His examples of practices such as chess, portrait painting, or cricket provide little opportunity to understand how practices are shaped by the world-structures of technology. *After Virtue*, pp. 188–191.

61. Taylor, *Sources of the Self*, p. 206.

62. Charles Taylor, "A Most Peculiar Institution," in *World, Mind, and Ethics: Essays on the Ethical Philosophy of Bernard Williams*, eds. J.E.J. Altham and R. Harrison (Cambridge: Cambridge University Press, 1995), 132–155, p. 153.

63. Taylor, *Social Theory as Practice*, p. 106.

64. Dana R. Villa, *Arendt and Heidegger: The Fate of the Political* (Princeton, NJ: Princeton University Press, 1996), p. 224.

65. See Martin Heidegger, "Letter on Humanism" in *Basic Writings*, trans. D. F. Krell, revised and expanded edn. (San Francisco: Harper & Row, 1993), 217–265, p. 255.

66. It is not that Heidegger was inattentive to Aristotle's account of moral rationality. We learn from Gadamer that early Heidegger was profoundly influenced by Aristotle's description of *phronesis* as a "type of knowing . . . that admits of no reference to a final objectivity in the sense of a science." Hans-Georg Gadamer, *Heidegger's Ways*, trans. J. W. Stanley (Albany, NY: State University of New York Press, 1994), p. 33. Gadamer extends this claim further to observe that "in the critique of Plato that crystallises the differentiation between *techne*, *episteme,* and *phronesis*, Heidegger took his first, decisive step away from philosophy as a rigorous science," p. 141. Yet the crucial thing to understand about

this step is that, beginning with his middle essay *On the Origin of a Work of Art*, Heidegger increasingly interprets *phronesis* as arising out of *Techne*.

67. See Flyvbjerg, *Sustaining Non-Rationalized Practices*, pp. 101–102.

68. Hubert Dreyfus, interviewed by Flyvbjerg in Ibid., p. 103.

69. Hubert L. Dreyfus and Charles Spinosa, "Highway Bridges and Feasts: Heidegger and Borgmann on How to Affirm Technology," *Man and World* 30 (1997): 159–177; Charles Spinosa, Fernando Flores, and Hubert L. Dreyfus, "Disclosing New Worlds: Entrepreneurship, Democratic Action, and the Cultivation of Solidarity," *Inquiry* 38 (1995): 3–63. The argument of the latter is developed more fully in Charles Spinosa, Fernando Flores, and Hubert L. Dreyfus, *Disclosing New Worlds: Entrepreneurship, Democratic Action, and the Cultivation of Solidarity* (Boston: MIT Press, 1997). It would seem that Dreyfus draws upon Heidegger in two distinct and somewhat contradictory modes. As we saw in chapter 5, Dreyfus has been at pains to protect later Heidegger's ontological concern about the technological understanding of being from mere practical concerns. However in the papers cited here, Dreyfus appears to have almost entirely discounted Heidegger's later explanation of technological ontology. What is more, he makes the assertion that at the very end of his professional life, Heidegger also abandoned his ontological account of the technological world as a unitary epochal destining in the history of being. According to Dreyfus, Heidegger came to see human worldhood as an ambiguously plural phenomenon. Drawing what seems to me to be a long bow from a short passage in Heidegger's *Building, Dwelling, Thinking*, Dreyfus and Spinosa, in *Highway Bridges and Feasts,* suggest that Heidegger thus did not caution against modern technology as a systemic, paradigmatic form of truthlessness, but rather embraced modern technology as simply another, inherently ambiguous, form of local disclosure. In this account, humans exist not within an epochal, unitary world, but within an almost limitless plurality of such local worlds, pp. 169–170.

70. Dreyfus and Spinosa, *Highway Bridges and Feasts*, p. 171.

71. Spinosa, Flores, and Dreyfus, *Disclosing New Worlds*, pp. 14–15.

72. Ibid., p. 3.

73. Albert Borgmann, "Theory, Practice, Reality," *Inquiry* 38 (1995): 143–156, p. 149.

74. Joanna Hodge, *Heidegger and Ethics* (London & New York: Routledge, 1995), p. 64.

75. Ibid., p. 28.

76. David Strong, *Crazy Mountains: Learning from Wilderness to Weigh Technology* (Albany, NY: State University of New York Press, 1995), p. 76.

77. Borgmann, *Theory, Practice, Reality*, pp. 148–149.

78. Stephen D. Ross, *Locality and Practical Judgement: Charity and Sacrifice* (New York: Fordham University Press, 1994), p. 213.

CHAPTER EIGHT: A WORLD WORTH CARING FOR

1. Albert Borgmann, *Crossing the Postmodern Divide* (Chicago: University of Chicago Press, 1992), p. 96.

2. Ibid., p. 144.

3. David Strong, *Crazy Mountains: Learning from Wilderness to Weigh Technology* (Albany, NY: State University of New York Press, 1995), p. 101.

4. Borgmann, *Crossing the Postmodern Divide*, p. 117.

5. Judith Wright, "Geology Lecture," in *A Human Pattern: Selected Poems* (Sydney: ETT Imprint, [1973] 1996), p. 172, (quotation from lines 1–4, 17–24).

6. With apologies to Wendell Berry's "Long-Legged House."

7. A fact still not fully acknowledged in the general Australian consciousness that continues to hold on to the aesthetics of pastoral Europe. See Tim Flannery, *The Future Eaters: An Ecological History of The Australasian Lands and People* (Sydney: Reed Books, 1994).

8. Isaac Scott Nind opened his description of the indigenous people of the Swan River, published in 1831, thus: "King George's Sound, the entrance of which is latitude 35°6'20" south, and longitude 118°1' east of Greenwich, is situated on the south coast, but very near the southwest extremity of New Holland." "Description of the Natives of King George's Sound (Swan River Colony) and Adjoining Country," in *Nyungar—The People: Aboriginal Customs in the Southwest of Australia*, ed. N. Green (Perth, Western Australia: Creative Research Publishers, [1981] 1979), 15–45.

9. A view set out in John K. Ewers, *The Western Gateway: 100 Years of Local Government in Fremantle* (Fremantle, Western Australia: Fremantle City Council, 1948).

10. The long-local words used here are from the language of the *Nyungar* peoples (also—*Noongar* or *Nyoongar*) who lived for tens of thousands of years, and who live still, throughout the Southwest corner of the Australian mainland. Where possible I use the dialect of the Swan River *Nyungar's*. The scientific names of the flora and fauna I refer to are (in order of appearance): *Jarrah—Eucalyptus marginata*; *Tuart—Eucalyptus gomphocephala*; *Marri—Eucalyptus calophylla*; *Condil—Allocasuarina fraseriana* (Common Sheoak); *Wonnil—Agonis flexulosa* (peppermint gum); *Kabbur—Jacksonia sternbergiana* (Green Stinkwood); *Magnite—Banksia grandis* (Bull Banksia); *Woorening—Phylidonyris novaehollandiae* (New Holland Honeyeater); *Dumbari—Santalum acuminatum* (Quandong); *Tjunguri—Thysantotus patersonii*; *Balga—Xanthorrhoea preissii*; *Kulbardie—Gymnorhina tibicen* (Magpie). Peter Bindon and Ross Chadwick, *A Nyoongar Wordlist from the South-West of Western Australia* (Perth, Western Australia: Western Australian Museum, 1992); Nick Thieberger and William McGregor, *Aboriginal Words* (New South Wales: The Macquarie Dictionary, 1994); Munyari (Ralph Winmar), *Walwalinj: The Hill that Cries, Nyungar Language, and Culture* (Perth, Western Australia: Dorothy Winmar, 1996). Many of the terms relating to plants and animals and their uses are also taken from Brad Daw, Trevor Walley, and Greg Keighery, *Bush Tucker Plants of the South-West* (Perth, Western Australia: Department of Conservation and Land Management, 1997); Robert Powell, *Leaf and Branch: Trees and Shrubs of Perth* (Perth, Western Australia: Department of Conservation and Land Management, 1990).

11. See, e.g., Maxine Fumagalli's poem, "Dreamtime Breezes" in *Southwest Noongar Woman* (Denmark, Western Australia: Denmark Environment Centre, 1992), p. 57.

12. Freya Mathews, "Mysticeti Testament," *The Trumpeter* 6, no. 3 (1989): 119–120 (quotation from last 9 lines).

13. The devastating ecological impact of the introduction of feral animals into Western Australia is well described in Gary Burke, *More than Pests: The Strategic Significance of Feral Animals and Invasive Plants for Nature Conservation and Landcare in Western Australia* (Perth, Western Australia: Institute for Science and Technology Policy, 1996), p. 9.

14. Nunzio Gumina cited in Trish Ainslie and Roger Garwood, *Fremantle: Life in the Port City* (Fremantle, Western Australia: Plantagenet Press, 1994), p. 38.

15. Ken Kelso, "Introduction," in *Fremantle: Life in the Port City*, by Trish Ainslie and Roger Garwood (Fremantle, Western Australia: Plantagenet Press, 1994), 8–16, p. 15.

16. Ibid.

17. Wendell Berry, *What Are People For?* (San Francisco: North Point Press, 1990), p. 9.

18. Nathaniel Ogle, *The Colony of Western Australia: A Manual for Emigrants 1839* (Sydney: John Ferguson, [1839] 1977), p. 9.

19. Ibid., pp. 44–45.

20. Ibid.

21. Ibid., p. 144.

22. Ewers, *The Western Gateway*, p. 46. The very first Fremantle jetty was built out of shipwrecked oak and was soon eaten by termites.

23. Bryn Griffiths, *Wharfies: A Celebration of 100 Years on the Fremantle Waterfront 1889–1989* (Perth, Western Australia: Platypus Press, 1989); Patricia M. Brown, *The Merchant Princes of Fremantle: The Rise and Decline of a Colonial Elite 1870–1900* (Perth, Western Australia: University of Western Australia Press, 1996).

24. Griffiths, *Wharfies*, pp. 17–22.

25. A dispute recorded in the pages of *The West Australian* newspaper of 15–22 April, 1998.

26. American Youth Commission, *Youth and the Future: A Report of the American Youth Commission* (Westport, CT: Greenwood Press, [1942] 1973), pp. 276, 278–279.

27. Wendell Berry, *Sex, Economy, Freedom and Community: Eight Essays* (New York: Pantheon, 1993), p. 35.

28. Judith Wright, "Habitat," in *A Human Pattern*, pp. 156–162 (quotation from Part III, lines 24–39).

29. Wendell Berry, *A Part* (San Francisco: North Point Press, 1980), p. 30.

30. Don McQuiston and Debra McQuiston, *Dolls and Toys of North America: A Journey Through Childhood* (San Francisco: Chronicle Books, 1995), pp. 39–47.

CHAPTER NINE: SUSTAINING TECHNOLOGY

1. Charles Taylor, "Social Theory as Practice," in *Philosophy and the Human Sciences: Philosophical Papers*, Vol. 2 (Cambridge & New York: Cambridge UP, 1985), 91–115, p. 111.

2. David Abram, *The Spell of the Sensuous: Perception and Language in a More-Than-Human World* (New York: Vintage Books, 1997), p. 28.

3. Ivan D. Illich, *Tools for Conviviality* (London: Calder & Boyars, 1973).

4. Abram, *The Spell of the Sensuous*, p. 202.

5. Hannah Arendt's crucial point is that, "[s]een from the viewpoint of man, who always lives in the interval between past and future, time is not a continuum, a flow of uninterrupted succession; it is broken in the middle, at the point where 'he' stands; and 'his' standpoint is not the present as we usually understand it but rather a gap in time which 'his' constant fighting, 'his' making a stand against past and future, keeps in existence." "Preface: The Gap Between Past and Future," in *Between Past and Future: Eight Exercises in Political Thought* (New York: Viking, [1954] 1968), 3–15, p. 11.

6. Wendell Berry, *The Unsettling of America: Culture and Agriculture* (San Francisco: Sierra Club Books, 1977), p. 56.

7. Albert Borgmann, *Crossing the Postmodern Divide* (Chicago: University of Chicago Press, 1992), p. 96.

8. The rampant optimism of technocratic discourses about sustainability stand in stark contrast to the increasing cynicism and anxiety expressed about environmental problems in the wider, lay community. A fact I think displayed, disturbingly, in Phil Macnaghten and John Urry's focus group survey of various sections of the Lancashire community in the UK on the theme of "sustainability discourse and daily practice," in *Contested Natures* (London, Thousand Oaks, CA & New Delhi: Sage, 1998), ch. 7, pp. 212–248.

9. Gregg Easterbrook, *A Moment on the Earth: The Coming Age of Environmental Optimism* (New York: Viking, 1995), pp. 675–676.

10. Abram, *The Spell of the Sensuous*, p. 272.

11. Borgmann, *Crossing the Postmodern Divide*, pp. 2–3.

12. A fact affirmed in Abram's description of the "flesh of language." See *The Spell of the Sensuous*, chs. 2–3, pp. 31–92.

13. David C. Toole, "Farming, Fly-Fishing, and Grace: How to Inhabit a Postnatural World Without Going Mad," *Soundings* 76, no. 1 (1993): 85–103, p. 90.

14. Wendell Berry, *What Are People For?* (San Francisco: North Point Press, 1990), p. 9.

15. Marcia Keegan, *Mother Earth, Father Sky: Navajo and Pueblo Indians of the Southwest* (London: Wildwood House, 1974), p. 62.

16. Bent Flyvbjerg, "Aristotle, Foucault, and Progressive Phronesis: Outline of an Applied Ethics for Sustainable Development," in *Applied Ethics: A Reader*, eds. E. R. Winkler and J. R. Coombs (Oxford & Cambridge: Basil Blackwell, 1993), 11–27, p. 12.

17. Albert Borgmann, *Technology and the Character of Contemporary Life: A Philosophical Inquiry* (Chicago: University of Chicago Press, 1984), p. 144.

18. See, once again, the focus group survey conducted by Macnaghten and Urry in *Contested Natures*, for evidence of this shrinkage of the present in the face of widely held hopes and fears.

19. Berry, *The Unsettling of America*, p. 223.

20. Adam B. Seligman, *The Problem of Trust* (Princeton, NJ: Princeton University Press, 1997), p. 153.

21. Berry, *What Are People For?*, p. 9.

22. Mary Oliver, "Home," *Aperture* 150, Winter (1998): 22, 25, p. 25.

23. Milan Kundera, *Slowness*, trans. L. Asher (London & Boston: Faber & Faber, 1996), p. 4.

24. Borgmann is right, modernism "can be met not through embargoes and prohibitions but through a genuine alternative." *Crossing the Postmodern Divide*, p. 116.

25. Berry, *The Unsettling of America*, p. 18.

26. Walt Whitman, "Song of the Open Road," in *Leaves of Grass: The 1892 Edition* (New York: Bantum Books, 1983), p. 122.

INDEX

Abram, David, 166–67, 201, 204, 254n. 26
Achterhuis, Hans, 72
Agenda 21: definition of sustainable development in, 11; funding of, 19, 24, 49; international implementation of, 18, 50; local approaches to, 21; national approaches to, 20–21; Northern bias in, 47; technological optimism in, 25–26. *See also* Earth Summit; Rio Declaration on Environment and Development; United Nations
anthropocentrism, 32, 106, 121
Arendt, Hannah, 87, 138, 151, 154, 202
Aristotle, 147, 152, 160, 161, 162, 164, 167–68, 169. *See also* practical reason
Australian National Strategy for Ecologically Sustainable Development, 20
automobile dependence, 104, 107–08
Ayer, Alfred, 144, 149

Bacon, Francis, 69–71, 110, 119, 121, 168
Bacon, Roger, 70
Baier, Kurt, 144
Barns, Ian, 141–42, 151
Barry, John, 61
Beck, Ulrich, 108
Beckerman, Wilfred, 14
Beney, Guy, 76
Benner, Patricia, 164
Bentham, Jeremy, 144
Berman, Morris, 79
Bernstein, Richard, 138–39, 160
Berry, Wendell, 1, 58, 106, 154–55, 188, 195, 202, 205, 211
Biosphere, 2, 75
Bookchin, Murray, 41, 72, 77, 79
Borgmann, Albert: device paradigm, 109–10, 111; dislocation of means and ends, 110, 112–13; focal practices, 112–13; on Heidegger, 109, 252n. 1, 254n. 26, 258n. 83; knowledge as reality, 97; philosophy of technology, 102–03; 162, 165, 168–69, 176, 181–82, 202, 204, 206, response to Dreyfus, 174
Browning, Robert, 166
Brundtland Commission, 15, 24
Brundtland Report: agriculture in, 29; definition of sustainable development, 11, 12, 39, 41; Enlightenment, legacy of, 67–68; ethics in, 32–33; Eurocentrism of, 42; failure of, 45; globalism and, 54; institutional reform and, 17–18, 19; technological optimism within, 24, 25, 66. *See also* sustainable development
Buckminster Fuller, R., 75–76, 98

Capra, Fritjof, 73
Carson, Rachel, 13, 40
Cartesian metaphysics. *See* dualism; Enlightenment
Cartesian skepticism, 154
Chatterjee, Pratap, 25, 48, 52, 54
Cheney, Jim, 163
Chipko movement (India), 14, 59
Club of Rome, 16, 28, 34. *See also* Limits to Growth
Code, Lorraine, 160
coevolutionary development, 58–59, 63
Cold War, 16, 30, 46, 50, 54

colonization, conceptual. *See* dualism
Commission on Global Governance, 11, 18, 35
Commoner, Barry, 13, 40, 41
Connolly, William, 60–62
consumption, culture of, 47, 112, 130–31, 140, 141–42, 143, 150, 157, 159, 171, 174, 175. *See also* Borgmann, Albert
cost-benefit analysis, 23
Council of Academies of Engineering and Technological Sciences, 26
Cubitt, Sean, 94
cyberspace, 102

Daly, Herman, 13, 67, 72
Derrida, Jacques, 138
Descartes, Rene, 69–71, 80, 87–88, 110, 120–21, 157, 168, 193. *See also* Enlightenment
Dessauer, Frederick, 68–70
development: crisis in, 2–5; discourses of, 45–47, 61–62; Eurocentrism of, 41–42, 47–48, 56–59; technology and, 24–26, 30–31, 209–10
Dewey, John, 161, 163
Dexter, Frank, 111–12
Diaz, Bernal, 3
Dobson, Andrew, 42–43
Dreyfus, Hubert: interpretation of Heidegger, 133–35; moral *praxis*, 160, 162; pretechnological practice 137; skill and technology, 173–74, 176
Dryzek, John, 15
dualism: feminist critique, 79–81; metaphysics underlying, 79, 121; of ends and means, 110–12; technology and, 85–89, 100–01, 248n. 41
Durbin, Paul, 93

Earth Summit: '92 Global Forum and, 47–48; *Earth Summit+5*, 18–19; hegemony, 46, 47, 48; non-governmental organizations and, 52–53; preparatory conferences, 18; technocratic managerialism and, 45, 47; World Business Council for Sustainable Development and, 21, 48–49. *See also Agenda 21*; United Nations
Easterbrook, Gregg, 16, 203
ecoefficiency, 23, 26, 35, 37, 54, 203
ecological crisis, 1–4, 13–14, 63
ecological modernization. *See* ecomodernism
ecological scarcity, 71–72
ecomodernism: economic assumptions of, 16, 22–24; effect on environmental movement, 41–42; ethical assumptions of, 33, 34–35, 38; explanation of term 15; history of, 65–67; institutional reform and, 19; role of business in, 21, 30–31; technocratic sustainability and, 39, 43–44, 53, 56, 62, 64, 82, 108, 202–04; technology and, 25, 27–30, 75, 93, 101, 203–04
economic growth, 12, 16, 17, 22–23, 30–31, 33, 34, 71–72. *See also* progress
Ehrlich, Paul, 13, 40
Ellul, Jacques, 98–99, 247n. 34. *See also* neo-Luddites
Engel, Ronald, 56
Enlightenment: Christianity and, 70–71; epistemology, 152–53; idea of scarcity, 68–70, 72; idea of sustainable development, 65–68, 154–55; ideal of authenticity, 155–57; ideal of disengaged agency, 120–21, 153–54, 170–71; mastery of nature and, 78–80, 86–88; promise of technology, 70, 110, 111, 150. *See also* Bacon, Francis; Descartes, Rene
environmental economics, 22–24
environmental ethics, 13, 31–34, 37, 41. *See also* radical ecophilosophy
environmental technology, 24–26
environmentalism: backlash against, 14–15, 16; confusion within, 41, 42–43, 73–74; corporate activism and, 15, 16; counterculture and 13–15; ecomodernism and, 15–16, 38; polarization of, 15; Promethean optimism within, 15, 26, 76–77; Southern concerns about, 14, 44–48, 50–51, 53, 55, 57–60; two post-war waves of, 13–16, 22, 40
epistemological proceduralism, 153–55

essentially contested concepts, 60–62
Esteva, Gustavo, 4, 53–54, 58
ethics. *See* moral life; moral science; sustainable development, ethics of
eudaimonia, 164
Eurocentric hegemony, 41–42, 44–53, 56–57, 60, 106

Finger, Matthias, 25, 48, 52, 54
Flores, Fernando, 173
Flyvbjerg, Bent, 205, 250, n. 67
fossil fuels, 28–29
Fremantle, Port of, 180–92, 197
Friedman, Marilyn, 148
Friends of the Earth, 14
future: consumption of, 192: orientation to, 201–05

G-8 (formerly G-7), 50
Gadamer, Hans-Georg, 151, 152, 157–58, 162
Gaita, Raymond, 162
Galileo Galilei, 87
gardening, 194–95, 210–11
General Agreement on Trade and Tariffs, 19
Giddens, Anthony, 94, 102
Gilligan, Carol, 160
global civil society, 34, 52–55
Global Environmental Facility, 18, 19, 40, 49. *See also* World Bank
global environmental politics, 18–19, 21, 44–46, 54–55
global images, 41, 54
globalization, 50, 51, 54–55
good life, the, 93, 149, 188
Gouinlock, James, 144–45, 149, 150, 156, 163–64, 167
Green Belt Movement (Kenya), 14
Greenpeace, 14

Habermas, Jürgen, 155, 160
Haraway, Donna, 112
Hardin, Garrett, 14
Hare, Richard, 144, 150
Hart, John, 143
Heidegger, Martin: account of human freedom 124–25; art as *techne*, 125–26; contrast between *poiesis* and production, 120–21; critique of metaphysics, 119; displacement of *praxis*, 138–40, 172–73; dualism of theory and practice, 136–38; essence of practice, 117; history of philosophy, 119–24; nostalgia of, 126–28; National Socialism and, 138–39; ontology of human existence, 117–18, 122–23; philosophy of technology, 116, 252n. 1; the technological world, 118–19, 129–32, 143, 202. *See also* Dreyfus, Hubert
Henderson, Hazel, 41, 67
Hildyard, Nicholas, 47, 53
Hill, Stephen, 109
Hobbes, Thomas, 71–72
Hodge, Johanna, 122, 132, 142, 175
Hölderlin, Friedrich, 127–287 131, 253n. 8
holism: conceptual ambiguity in, 73–74, 82; ecomodernism and, 74–75; Gaian theory, 76, 82; new physics and, 73–74
Holmberg, Johan, 39
Homer, 147
Hume, David, 144

Illich, Ivan, 39, 75, 98, 107, 201
individualism: modern ideal of freedom and, 155–58; self/world dualism and, 84, 121, 141–42; technology and, 88, 119
industrial ecology, 27–28: decarbonization, 28; dematerialization, 29
instrumental reason, 144–45, 153–56
instrumentalism, of technology, 85–87, 94, 96, 100–02, 106, 108–09
intergenerational equity, 17, 22–23, 32–33, 35, 37, 39, 59
International Chamber of Commerce, 21, 48
International Commission on Disarmament and Security Issues, 15
International Commission on International Development Issues, 15
International Monetary Fund, 19, 51
International Union for the Conservation of Nature (IUCN), 11, 37
Ismail, Razali, 18

Japan, as a technological culture, 133–34
Jarrah (Eucalyptus marginata), 183–85, 187, 192–93, 198
Jonas, Hans, 95, 143

Kant, Immanuel, 35, 144, 152–53, 166
Kateb, George, 102
Keynes, John Maynard, 71, 72
Khor, Martin, 19
Kuhn, Thomas, 73, 106
Kundera, Milan, 209
Kyoto International Summit on Climate Change, 20

Leiss, William, 78–79
Lélé, Sharachchandra, 4, 37
lifeboat ethics, 14
Limits to Growth, 13–14, 40
Lipietz, Alain, 52
Locke, John, 69
logos. *See* Heidegger, Martin
Lovekin, David, 99
Lovelock, James, 76
Lovitt William, 117, 118, 120, 254n. 22, 256n. 58, 259n. 104
Luddites, 99, 247n. 33
Luijf, Reginald, 67–68, 71

Maclntyre, Alasdair, 146–47, 160, 164, 169
Maddox, John, 14
Manne, Anne, 146
Marcuse, Herbert, 64, 73, 78, 85, 98
Margulis, Lynn, 73, 76
Mathews, Freya, 89, 185
McKibben, Bill, 66, 74
Mehmet, Ozay, 41–42
Merchant, Carolyn, 78, 79
Mexico City, 3–4
Mies, Maria, 51–52
Mill, John Stuart, 71, 72
Miller, Marian, 50, 54–55
Mitcham, Carl, 63, 96–97, 248n. 41
Moore, George, 144, 146
moral life, 143–44, 148–49, 168, 205
moral ontology, 149–51, 165
moral relationality, 83–84, 147–49, 161–67, 168–69, 172–73, 175
moral science, 144–47, 152–53, 163
moral subjectivism, 146–47
moral *telos*, 146–47, 150
Multilateral Agreement on Investment, 51. *See also* neocolonialism
Mumford, Lewis, 71, 78, 98, 106

Naess, Arne, xi, 13, 41, 77
neocolonialism: corporate, 51, 75; environmental, 50–51
neo-Luddites. *See* Luddites
New Age, 73
Nietzsche, Friedrich, 66, 131, 138, 147
Noble, David, 70, 79
Noddings, Nel, 160
Norgaard, Richard, 38, 58–59. *See also* coevolutionary development
Nyungar people, South-Western Australia, 184–85, 186, 188, 190, 194

Ogle, Nathaniel, 188–89
Oliver, Mary, 167, 207
OPEC, 15
Ophuls, William, 67
O'Riordan, Timothy, 48
Orr, David, 43–44, 62, 93
Orwell, George, 98
Osborn, Fairfield, 13
Our Common Future. *See* Brundtland Report
Overseas Development Assistance, 19, 49

Pacey, Arnold, 99
parenting, 195–98, 208–09
Pascarella, Perry, 14
Pearce, David, 22–24, 33, 41, 43. *See also* environmental economics
Peet, John, 60
Peterson, Tarla, 41, 57
philosophy of technology, 93–94, 243–44n. 6. *See also* Borgmann, Albert; Winner, Langdon
phronesis. *See* practical reason
Piaget, Jean, 145
Pinchot, Gifford, 66
Pirages, Denis, 45
planetary science, 30
Plato, 80, 119
Plautus, 127

Plumwood, Val, 79–86, 120
poiesis. *See* Heidegger, Martin
polis, 164, 167
Postman, Neil, 69
practical reason: Aristotle's account of *phronesis*, 152, 161, 162, 164, 167–68, 169, 196; distinct from technical reason 151–52, 163; internal to moral practice, 162–64; Kantian proceduralism, 152–53, 166; the rationality of relationality, 165–67, 169; social theory and, 169–72; technology and, 157–58, 167–68, 169–72; world-affirming, 162, 163, 165–67
Prakash, Madhu Suri, 53–54, 58
praxis, 137–40, 141, 158, 160–61, 164, 167–68, 171–73, 175–76
precautionary principle, 17
progress, cultural project of, 4, 16, 70–72, 73–74, 108, 113, 150, 156, 188–91

Radford Reuther, Rosemary, 13, 77
radical ecophilosophy, 41, 42, 77–78, 88–89
Raghavan, Chakravarthi, 51
Rahnema, Majid, 53
Rawls, John, 144, 146
Redclift, Michael, 21, 44–45, 57, 59
relativism. *See* moral subjectivism
Rescher, Nicholas, 72
Rio Declaration on Environment and Development, 33, 46–47. *See also* Earth Summit
Ross, Stephen, 176
Rostow, Walter, 22
Roszak, Theodore, 13
Rotterdam Charter for Sustainable Development. *See* International Chamber of Commerce

Sachs, Wolfgang, 45–46, 55
Saint Augustine, 70
Sandbrook, Richard, 39
Sandel, Michael, 147, 148
Schmidheiny, Stephen, 21
Schmitz, Gerald, 45, 53
Schumacher, Fritz, 14, 41, 236n. 112
Shiva, Vandana, 47, 54, 59, 79
Sidgwick, Henry, 144, 145
Simon, Julian, 15, 66
Slyvan (Routley), Richard, 13
Smith, Richard, 60
Socolow, Robert, 27, 28
Socrates, 63, 139
spaceship earth, 67
Spinosa, Charles, 173
stability, ideal of, 67, 68–69
Stevenson, Charles, 146
Strong, David, 93, 176, 182
Strong, Maurice, 21, 53. *See also* Earth Summit
sustainability: basic questions of, 64, 181, 213; definition, 38, 63–64, 66; distinction between cultural and technocratic discourses of, 5, 40, 43–44, 62, 64, 177, 201, 202–04; evidence of unsustainability, 2, 47–48; morally contested concept, ix, 4–5, 60, 62; Southern perspectives on, 47, 50–52, 53–54, 57–60; technocratic forms of, 5, 43–44, 47, 77, 81, 154, 177, 201, 202–04; weak versus strong distinction, 23, 43
sustainable development: agriculture and, 29–30, 58, 59; Bretton Woods organizations and, 19; definitions, 11; discourse analysis of, 41; emergence of concept, 1–2, 11–12; environmental economics and, 23–23; ethics of, 31–30, 50, 60; global economic liberalization and, 23, 30–31, 44–45, 46–49, 50–51; global *mythos* and, 53–54; hegemony of, 44–45; indicators of, 42–43; North South politics and, 47; population and, 17; precautionary approach, 17; principles of, 17; role of United Nations in, 17–18 (*see also Agenda 21*; Brundtland Commission; Brundtland Report; Earth Summit); strategic aims of, 17; technological optimism and, 24–26, 30, 43–44; trade liberalization and, 23
sustenance and sustainability, 64; as distinct from consumption, 181, 198, 199–201, 207; as essence of everyday practice, 196, 201, 212; as inherently public, 206–07, 209; as practical craft,

sustenance and sustainability (*continued*) 210–11; practical reason and, 213; reproduction and, 208–09, 210; temporality and, 204

Taylor,Charles, 148, 149–50, 153, 155–56, 169–72, 200. *See also* moral ontology
teche. *See* Heidegger, Martin
technocratic sustainability. *See* ecomodernism; sustainability
technological determinism, 98–99, 100–02
technological ontology, 119–23, 156, 168. *See also* Heidegger, Martin
technological optimism, 22
technological society, the, 2, 12, 93
technological systems (technosystems) analogy to ecosystems, 105; as background of practice, 106–08, 110–11, 141–43, 150–51; moral significance of, 168, 170–76
technology: as correlational practice, 95, 113, 143, 158, 167–69, 172, 175–77, 182; as cultural experiment, 104–05, 211; as cultural trajectory, 107–09; as moral *telos*, 150–51; as world-building 104–09, 176, 181 (*see also* Winner, Langdon); Christianity and, 70; determinism (*see* technological determinism); difficulty in defining, 96–97; fallacy of externalization and, 96, 143; economic progress and, 40; first order and second order explanation of, 95–96; history of, 106, 250n. 63; irony of, 111; means and ends in (*see* Borgmann, Albert); neglected by philosophers, 86, 88–89; non-neutrality of (*see* instrumentalism); ontology and (*see* technological ontology); risk and, 97–98; social constructivism, 99–100, 247n. 30. *See also* philosophy of technology
technophilia/technophobia. *See* technological determinism
Tenochtitlán, Aztec City, 3–4
Thiele, Leslie, 125, 131
Tijmes, Pieter, 67–68, 71
time, technologization of, 70–71, 87, 127–28, 130, 201–02
Todd, John, 41
transnational corporations: control of media, 50; influence at Earth Summit, 48–49; neocolonial power of, 51, 59; use of whole-earth images, 40
Truman, Harry, 31
trust, 207

unintended consequences of technology, 107–09
United Nations (UN): Center for Transnational Companies, 48–49; Commission on Sustainable Development, 18–19; Conference on Environment and Development (*see* Earth Summit); Convention on Biological Diversity (Cartagena), 48; Development Decade, 22; Development Program, 18, 35, 49–50; Division for Sustainable Development Environment Program, 11–12, 23, 32, 49–50; Financial crisis and, 49; General Assembly, 17, 18, 182; internal conflict within, 49–50; Stockholm Declaration, 11, 32; World Charter for Nature, 32
United States Government, 31, 48, 49, 222–23n. 60

van Gogh, Vincent, 126–27
Vico, 154
Villa, Dana, 172
Vogt, William, 13
von Weizäcker, Ernst, 16, 30, 34

Wertheim, Margaret, 102
Weston, Anthony, 163
White, Lynn, Jr., 87
Whitman, Walt, 212
Will, Frederick, 153
Williams, Bernard, 152, 160, 161, 169
Winner, Langdon: critique of efficiency, 38; philosophy of technology, 94, 97; technological autonomy, 99–100; technological somnambulism, 100–01; technology as forms of life, 105; technology and moral theory, 142–43; technology and reverse adaptation, 107–08; world-building, 103–04

Wise-Use Movement, 16
World Bank, 19, 31, 46, 51
World Business Council for Sustainable Development, 21, 76
World Commission on Environment and Development. *See* Brundtland Commission
World Resources Institute, 24, 26, 75
World Trade Organization (WTO), 19, 52
World Wide Fund for Nature, 12
Worster, Donald, 39, 47, 65
Wright, Judith, x, 182–83, 191–92

Zapatista movement (Mexico), 14
Zimmerman, Michael, 61, 78, 122, 139–40
Zukav, Gary, 73